Principles of
Hydrology

Second Edition

R.C.Ward

Senior Lecturer in Geography
University of Hull

London · New York · St Louis · San Francisco · Düsseldorf
Johannesburg · Kuala Lumpur · Mexico · Montreal · New Delhi
Panama · Paris · São Paulo · Singapore · Sydney · Toronto

Published by
McGraw-Hill Book Company (UK) Limited
MAIDENHEAD · BERKSHIRE · ENGLAND

Library of Congress Cataloging in Publication Data

Ward, R. C.
Principles of hydrology.
Second edition.

Includes bibliographies.
1. Hydrology. I. Title.
GB661.W3   1975   551.4   74-32444
ISBN 0-07-084055-5

Principles of Hydrology, 2/e

*To Charles and Hilda
and to Kay*

Preface

The year 1974 marked not only the end of the International Hydrological Decade but also the tercentenary of the founding of scientific hydrology by Perrault and Mariotte in France. Such occasions prompt a mood of contemplation and review which in this instance culminated in the preparation of a second edition of *Principles of Hydrology*. I am grateful for the kind way in which the first edition was received and gratified to know that so many people found it useful. Since its preparation, however, important advances have been made, particularly in the fields of soil moisture, groundwater, and runoff, which called for extensive revision. In fact much of the book has been rewritten although some original material has been retained, especially in the earlier chapters. A chapter on the drainage basin has been added by way of conclusion and the chapter bibliographies have been revised and updated.

Once again I am sincerely grateful to all who have helped in the preparation of this book by granting permission to quote material and to copy or adapt a large number of the many diagrams; individual acknowledgements are made in each case. Particular thanks are also due to Dr Keith Smith for his valuable advice, to Mr K. Scurr and his colleagues for the preparation of all the diagrams, to Miss M. Micklethwaite for typing most of the manuscript, and finally to my wife, Kay, for her encouragement, criticism, and practical help.

R.C. WARD
Hull, 1974

Preface to First Edition

This book presents a non-mathematical treatment of 'pure' as opposed to 'applied' hydrology. It is intended to fill a long-felt need at university level, among geographers in particular and earth scientists in general, for a straightforward, systematic analysis of the distribution and movement of water in the physical environment. It is, therefore, the first British textbook covering the general field of hydrology that is not aimed exclusively at the engineer. Nevertheless, it will probably be found useful as an elementary text for students in engineering faculties in which hydrology forms a compulsory part of the course, and for technical staffs in water boards and river authorities. In addition, the book is written in a way that should make it quite acceptable to the intelligent layman seeking background information on the pressing problems of water supply, and on the factors underlying the initiation in 1965 of the International Hydrological Decade.

Inevitably, its general scope and framework are similar to those of existing engineering texts. The essential differences lie in the selection, emphasis, and treatment of material and, particularly, in the inclusion of numerous examples from the British Isles. A conscious effort has been made to reduce engineering and other 'applied' aspects to a minimum, although in many cases natural events have been so modified by man's activities that references to the latter are unavoidable. On the other hand, since many of the problems of hydrology are geographical problems of spatial distribution and of climatic and regional differences, the application of geographical methods and techniques should contribute positively to their solution.

The arrangement of the text follows the conventional, although logical, systematic approach, whereby the concept of the continuous natural movement of water in the hydrological cycle is very briefly introduced in chapter 1, and, in the remaining seven chapters, the main phases of the cycle, i.e., precipitation, interception, evaporation, evapotranspiration, soil moisture, groundwater and runoff, are examined in detail. There was a temptation to write a ninth, concluding chapter but it was felt that the discussion of runoff in chapter 8 provides a natural conclusion, emphasizing, as it does, the effects and interactions of many of the components of the hydrological cycle that have been previously discussed. In view of this arrangement, I have necessarily drawn quite heavily in places on the existing standard hydrological texts, and I hope that this debt is adequately acknowledged in the bibliographies. These have been placed at the end of each chapter, despite the fact that this has resulted in some obvious repetitions, in order to avoid a large and cumbersome bibliography at the end of the book. It will be noted that reference has been made to a large number of articles and other publications throughout this book in the hope that the interested reader will be encouraged to extend his study of particular topics. For this reason, the articles referred to are not necessarily always the normally accepted 'classics'. They have been selected largely because they represent a modern statement of ideas and facts; or they themselves contain an extensive bibliography that will lead the reader back through the development of thought on a given topic; or the journals in which they appear are comparatively easily accessible in this country.

I am sincerely grateful to all who have helped in the preparation of this book. Particular thanks are due to the numerous individuals and publishers for permission to copy or adapt a large number of the many diagrams; individual acknowledgements are made in each case. I would also like to thank Professor H. R. Wilkinson, of the Department of Geography at Hull University, and other colleagues for reading parts of the manuscript, although they are in no way responsible for the deficiencies that still remain; Mr R. R. Dean and the staff of the Geography Department drawing office for the preparation of all the diagrams; Miss J. M. Bailey and Miss P. A. Ashcroft, who bore the brunt of the typing; finally, my wife, Kay Ward, for her commendable assistance during many hours of proofreading and checking.

R. C. WARD
Hull, 1967

Contents

CHAPTER 4 EVAPORATION

CHAPTER 5 EVAPOTRANSPIRATION

CHAPTER 7 SUBSURFACE WATER—GROUNDWATER

The Occurrence of Groundwater

Groundwater Storage

CHAPTER 8 RUNOFF

(C) *Minimum Flows*

CHAPTER 9 THE DRAINAGE BASIN

1. Introduction

1.1 Definition and scope of hydrology

Although hydrology can be simply defined as the science of water, such a broad definition is rather misleading. Usage has tended to restrict hydrology to the study of water as it occurs on, over, and under the earth's surface as streamflow, water vapour, precipitation, soil moisture, and groundwater but in recent years two trends in particular have resulted in important modifications to this generalized view. The first trend has been the development of the system concept and the resulting improved understanding of the hydrological cycle on a more sophisticated and higher conceptual level [7]. Thus not only may we recognize that the physical processes, which together constitute *physical hydrology,* can be investigated and explained by modern systems analysis techniques but also that these physical processes and subsystems can be simulated mathematically. Numerous mathematical and statistical techniques are becoming available to the hydrologist and the system concept has opened up new possibilities in the fields of *theoretical hydrology,* e.g., systems hydrology, parametric hydrology, and stochastic hydrology.

The second trend has been that towards relevance, i.e., the extent to which disciplines, including hydrology '. . . bear upon, relate, or are germane and applicable to the problems of society' [1]. Within hydrology the quest for relevance has resulted in the growth of interest in man's impact on hydrological conditions, e.g., urban hydrology, the hydrology of vegetation and land-use manipulation and the long-overdue recognition of major omissions such as water quality which has in the past been virtually excluded as a parameter of water science in favour of almost total attention to quantitative aspects [1].

The scope of hydrology is thus wider now than it has ever been. Discussion of the principles of hydrology, however, involves a much more restricted field of study. Principles are concerned with the basic physical processes, i.e., with an accurate knowledge and understanding of the occurrence, distribution, and movement of water over, on, and under the surface of the earth, and with the recognition that water is an element in the physical environment, just as soil, vegetation, climate and rock type are and that it is in this role that water is studied by the environmental scientist. Application of these principles to water resources management, to the prediction and forecasting of streamflow, and to the modification of drainage basin and channel characteristics, takes one, in theory, into the field of *applied hydrology,* although it will become increasingly clear in later pages that, in fact, this distinction is very blurred. As far as possible, however, it is intended that this discussion of the principles of hydrology should be predominantly concerned with introducing the physical basis of hydrology.

1.2 A brief history of hydrology

That water is essential to life and that its distribution and availability are closely associated with the development of human society seems so obvious as to be a fundamental truism. This being so it was almost inevitable that the development of water resources preceded any real understanding of their origin and formation. Archaeological discoveries and later documentary evidence emphasize the significant part played by the location and magnitude of water supplies in the lives of, for example, the Old Testament peoples, the ancient Egyptians, and later of the Greeks and Romans. During these periods, throughout the Middle Ages, and indeed until comparatively recent times, the search continued for an explanation of springs, streamflow, and the occurrence and movement of groundwater. But the hypotheses put forward were either based on guesswork or mythology or else were biased by religious convictions; few, if any, of the hypotheses, were based on the scientific measurement of the relevant hydrological factors. And yet some of the ideas developed by the ancient writers were remarkably close to the truth as we now know it. Thus Aristotle (384-322 B.C.) explained the mechanics of precipitation, Vitruvius, three centuries later, believed in the pluvial origin of springs, da Vinci (1452-1519) had somewhat confused ideas about the hydrological cycle but a much better understanding of the principles of flow in open channels than either his predecessors or contemporaries, and Palissy (1510-1590) stated categorically that rainfall was the only source of springs and rivers [2].

It was not until near the end of the seventeenth century, however, that plausible theories about the hydrological cycle, based on experimental evidence, were advanced. The greatest advances came largely through the work of three men; Pierre Perrault and Edmé Mariotte, whose work on the Seine drainage basin demonstrated that, contrary to earlier assumptions, rainfall was more than adequate to account for river flow; and the English astronomer, Edmund Halley, who showed that the total flow of springs and rivers could be more than accounted for by oceanic evaporation. Because Perrault, Mariotte, and Halley undertook hydrological research of the modern scientific type they may well be regarded as the founders of hydrology [13].

After a period of modest consolidation during the eighteenth century there was a remarkably rapid growth of knowledge in hydrology during the nineteenth century, which saw the beginnings of systematic river flow measurement, e.g., on the upper Rhine near Basle in 1809, the Tiber at Rome in 1825, and on the Garonne in 1837 [2], the derivation of a universal discharge formula by Manning and the development of the 'rational' method for discharge estimation. In the United States the earliest records of river flow were made on the Ohio River at Wheeling, West Virginia between 1838 and 1848, but in Britain, despite the early start in the organized recording of rainfall by Symons, the first regular measurements of streamflow were made by Captain W. N. McClean in 1912 [3] in Scotland and even by 1936 there were still only twenty-seven regular gauging stations in Britain.

2

Groundwater hydrology also developed substantially during this period. For example, William Smith, the 'Father of English Geology', was, in 1802, the first person to apply stratigraphic principles directly to groundwater exploitation [22], while in 1856 Darcy laid the foundation of groundwater theory in his report on the public water-supply system for Dijon [2]. Darcy's work was later extended by a fellow Frenchman Dupuit, the German Thiem, who analysed problems concerning the flow of water toward wells, and the Austrian Forchheimer, whose major contribution was the introduction of complex mathematical analyses to the solution of groundwater problems.

The nineteenth century also saw the publication of the first textbook in hydrology. This was Nathaniel Beardmore's *Manual of Hydrology* published in 1862 which was itself a revision of an earlier work, *Hydraulic Tables,* of 1850 [2]. In 1904 Daniel W. Mead, of the University of Wisconsin, published his *Notes on Hydrology* as the first American text and in fact his later texts [16] [17] are still widely used today [13].

As would be expected, enormous advances in the subject have been made during the twentieth century, particularly in the United States. In the thirties the US Government provided large sums of money for studies and projects in the fields of conservation, irrigation, and flood control and Eagleson [11] somewhat facetiously referred to this period as the 'First Golden Age of Hydrology'. Military technological developments during and immediately after World War II, again particularly in the United States, created new equipment and techniques, including greatly improved remote sensing equipment, which were subsequently adapted for research in the earth sciences and Eagleson [11] suggested that hydrology's 'Second Golden Age' began in the 'sixties as this new phase of research began to be specifically applied to the rational solution of problems in water resources. At the present time, as has been suggested earlier, hydrology is in a continual state of flux as new trends and new concepts develop. Because of this we perhaps tend to over-emphasize recent developments and thereby give the impression that hydrology is a new subject, with the result that few hydrologists have a good knowledge of their heritage or even a broad understanding of how the science has developed even to its present imperfect state [2].

1.3 Occurrence of water

Water occurs in the atmosphere and above and below the surface of the earth as a liquid, a solid, and a gas. As a liquid, it is of direct importance to the hydrologist in the form of falling rain in the atmosphere; as streams, lakes, and oceans on the surface; and as soil water and groundwater below the surface. In its solid state it occurs in such forms as snow, hail, and ice. Finally, water vapour occurs abundantly in the lower layers of the atmosphere and, to a limited extent, within the immediate surface layers of the earth's crust.

Hydrologically, a most significant factor is that not only does water occur in these various forms, but also that in no case does it occur independently of all the other forms. Water is, in fact, continually changing in both state and place. Thus, it changes from solid to liquid as a result of the melting of snow, hail, and ice, or from a liquid to a solid as a result of freezing. It changes from a liquid to a gas as a result of evaporation, or from a gas to a liquid by means of condensation. Such changes take place all the time and are partly responsible for, and partly caused by, the continuous movement of water from one place to another. Water moves rapidly as falling precipitation in the atmosphere, or as flowing streams across the ground surface. It moves more slowly beneath the ground surface, as groundwater gradually flows to streams or the oceans, or as water percolates through the surface layers of the soil and subsoil towards the underlying groundwater, or moves upwards by capillary action towards the surface.

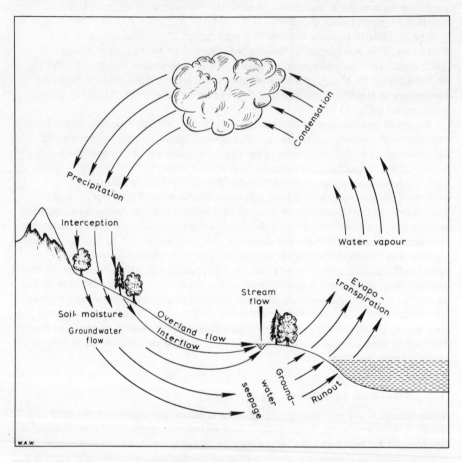

Fig. 1.1. *Simplified diagram of the hydrological cycle*

1.4 The hydrological cycle

The interdependence and continuous movement of all forms of water provide the basis for the concept of the hydrological cycle, which Meinzer [18] regarded as the central concept in hydrology, and which is illustrated simply and diagrammatically in Fig. 1.1. Essentially, the hydrological cycle envisages that all water is involved in a cyclical movement which continues indefinitely. Some of the *water vapour* in the

4

atmosphere condenses and may give rise to *precipitation.* Referring specifically to the land areas, not all of this precipitation will reach the ground surface because some will be evaporated while falling and, more important, some will be caught or *intercepted* by the vegetation cover or by the surfaces of buildings and other structures, and will from there be evaporated back into the atmosphere.

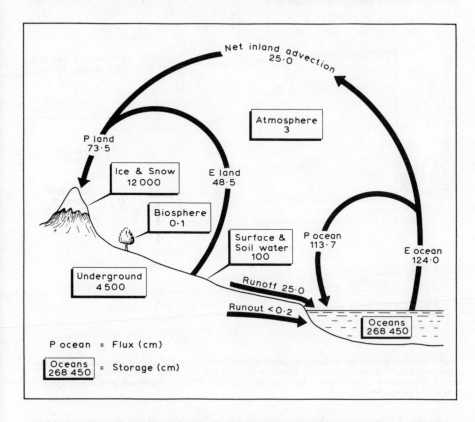

Fig. 1.2. *Approximate values of the fluxes and storages in the global hydrological cycle. (Based on data from Lvovitch [15], Nace [20], and Sutcliffe [23]).*

   The part of the precipitation which reaches the ground surface may then follow one of three courses. It may, first, remain on the surface as *surface storage* in the form of pools, puddles, and surface moisture which are eventually evaporated back into the atmosphere. Secondly, it may flow over the surface as *overland flow* into depressions and channels to become *surface runoff* in the form of streams and lakes, from which it will move either by evaporation back into the atmosphere, or by seepage towards the groundwater, or by further surface flow into the oceans. Following the third course, falling precipitation may *infiltrate* through the ground surface to join existing *soil moisture.* This may be removed either by *evaporation* and *transpiration* from the soil and vegetation surfaces, or by *interflow* towards

5

stream channels, or by downward *percolation* to the underlying *groundwater,* where it may be held for weeks or months or even longer. The groundwater component will eventually be removed either by upward capillary movement to the soil surface and vegetation cover, whence it will be returned by evaporation and transpiration to the atmosphere, or by groundwater seepage and flow into surface streams and by runout to the oceans.

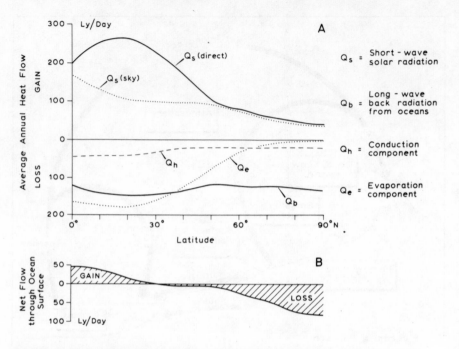

Fig. 1.3. *Average annual latitudinal values of heatflow across the ocean-atmosphere interface. Gain and loss in respect of oceanic values in langleys per day. (1 ly = 1 (small) calorie per cm²).* (From an original diagram by G. L. Pickard [21], Descriptive Physical Oceanography, © 1964, Pergamon Press Ltd.)

Figure 1.2 indicates very approximate values for some of the major fluxes and storages involved in the hydrological cycle. The excess evaporation from the oceans is transported atmospherically to the land areas whilst the excess of land precipitation over evapotranspiration is evacuated to the oceans as runoff and runout (i.e., groundwater flow directly into the ocean basins). Apparent discrepancies between oceanic and terrestrial figures result from the fact that these values are depth equivalents calculated for vastly different areas. Storage values emphasize the dominant role of the ocean basins and the major ice caps and the very small amount of water stored in the atmosphere from which all precipitation must obviously be derived. Such storage values can, however, only be meaningfully considered in relation to residence times which may vary from a few days in the case of atmospheric moisture, weeks or months in the case of soil moisture through to hundreds and even thousands of years in the case of groundwater and icecap storage and an indefinitely long period in the case of much oceanic water for which the average residence time is about 3000 years, although of course some molecules may reside only an instant before being evaporated into the atmosphere [19].

Although somewhat obscured in the simplified diagrammatic representation in Fig. 1.1 the dominant role of the oceans in the hydrological cycle (and by inference in the global energy budget) is well emphasized in Fig. 1.2. Although oceanic waters comprise only about 0.023 per cent of the earth's total mass they nevertheless comprise 97.6 per cent of its total water [19] and cover 71 per cent of its surface. Nace [19] remarked that the system ocean-atmosphere-continents is a great heat engine that drives the hydrological cycle and that the oceans are the principal heat reservoir. Since the heat content of that reservoir does not vary very much from year to year it follows that the large income of energy is balanced by equally large losses. As Fig. 1.3 shows, however, the magnitudes of the components of the heat balance vary considerably from one latitude to another with the result that the oceans gain heat across the earth-atmosphere interface from the equator to 30° N and lose it beyond this. As a result there is a net advective flow of heat polewards both within the oceans themselves (e.g., currents) and in the overlying atmosphere. Clearly this latitudinal adjustment and the changes in its magnitude from year to year have important implications for the global hydrological cycle.

Inevitably, simplifications and generalizations such as those involved in the concept of the hydrological cycle, are liable to mislead unless treated with caution. Thus, the implication of a smooth, uninterrupted, sequential movement of water is belied by the complexity of natural events. The cycle is short-circuited when, for example, water precipitated from the atmosphere and falling to the ground surface is immediately returned to the atmosphere by evaporation without becoming involved in streamflow, soil moisture, or groundwater movement, or the oceans. Similarly, precipitation may fall upon a lake and be evaporated from there without touching the land surface at all, or it may fall upon the land and percolate down to the main groundwater body within which it slowly moves towards a discharge point such as a spring, which it may not reach for a thousand years or more.

The irregularity of water movement within the cycle is further illustrated by conditions in hot deserts and subpolar regions. In the former, rainfall is spasmodic, occurring perhaps once in ten or twenty years. Other phases of the cycle, such as evaporation and surface runoff, which can take place only for a short period after rainfall, are, therefore, equally spasmodic, so that a short burst of hydrological activity for a week or so may be followed by a long period of virtual inactivity, apart, perhaps, from a slow redistribution of groundwater at some depth below the surface. In cold climates, where most precipitation is in the form of snow, there may be an interval of several months between precipitation and the active involvement of the precipitated moisture, after melting, in the subsequent phases of the hydrological cycle.

Systems approach to the hydrological cycle

1.5 Introduction

Although the simplified, descriptive cycle discussed so far provides a useful introductory concept it is of little practical value to the hydrologist concerned with a detailed, quantitative study of water occurrence, distribution, and movement in a specific area. There is clearly a need to replace the abstraction of the interminable cycle with a more positive approach. At first this more direct approach took the form of a simple application of the continuity equation in the form of the hydrological or water balance equation:

$$\text{Inflow} = \text{Outflow} \pm \Delta \text{ Storage}$$

If this equation can be solved, a quantitative assessment of the movement of water over, through, and across the land is possible because the equation must be applied to a specific area for a specific period of time.

The size of the area may vary from a small experimental plot of a few square metres to a large continent, or even the globe, but for most hydrological purposes it will comprise a relatively watertight river catchment or group of catchments. At the same time, the components of the equation will tend to vary according to the size of the area being considered, as also will the time period to which it is applied. A ten-minute run of data may suffice for a small plot, whereas data for a year, or long-term averages for several decades may be necessary for larger areas.

More recently the growing interest in systems analysis has led to a more rigorous application of the continuity equation in a systems approach to the hydrological cycle.

1.6 Systems and subsystems

Although the word 'system' has been defined in a number of ways we can accept with Chow [5] that a system is '. . . an aggregation or assemblage of parts, being either objects or concepts, united by some form of regular interaction'. The hydrological cycle can therefore be regarded as a system comprised of a number of parts such as precipitation, interception, evaporation, etc. There are many ways of classifying systems and it will suffice here to draw attention to some of the main types.

Again using Chow's terminology [9] , a *physical system* is a system in the real world and a *sequential system* is a physical system which consists of input, output, and a throughput of some working medium (matter, energy, or information) which passes through the system as is illustrated schematically in Fig. 1.4. A *dynamic system* is a physical system which receives certain quantitative inputs and accordingly acts concertedly under given constraints to produce certain quantitative outputs.

When the way in which a system operates is independent of the output produced the system is regarded as being *open* whereas if its operation depends upon the feedback of all or part of the output the system is regarded as *closed*. Chow [9] pointed out that most local or regional hydrological systems are open in the sense that their output exerts little significant control on the system, while the ground-water system on an island or the global hydrological cycle may be considered as closed in the sense that the output of these systems will have a substantial effect upon the balance of the system.

If each of the component parts of a system itself has the qualities of a system it may be regarded as a *subsystem* or as a separate system depending upon the objec-

8

tive of the investigation. For example, within the hydrological cycle system one can define various subsystems such as the vegetation, ground surface, soil moisture, groundwater, and channel subsystems. To the groundwater hydrologist, however, groundwater flow is a hydrological system in its own right.

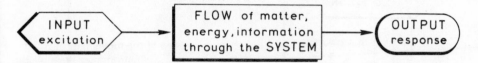

Fig. 1.4. *Diagrammatic representation of a sequential system.*

1.7 The hydrological cycle as a system

In terms of the foregoing discussion it will be apparent that the hydrological cycle is a physical, sequential, dynamic system which operates within a set of constraints or physical laws that control the movement, storage, and disposition of water within the system and which derives its energy from the spatial imbalance between incoming and outgoing radiation [12]. Furthermore, as we have seen, when considered with respect to the storage and movement of water within the system the global hydrological system is a closed system and therefore conforms to the principle of conservation of mass [12].

Figure 1.5 is an elegantly simple systems diagram of the hydrological cycle, at a local or regional scale, constructed by J. Lewin, which at once illustrates both the essential simplicity and also the great complexity of the system. At a local or watershed scale the system will normally have a single input, precipitation and two major outputs, evapotranspiration and runoff. Throughput consists of the varied transfer of water through the system, from one storage to another by means of the processes labelled from (a) to (j). In some cases a third output, leakage, may also occur from deeper subsurface water so that for other basins leakage will form an additional input of water, in which case the shape of the box and the direction of the arrow would have to be changed appropriately.

Clearly, each of the five storages shown has the qualities of a system and may therefore be considered subsystems of the basin cycle. Instead of being shown as a shaded box, therefore, each would be more appropriately represented by an internal diagram of storages and transfers although for the sake of clarity this has not been done.

With a small number of modifications Fig. 1.5 becomes a systems diagram of the global hydrological cycle as is shown in Fig. 1.6. With this closed system there are no inputs or outputs of water, only of energy, so that the global hydrological cycle consists of a series of transfers of water from one storage to another, including now the atmosphere and of course, the ocean basins.

1.8 The man-modified system

A systems approach to the hydrological cycle will not of itself be adequate if it perpetuates traditional weaknesses in the study of the hydrological cycle to which

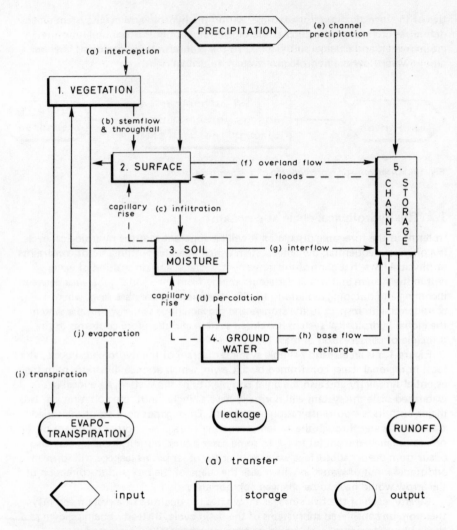

Fig. 1.5. *Systems diagram of the hydrological cycle at a local or regional scale. (From an original diagram by Dr J. Lewin, Department of Geography, UCW, Aberystwyth.)*

some references have already been made. Thus in the past hydrologists have been much concerned with quantitative aspects of the terrestrial cycle and little concerned with the role of the oceans, the global atmospheric circulation, or man-induced modifications in the cycle.

With specific reference to the latter, Ackermann hyperbolically pointed out that hydrology is unique among geophysical sciences in that it must deal increasingly with alterations in the natural order of things [1]. There is no need to labour this point but using the basic framework of the systems diagram shown in Fig. 1.5, Fig. 1.7 summarizes the main categories of man's attempts to modify thè natural

10

system. As can be seen, human modifications may be made to virtually every component of the system. At the present time, however, the most important of these modifications relate to (a) large-scale changes of channel flow and storage, e.g., by means of surface changes which affect surface runoff and the incidence or magnitude of flooding, (b) the widespread development of irrigation and land drainage, and (c) the large-scale abstraction of groundwater and surface water for domestic or industrial uses. Other modifications which are locally of great importance and which will undoubtedly become more generally important in the near future include artificial recharge of groundwater, and interbasin transfers of surface and groundwater. Still other modifications, such as the artificial stimulation of precipitation or the use of transpiration and evaporation suppressants, are at present in the early stages of development but may become important in due course.

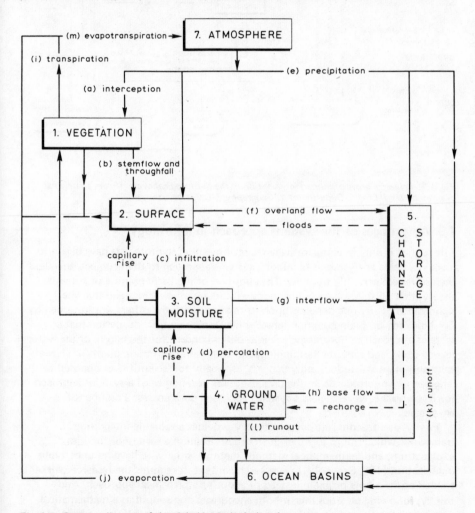

Fig. 1.6. *Systems diagram of the global hydrological cycle.*

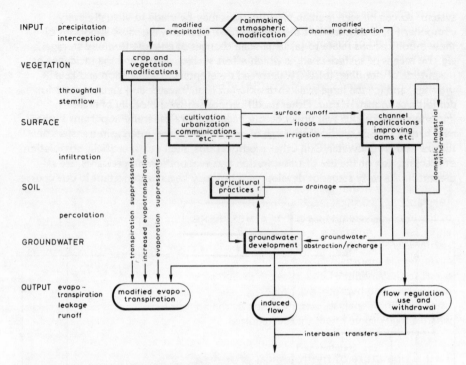

Figure content labels:

INPUT precipitation interception

VEGETATION

throughfall stemflow

SURFACE

infiltration

SOIL

percolation

GROUNDWATER

OUTPUT evapo-transpiration leakage runoff

modified precipitation

rainmaking atmospheric modification

modified channel precipitation

crop and vegetation modifications

cultivation urbanization communications etc.

surface runoff
floods
irrigation

channel modifications improving dams etc.

transpiration suppressants
increased evapotranspiration
evaporation suppressants

domestic industrial withdrawals

agricultural practices

drainage

groundwater development

groundwater abstraction/recharge

modified evapo-transpiration

induced flow

flow regulation use and withdrawal

interbasin transfers

Fig. 1.7. *Principal areas of human intervention in the hydrological cycle. (From an original diagram by Dr J. Lewin, Department of Geography, UCW, Aberystwyth.)*

1.9 Advantages of the systems approach in hydrology

Although it is only in comparatively recent times that hydrologists have begun to invoke systems analysis and to adopt a systems approach to their subject, the ideas themselves are certainly not new. The adoption of the basic premise of systems theory, that all things have connections with many other things and that the significance of any one depends upon its relations with the others, was regarded by Chisholm [4] as 'plain common sense' while Domenico [10] suggested that a systems approach to hydrology '. . . is as old as concern for the nature of the water regime itself, and only the terminology has changed'. Both these points of view are valid and there are undoubtedly dangers, especially for 'scientific' as opposed to 'engineering' hydrologists, in the unquestioning acceptance of a systems approach for its own sake. Nevertheless a systems approach does possess a number of advantages.

Firstly, as a teaching aid systems theory provides a valuable integrating framework which brings together in an orderly fashion a variety of theories, explanations, and mathematical methods that may otherwise have no underlying organization [10]. Secondly, in a research context, it requires the research worker to look at the whole problem instead of adopting a piecemeal approach. And thirdly, following on from this, it both encourages and simplifies a mathematical problem-simulation whose objective will be to formulate models whose behaviour approximates the real system and which are amenable to a mathematical analysis and solution [9]. It is this third and most important factor which has opened up a whole new field of hydrological investigation whose results have figured prominently in recent hydrological literature.

Mathematical models of hydrological systems may be either *lumped-system* models or *distributed-system* models [6]. The former is illustrated by the commonly encountered 'black-box' approach whereby the system has input and output but where the investigator has no knowledge of (or interest in) the internal workings of the system. In this type of model, exemplified by statistical analyses of drainage-basin rainfall-runoff relations, all parts of the system being simulated are regarded as being located at a single point in space whereas in a distributed-system model the internal processes of the system are considered and various distributed points or areas within the internal space of the system must therefore be considered [9]. Most hydrological models which have been developed for systems analysis are of the lumped-system or black-box type. Mathematical models may also represent either *linear* or *non-linear* systems. In a linear system the input is directly proportional to the output so that in the case of the unit-hydrograph method, for example, the input rainfall excess falling on the drainage basin is assumed proportional to the amount of direct runoff [8]. In actual drainage basins, however, this is not the case and hydrologists are now giving more attention to the possibility of using non-linear analysis in the solution of some hydrological problems, particularly where a high degree of accuracy is required.

1.10 The nature of hydrological processes

Some characteristics of hydrological processes are illustrated in Fig. 1.8. From this it can be seen that hydrological processes may be regarded as *deterministic* (i.e., chance-independent) or *stochastic* (i.e., chance-dependent). All hydrologic phenomena are subject to laws which govern their evolution and behaviour so that at a certain scale of investigation all hydrologic processes are deterministic. However, as Leopold and Langbein [14] recognized, although on a microscopic scale each minute event in landscape evolution is surely deterministic, the large-scale processes which comprise the combined effect of the individual

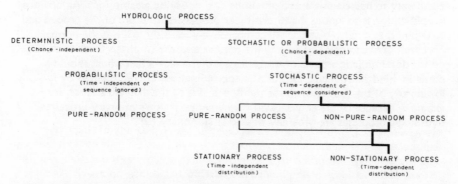

Fig. 1.8. *Some characteristics of hydrological processes. (From an original diagram by V. T. Chow (Ed.),* Handbook of Applied Hydrology, *McGraw-Hill, New York, 1964.)*

miscroscopic events may be stochastic in nature. Todorovic and Yevjevich [24] suggested that in general a physical phenomenon is assumed to be deterministic if, on the basis of its present state, its future characteristics may be predicted with certainty. Thus the Newtonian laws of motion are deterministic in the sense that the present state of a moving body uniquely determines its future states and similarly, in hydrology, the outflow hydrograph from an impermeable surface is a deterministic process if the characteristics of the rainfall input are known and the evaporation is negligible. On the other hand, the evolution of rainfall phenomena in time are stochastic in the sense that, although each micro-event is deterministic, in macro terms only probabilities of future conditions may be determined from the present situation.

Secondly, chance-dependent processes may be stochastic or probabilistic. A *probabilistic* process is one in which the sequence of occurrence of the variables involved is ignored, as in many analyses of flood records, and the chance of their occurrence is assumed to follow a definite probability distribution in which the variables are considered to be pure-random variables [5]. A *stochastic* process is time-dependent involving both probability and a sequential relationship between the variates. In this case the variables may be either pure-random or non-pure-random but the probability distribution of the variables may or may not vary with time [5].

If the process is *pure-random,* the members of the time series are mutually independent and thus constitute a random sequence whereas in a *non-pure-random* process the members of the time series are dependent amongst themselves, possess both a deterministic and a pure-random component and thus constitute a non-random sequence.

Finally, if the probability distribution of a random variable remains constant throughout the process, the process and the time series are said to be *stationary.* On the other hand, for example, streamflow from a drainage basin which had been affected by human activity or climatic change would be *non-stationary* in character.

Figure 1.8 indicates that actual hydrological processes, depicted by the heavy line, are stochastic or probabilistic, non-pure-random, and non-stationary. This means that natural hydrological processes are extremely difficult to treat mathematically because of the complexities involved in the analyses and as a result it is customary to make some approximations or simplifying assumptions. An obvious simplification is to ignore the stochastic or probabilistic nature of the process and to regard it as deterministic. Until comparatively recently this was the normal approach in hydrology and even in the present text the emphasis tends to be largely on the deterministic aspects of the processes discussed—a fact which should be borne in mind throughout. Less drastic approximations are represented by following any of the thin lines on the right-hand side of the diagram so that, for example, because of the complicated mathematics of non-stationary processes, hydrologic processes are generally treated as stationary [5].

References

1. ACKERMANN, W. C., Scientific hydrology in the United States, *The Progress of Hydrology,* University of Illinois. Urbana, Illinois, 1: 50-60, 1969.

2. BISWAS, A. K., A short history of hydrology, *The Progress of Hydrology,* University of Illinois, Urbana, Illinois, **2**: 914-936, 1969.
3. BOULTON, A. G., Surface water survey and modernization, Informal discussion of the Hydrological Group, Institution of Civil Engineers, London, 1966.
4. CHISHOLM, M., General systems theory and geography, *Trans. IBG,* **42**: 45-52, 1967.
5. CHOW, V. T., Statistical and probability analysis of hydrologic data, Part 1. Frequency analysis, in *Handbook of Applied Hydrology* (Ed. V. T. Chow) pp. 8-1—8-42, McGraw-Hill, New York, 1964.
6. CHOW, V. T., Laboratory study of watershed hydrology, *Proc. Intern. Hydrol. Sympos.,* Fort Collins, Colorado, **1**: 194-202, 1967.
7. CHOW, V. T., New trends in hydrology, *Nature and Resources,* **3**, No. 2, 1967.
8. CHOW, V. T., Hydrologic systems for water resources management, *Hydrology in Water Resources Management Conf. Proc.,* Water Resources Research Institute, Clemson Univ., Clemson, S.Carolina, pp. 8-22, 1968.
9. CHOW, V. T., Systems approaches in hydrology and water resources, *The Progress of Hydrology,* University of Illinois, Urbana, Illinois, **1**: 490-509, 1969.
10. DOMENICO, P. A., *Concepts and Models in Groundwater Hydrology,* McGraw-Hill, New York, 405, pp., 1972.
11. EAGLESON, P. S., *Dynamic Hydrology,* McGraw-Hill, New York, 462 pp., 1970.
12. FREEZE, R. A., and R. L. HARLAN, Blueprint for a physically-based, digitally-simulated hydrologic response model, *J. Hydrol.,* **9**: 237-258, 1969.
13. JONES, P. B., G. D. WALKER, R. W. HARDEN, and L. L. McDANIELS, The development of the science of hydrology, *Texas Water Commission, Circ.* 63-03, 35 pp., 1963.
14 LEOPOLD, L. B. and W. LANGBEIN, The concept of entropy in landscape evolution, *USGS Prof. Pap.* 500A: A1-A20, 1962.
15. LVOVITCH, M. I., World Water balance (General Report), *Symposium on World Water Balance, IASH Publ.* **93**: 401-415, 1970.
16. MEAD, D. W., *Hydrology, the fundamental basis of hydraulic engineering,* McGraw-Hill, New York, 626 pp., 1919.
17. MEAD, D. W., *Hydrology, the fundamental basis of hydraulic engineering* (2nd Ed. 2nd impression by Mead and Hunt, Inc. Consulting Engineers), McGraw-Hill, New York, 717 pp. 1950.
18. MEINZER, O. E., Chapter 1, Introduction, in *Hydrology* (Ed. O. E. Meinzer), McGraw-Hill, New York, 1942.
19. NACE, R. L., World water inventory and control, in *Water, Earth and Man,* (Ed. R. J. Chorley), pp. 31-42, Methuen, London, 1969.
20. NACE, R. L., World hydrology: status and prospects, *Symposium on World Water Balance, IASH Publ.* **92**: 1-10, 1970.
21. PICKARD, G. L., *Descriptive Physical Oceanography,* Pergamon, Oxford, 199 pp., 1964.
22. SMITH, K., *Water in Britain: a study in applied hydrology and resource Geography,* Macmillan, London, 232 pp., 1972.
23. SUTCLIFFE, R. C., World water balance—A geophysical problem, *Symposium on World Water Balance, IASH Publ.* **92**: 19-24, 1970.
24. TODOROVIC, P. and V. YEVJEVICH, Stochastic process of precipitation, *Colorado State Univ. Hydrol. Papers,* Fort Collins, Colorado, **35**: 61 pp., 1969.

2. Precipitation

2.1 Introduction

All the water moving in the land-bound portion of the hydrological cycle derives either directly or indirectly from precipitation. The study of precipitation is, therefore, of fundamental importance to the hydrologist but is, nevertheless, in its entirety, the domain of the meteorologist and climatologist. This chapter will, therefore, emphasize those aspects of precipitation which are of direct interest to the hydrologist, and it will be assumed that the reader is familiar with the systematic treatment of the subject as it is normally presented in standard meteorological and climatological texts.

The total amount of water in the atmosphere at any given moment represents only a minute proportion of the world's total water budget. Nace [83] in fact, estimated that atmospheric moisture represents 0.0001 per cent of the world's total supply of land, oceanic, and atmospheric water, and yet this small amount serves as a continuing source of supply in the form of precipitation. Bernard [10] expressed this rather differently by noting that if all the moisture in the atmosphere were precipitated at the same time it would result in an average rainfall over the entire surface of the earth of only 25 mm. Kuenen [67] estimated the capacity of the atmosphere over Europe in summer as being the equivalent of 38 mm of rainfall. Annual totals of precipitation are, of course, many times larger than these amounts, thus emphasizing the short residence time of atmospheric moisture. Frequently, however, these amounts of precipitation are exceeded many times in a single storm, indicating that large-scale lateral movements of moist air must be a major feature in the distribution of precipitation. It is, however, the task of the meteorologist to analyse and explain the mechanism of, and the reasons for, this distribution. The hydrologist is interested in the distribution itself, in how much precipitation occurs, and in when and where it occurs. Thus, the hydrological aspects of precipitation study are concerned with the form in which precipitation occurs, its variations in both time and space, its measurement, and the problems associated with correct use and interpretation of the measured data.

Types of precipitation

Before discussing spatial patterns and regimes of precipitation, it will be helpful briefly to summarize the types of precipitation. This can be done from two main points of view: a classification may be made either on the basis of the form or appearance of the precipitation, or on the basis of genesis, each method emphasizing aspects of hydrological significance.

2.2 Classification on basis of form or appearance

A simple but fundamental distinction can initially be made between *liquid* and *solid* precipitation. The former, comprising drizzle, rain, and dew, tends to play an immediate part in the movement of the hydrological cycle upon reaching the ground, while the latter, comprising mainly snow, glaze, and frost, may remain upon the ground surface for some considerable time after precipitation has occurred until the temperature rises sufficiently for the ice to melt into water. The hydrological implications of this are such that solid precipitation, particularly snow, will be discussed separately at the end of this chapter. Hail is a rather special case since, although it falls as a solid, it normally does so in temperature conditions that favour rapid melting at the ground surface, and so it tends to react hydrologically like a heavy shower of rain.

2.3 Classification on basis of genesis

Two main factors are necessary in order for precipitation to occur: a body of moist air, and a means of lifting this body of air sufficiently for condensation and then precipitation to take place. The presence of moist air over a land surface normally results from the lateral movement, sometimes over many thousands of kilometres, of maritime air masses and, in the British Isles, for example, this reflects the frequent movement of air from the Atlantic ocean. The uplift of this air may then be of the cyclonic, orographic, or convectional type.

Cyclonic precipitation may be either non-frontal or frontal in origin. The non-frontal type results from the convergence and subsequent uplift of air within a low-pressure area. Tropical precipitation of this type may deliver as much as 380 mm of rain within 12 to 24 hours [36] while in extratropical cases non-frontal cyclonic precipitation produces rains (or snows) of moderate intensity and longer duration, e.g., 50 to 150 mm in 24 to 72 hours [36] . In the case of frontal precipitation, uplift results when warm, moist air slides up and over a wedge of denser, cold air at either a warm or cold front. This occurs very commonly in the British Isles, particularly in western districts. The gradient of the frontal surface considerably influences the type of precipitation received. Thus the very shallow warm front gradients produce gradual lifting and cooling and result in widespread rains, occurring in belts several hundreds of kilometres in width. These rains are usually of moderate to low intensity and may last for two or three days at a time. The steeper cold front gradients produce more rapid uplift and cooling over shorter horizontal distances and may give rise to short-duration, relatively intense rainfall.

Orographic precipitation is best developed when a deep layer of moist air [35] is forced to rise over a range of hills or mountains. The effect is largely dependent on the size of the relief barrier and on its alignment with respect to air movement. Frequently, more rain falls on the windward than on the leeward slopes, with the consequent development of a rainshadow area on the latter. This effect can be seen in the northern and western highland areas of the British Isles. Over narrow uplands, however, precipitation may be carried over the crest-line by the wind, causing a lee-side maximum [6]. In other cases maximum precipitation occurs not at the crest-line but at some lower level on the windward slopes. It is thus clear that simple concepts '. . . of rainfall increase with height and lee-side rain shadow need to be replaced by more realistic models for a variety of synoptic situations in each mountain area' [6]. The intensity of orographic precipitation tends to vary with the depth of the uplifted layer of moist air. A deep layer produces heavy rainfall while the uplift of a shallow layer may result only in a light drizzle [35].

Finally, in the case of *convectional precipitation,* instability resulting from the heating of the earth's surface leads to the ascent of pockets or bubbles of air that cool adiabatically to below the dewpoint. Spatially, the occurrence of these convection cells may be organized or disorganized. The latter case is typified by summer heating of the ground surface resulting in scattered showers of high intensity and short duration. In the British Isles this occurs most frequently with *mP* air in inland areas during the summer months. An organized distribution of convection cells may occur, for example, where a cold, moist airstream moves over a warmer surface and the cells tend to travel with the wind producing a streaky distribution of precipitation parallel to the wind direction [9] or in tropical cyclones, where cumulo-nimbus cells become organized about the vortex in spiralling bands of cloud mass giving rise to heavy, prolonged rainfall affecting large areas [6].

It will be clear from this discussion that the three forms of precipitation, cyclonic, orographic, and convectional, do not necessarily occur independently of each other. Furthermore, each of the three types occurs in a number of forms. It is, therefore, very difficult to relate rainfall characteristics to genetic origins except in the general manner of the foregoing discussion.

Variations of precipitation

Two of the most important aspects of precipitation for the hydrologist are, on the one hand, its geographical distribution, the areal pattern of rainfall, and the magnitude of the contrasts between areas of high and low rainfall; and, on the other hand, the variations of rainfall in time, and the extent to which these variations are random or may follow a recurring, predictable regime.

2.4 Geographical variations

There is no reason to suppose that precipitation does not fall on every part of the earth's surface but, although the average annual amount for the land areas of the world has been estimated to be about 700 mm [67], there is, in fact, a very great contrast between some of the driest desert areas, which may receive rainfall only once in twenty years, and places like Mount Waialeale in the Hawaiian Islands, which receives more than 12 000 mm annually [67] and Cherrapunji where more than 26 400 mm have been recorded in one year [85]. Examples such as these emphasize the fact that the amount of rain falling at a given location is almost entirely independent of the amount of evaporation taking place there [81], so that large-scale horizontal and vertical movements within the atmosphere must be invoked to explain the transfer of large masses of moist air from areas of high evaporation to areas of high precipitation.

World pattern

Consideration of the main mechanisms of mass transfer within the atmosphere leads initially to a simple concept of the zonal distribution of precipitation over the earth. Thus, in polar and subtropical latitudes, the general tendency is for the gradual subsidence of air from higher levels in the atmosphere leading to its warming and relative desiccation. The mid-latitude and inter-tropical zones, however, are areas of convergence, disturbance, and uplift and it is, therefore, in these two zones that precipitation is most likely to occur. Reference to maps showing the world distribution of precipitation indicates that although this simple explanation holds the elements of truth, considerable modification to the pattern results from a number of factors in addition to, as yet, unexplained and/or random variations in the global atmospheric circulation.

Since evaporation from the oceans forms the main source of precipitable moisture in the atmosphere, it follows that precipitation will tend to decrease with increasing distance from the sea, resulting in areas of extremely low rainfall near the centres of most of the major landmasses. A further modification, however, results from the disposition of land and sea in relation to the prevailing air movement, and means that coastal areas with predominantly onshore winds receive far more precipitation than coastal areas where the dominant wind direction is offshore. Although orographic influences are more noticeable on a smaller scale, nevertheless, on a larger scale, mountain ranges tend to accentuate precipitation amounts, particularly in areas where the prevailing air movement is onshore.

Regional patterns

When smaller areas, such as the United States or the British Isles, are considered in detail, the orographic influence is far more apparent, dominating the monthly and, even more markedly, the annual distributions [35] [95].

In terms of the latter, precipitation in the British Isles (see Fig. 2.1A) increases from a minimum of about 500 mm near the Thames estuary in the east, to more than 2500 mm in parts of North Wales, the Lake District, and the mountains of Scotland, reaching a maximum of 5000 mm on Snowdon and at the head of the River Garry in Invernesshire [41]. In the United States (see Fig. 2.1B) the influence of the western cordillera and of the Appalachian Mountains on the annual distribution of precipitation can be distinguished but because of the greater size of the country other major influences, e.g., latitudinal, are also apparent. Monthly distributions are inevitably less static and, while some for the British Isles show the orographic influence quite clearly (see Fig. 2.2A), others, especially during the summer months, seem to lack any correlation with relief, reflecting instead the important part played by convectional activity in the more 'continental' districts of the south and east (see Fig. 2.2B).

Storm patterns

When even shorter term totals are considered, the orographic influence often ceases to be a dominant factor; instead, the distribution of precipitation tends to be more closely related to the prevailing weather systems and pressure patterns. Thus, the precise path of a depression will influence the location of areas affected by frontal

19

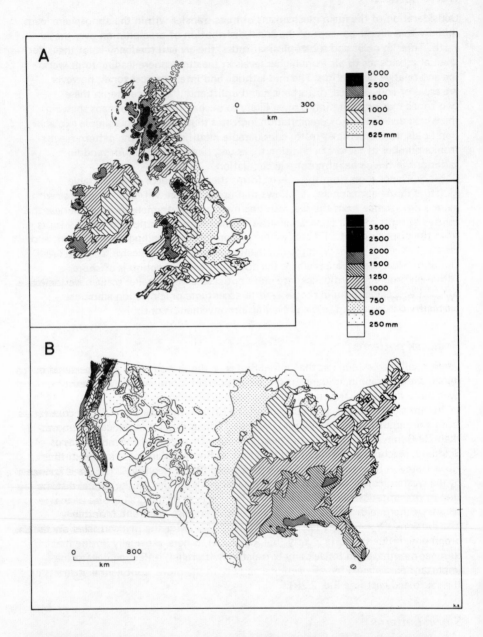

Fig. 2.1. *Mean annual precipitation over (A) the British Isles, 1901-30, and (B) the conterminous USA. (From (A) an original map in* Climatological Atlas of the British Isles, *HMSO, 1952. By permission of the Controller of Her Majesty's Stationery Office; and (B) an original diagram by J. J. Geraghty, D. W. Miller, F. van der Leeden and F. L. Troise,* Water Atlas of the United States, *Water Information Center, Inc., New York, 1973.)*

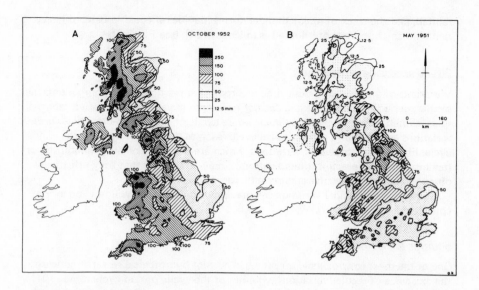

Fig. 2.2. *Monthly distributions of precipitation showing (A) a strong orographic influence and (B) negligible orographic influence. (From original maps in Meteorological Office,* Monthly Weather Report *for 1951 and 1952, HMSO. By permission of the Controller of Her Majesty's stationery Office.)*

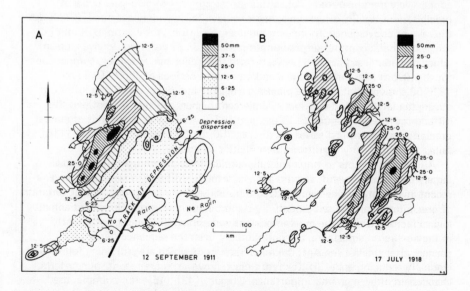

Fig. 2.3. *Rainfall distributions influenced by (A) the track of a depression and (B) thunderstorms. (From original diagrams by Salter [94].)*

rainfall, and the track of major thunderstorms may be reflected in belts of heavy rainfall many hundreds of kilometres in length [94] (see Fig. 2.3).

2.5 Variations with time

Variations of precipitation with time incorporate three time series components, i.e., secular, periodic, and stochastic. *Secular* or long-term variations are often referred to as 'climatic changes' and may incorporate cyclic or trend characteristics. *Periodic* variations are related to astronomical cycles such as the diurnal cycle or the annual cycle. Finally, the *stochastic* variations, which are the result of the probabilistic or random nature of precipitation occurrence, may be so great that they effectively obscure most of the secular and periodic variations.

Some of the principal variations of precipitation with time will now be briefly considered.

Secular variations

One of the most controversial questions in current hydrological research concerns the existence of secular variations, whether of the cyclic or non-cyclic type [101]. Many attempts have been made to determine the existence of *cyclic variations* of precipitation from studies of annual totals. Accurate information about such variations would be an important aid in forecasting the possibility of floods or the amount of water available for supply purposes, and in determining the advisability of establishing settlements in low-lying areas near rivers. The information is also needed in order to establish the length of record to be used in deriving accurate values for mean precipitation. Many cycles have been reported but few have been conclusively demonstrated [72], although the concept of a 35-year cycle of precipitation, the Bruckner cycle, has long been fairly widely accepted, if only for the sake of convenience. In the late nineteenth century, for example, Binnie [12] suggested that a continuous precipitation record of 35 years would give a mean which deviated less than two per cent from the true mean. However, comparisons for the British Isles between the standard rainfall periods of 1881-1915 and 1916-50 showed that, in some places in England and Wales, mean precipitation during the latter period was some 10 per cent higher, and that in Scotland the differences rose to as much as 14 per cent [17]. Analysis of thousands of annual precipitation and runoff series for the last 100 to 150 years by Yevjevich [116] failed to indicate any significant periodicities or trends.

Although attempts to find generally applicable cycles of precipitation have largely been unsuccessful, and are almost certainly destined to remain so, several recent investigations have emphasized that many of the *non-cyclic* secular variations of precipitation are caused directly by a combination of geographical and climatological factors. One of the most important widespread influences, particularly in the northern hemisphere, has undoubtedly been the sequence of climatic changes after the last Ice Age, the repercussions of which are still being felt on a gradually reducing scale. To the hydrologist, however, more recent shorter-term changes are of even greater importance. Gregory [43] [45], for example, used 40-year and 60-year periods to delimit areas in the British Isles which had a characteristic direction and degree of fluctuation of the annual rainfall. In a series of larger-scale studies, Kraus [64] [65] [66] examined secular rainfall variations in some temperate east coast and tropical areas, and concluded that these were largely the result of dynamic changes in the general circulation of the atmosphere at these localities. Similarly, Whitmore [108] investigated the phenomenon of 'persistence', the tendency for years of high rainfall or low rainfall to be grouped together, and suggested that, in South Africa, the boundary between areas where persistence was strong and other areas where it was weak seemed to coincide with the Great Escarpment, and with the associated discontinuity in the atmospheric circulation.

These and similar investigations help to emphasize the need for further research into problems of the secular variation of rainfall, but at the same time clearly indicate that local geographical, as well as general climatic factors, need to be considered.

Periodic variations

For time intervals of less than one year (i.e., seasonal, monthly, daily, hourly, etc)

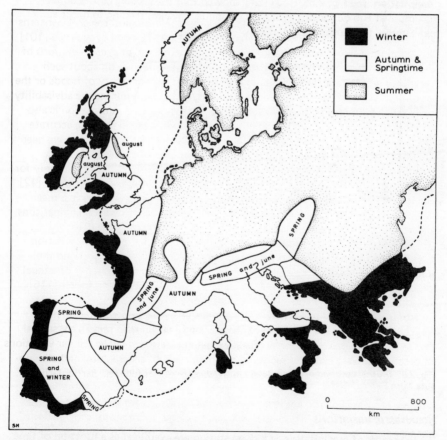

Fig. 2.4. *Season of highest precipitation over Europe. (From an original map by P. Thran in Agro-Climatic Atlas of Europe,* Elsevier, Amsterdam, 1965.)

continuous time series of precipitation or other hydrological phenomena, exhibit both periodic and stochastic components [101]. In some situations the periodic component may dominate whilst in others it may be obscured by the stochastic component. Examples of periodic components which are frequently dominant are seasonal and diurnal variations.

Seasonal variations in the precipitation regime are inevitably closely related to the general rhythm of climate in a given area, so that, for example, near the equator latitudinal movements of the inter-tropical convergence zone usually result in two maxima, whereas in tropical areas, there is often a distinct summer rainfall maximum. Western Europe provides an excellent example of a transitional area in which clear seasonal patterns are often difficult to discern (see Fig. 2.4).

Because convectional activity varies with temperature, and therefore also with insolation, some *diurnal variation* of precipitation occurs in certain areas. This is typified by the afternoon thunderstorms in parts of the equatorial zone, or by the frequent afternoon or early evening maximum of rainfall in continental interiors, or in other areas where thunderstorms contribute a significant proportion of the total precipitation (see Fig. 2.5). Hogg [54] showed that the evening convectional maximum at Rio de Janeiro exhibited an annual rhythm, tending to occur earlier in mid-summer than in spring or autumn.

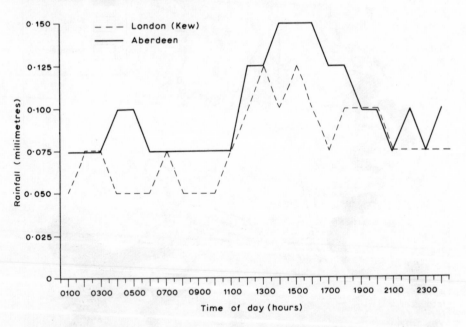

Fig. 2.5. *Diurnal variation of precipitation at London (Kew) and Aberdeen in July. (Based on data from Salter [94].)*

Stochastic variations

When records of precipitation at a given station are examined as a function of time they are found to be 'noisy', i.e., much of the variation results from a random component. This is clearly illustrated in Fig. 2.6A in which 90 years of June rainfall at Albury, N.S.W. are presented. The random variations in this record are so great that they largely obscure any long-term periodicity or trend which may exist. Similarly, when daily station values are examined in Fig. 2.6B, each storm event appears to occur randomly and also to vary randomly in terms of total depth and duration. When given storm events are examined in detail, however, as in Fig. 2.6C, it will be seen that although there is still a small random component there is now a considerable similarity in the time-distribution of rainfall intensity through each storm period.

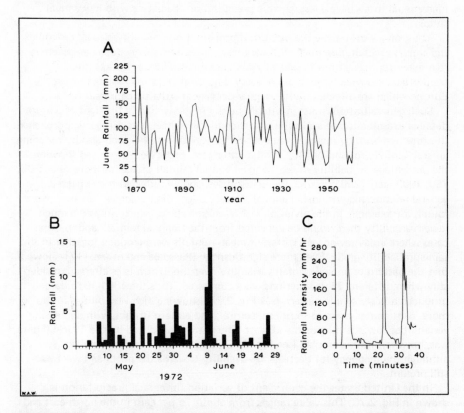

Fig. 2.6. *Variations of rainfall with time: (A) Variation of June rainfall at Albury, NSW, Australia, (B) Daily rainfall at Withernwick, Humberside County, England, and (C) Instantaneous rainfall intensity during a storm on 27 May, 1968 at Holmdel, New Jersey. (From: (A) an original diagram by P. S. Eagleson,* Dynamic Hydrology, McGraw-Hill, New York, 1970 *and (C) an original diagram by R. A. Semplak,* Rev. Sci. Instrum., **37**: 1554-1558, 1966.)

In any analysis of station precipitation, therefore, it is necessary to take into account both the stochastic or random component and the non-random or deterministic component. The timing and sizing of individual events is largely stochastic in nature while the internal structure of a given storm may be largely deterministic [36].

Variability of precipitation

In view of the complex nature of the precipitation record and particularly the combination of secular, periodic, stochastic, and internally deterministic components, it is normally extremely difficult to estimate the reliability or variability of precipitation, i.e., the likelihood of a given pattern, or regime, or amount, being repeated each year or month. Generally speaking, the reliability of precipitation increases first, with an increase in the time interval under consideration, so that annual totals and patterns are more stable than those for individual months or seasons; and second, with the amount of precipitation, so that areas with high annual totals have a less variable precipitation than areas with low annual totals.

Thus, on a world scale, the pattern of precipitation reliability closely resembles the annual precipitation map, with very low values in the large desert areas where rain may occur only once in several years, and high values in, for example, equatorial areas where rain may fall every day, or in areas such as north-west Europe which are affected by the regular passage of Atlantic depressions.

Such generalizations must inevitably be slightly amended in the light of a more detailed examination of a specific area such as the British Isles. Gregory approached this problem-both by mapping annual rainfall variability [44], and also by mapping annual rainfall probability [46]. In the latter case, maps were produced showing the percentage probability of receiving an annual rainfall of less, or more, than 500, 750, 1000, and 1250 mm over the British Isles, and all reflected the expected general relationship with mean annual rainfall. Slight discrepancies, however, were found, for example, in the peninsulas of western Ireland, which showed a much lower variability than would be expected from their annual rainfall, and in other cases where areas having very different mean rainfalls were grouped together in the same probability grade. Even more significantly, Bleasdale and others [17] showed that the pattern of annual rainfall variability over the British Isles altered considerably when different 35-year periods were compared. Thus the 1881-1915 data produced a fairly simple pattern (see Fig. 2.7A) showing high variabilities in the more 'continental' areas of southern, central, and eastern Britain, with decreasing variabilities towards the north and west. The pattern produced by the 1916-50 data was, however, more complex (see Fig. 2.7B), and more difficult to explain, although the tendency for decreasing variability towards the north and west was still present.

In the United States the coefficient of variation for annual precipitation is shown in Fig. 2.7C. This value ranges from about 15 per cent in the north-east and in southern Florida to more than 50 per cent in the arid south-west.

Such comparisons again emphasize how little is known about the trends and variations of precipitation and lend support to the increasing attention now being given to the stochastic nature of the precipitation record.

Rainfall intensity

The frequency with which rainfalls of high intensity occur is likely to have an important bearing on the susceptibility of a catchment area to flooding, or to sudden increases in stream flow and is, therefore, of considerable significance to the

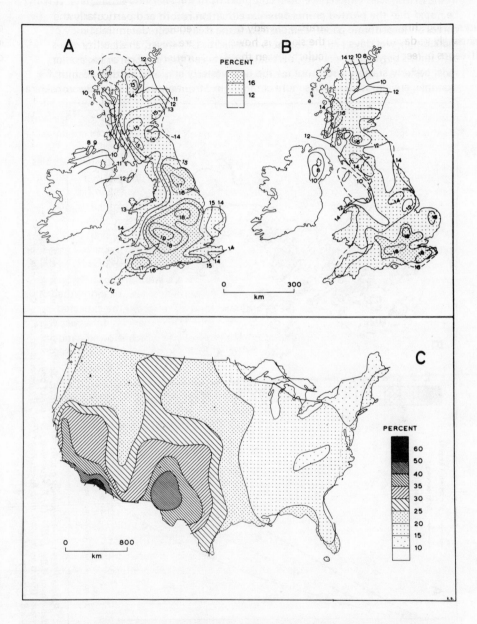

Fig. 2.7. *Percentage coefficient of variation of annual precipitation in (A) the United Kingdom for the period 1881-1915 and (B) 1916-50 and (C) in the conterminous USA for the period 1931-60. (From original diagrams by A. Bleasdale and others, in* Conservation of Water Resources, *pp. 121-136, Institution of Civil Engineers, London, 1963 and from an original map by Hershfield [51].)*

hydrologist. Figure 2.8A shows *rainfall intensity frequency graphs* for three stations in the British Isles which have been compiled from long-period rainfall data. It will be noted that the plotted points describe either straight lines or very shallow curves on logarithmic paper, demonstrating the steady decrease in frequency of increasingly wet days [79]. Despite the similarities, however, there are obvious differences between the curves, particularly between that for Scarborough, the most easterly station, and those for the two westerly stations. Falls of 25 mm, for example, are almost ten times more common in Strontian than in Scarborough.

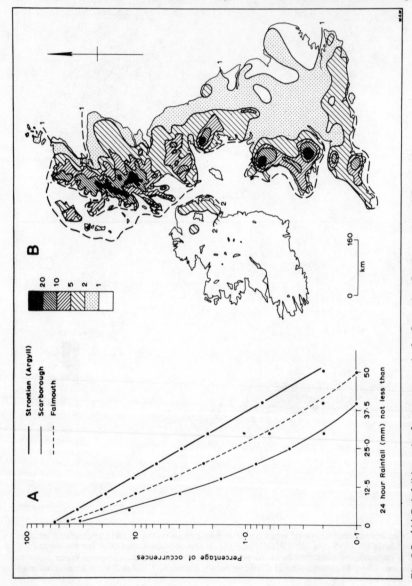

Fig. 2.8. *(A) Rainfall intensity-frequency graphs for Strontian, Scarborough, and Falmouth (Based on data from Miller [79]); (B) Frequencies per year of daily rainfalls of at least 25 mm in Great Britain. (From an original map by Holland [56]. By permission of the Controller of Her Majesty's Stationery Office.)*

This trend is confirmed when the frequency per year of daily rainfalls of at least 25 mm is plotted from the 1916-50 data for Great Britain. The resulting map (see Fig. 2.8B) shows that the heavy falls are most frequent in the highland areas of the west, and least numerous along parts of the east coasts of both England and Scotland.

In terms of rainfall intensity, daily rainfall totals may be very misleading and it is, therefore, relevant to consider briefly the frequency of falls of much shorter duration. Holland [55] presented a useful summary of some of the work on this difficult problem which had been done in Britain, and at the same time produced graphs based not only on data from the standard Meteorological Office rain

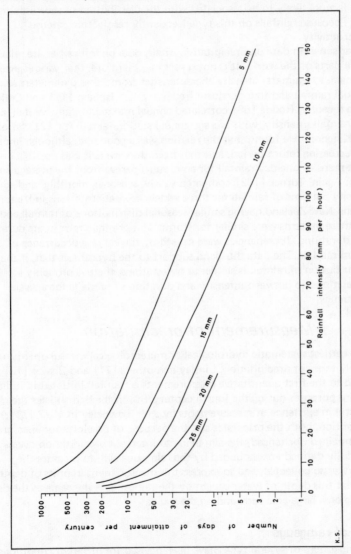

Fig. 2.9. Average intensity-frequency graphs for specific amounts of rainfall. (From an original diagram by Holland [55]. By permission of the Controller of Her Majesty's Stationery Office.)

recorders, but also from a number of open-scale recorders, allowing minute by minute totals to be read quite accurately. Average intensity frequency graphs for a number of stations in England are shown in Fig. 2.9. These indicate that the very high intensities occur only rarely, and also that they are less commonly experienced in connection with high totals of rainfall. Thus, for example, intensities of 60 mm/h were experienced on nine occasions per century, with total falls of 15 mm, on eighteen occasions with total falls of 10 mm, and on fifty-one occasions with falls of 5 mm. In contrast to the twenty-four hour rainfalls referred to earlier, a high frequency of the heavier short-period rainfalls represented in Fig. 2.9 is less easily correlated with areas of high annual rainfall, and with upland areas in particular, and indeed, is more likely to be associated with the distribution of severe thunderstorms [55], because rainfalls on this type frequently result from intense convectional activity.

In comparison with data on precipitation totals, data on intensities are relatively sparse for all parts of the world. It is not surprising, therefore, that various attempts have been made to estimate intensity characteristics from other parameters such as average annual rainfall and thunderstorm frequency (c.f. Bilham [11] and Collinge [29]). More recently, Rodda [89] correlated annual maximum daily rainfall, as a measure of rainfall intensity, with average annual station rainfall for 121 stations in the United Kingdom. He found that the relation was a good one, although improving with a decreasing return period, and that from the results it was possible to construct maps of one-day rainfall for any return period from the average annual rainfall map. Again, Turner [102] compared yearly, seasonal, monthly, and storm-intensity patterns of rainfall for three widely separated stations in Victoria, Tasmania, and New Zealand having similar seasonal distributions of rainfall and found that these patterns were similar for storms of more than four hours duration. For shorter durations, discrepancies were caused by the varying occurrence of thunderstorm rainfall. The data thus lend support to the hypothesis that, if the seasonal distribution of rainfall is similar at two stations, there is probably a consistent relationship between intensity and duration of rainfalls for a given average frequency.

Measurement of precipitation

Some of the earliest systematic hydrological or meteorological measurements made were probably those of precipitation. Linsley and others [71] and Biswas [13] [14] referred to the first quantitative measurements of rainfall initiated in India for tax assessment purposes during the fourth century BC. In the British Isles the records are still in existence of measurements by a Mr Towneley in 1677 [78]. Since its inception, both the principles and the purpose of precipitation measurement have remained unchanged, the aim being to intercept precipitation over a known, carefully defined area bounded by the raingauge rim, to measure the amount of water so collected, and to express this measurement in units of depth. It is assumed that this depth of water caught by the raingauge is the same as the depth of rain falling on a large area surrounding the gauge.

2.6 Types of raingauge

Non-recording gauges of several types have been devised for different conditions in

30

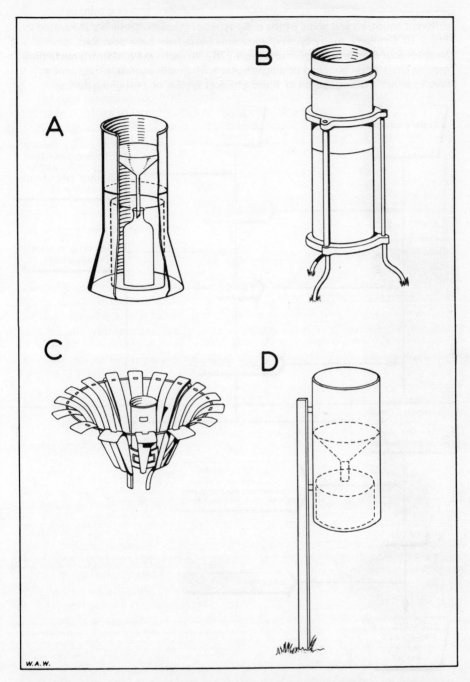

Fig. 2.10. *Four types of standard raingauge: (A) British Meteorological Office, (B) US Weather Bureau, (C) USSR Tretyakov gauge, and (D) German Hellman gauge. (From an original diagram by J. C. Rodda in R. J. Chorley (Ed.),* Water, Earth and Man, *Methuen, London, 1969.)*

different countries and some of the main types of standard gauge are illustrated in Fig. 2.10. Similarly, many types of recording gauge have been described, c.f. *Handbook of Meteorological Instruments* [78]. In each case a record is made on a moving chart or on punched or magnetic tape. The main operating mechanisms involve either a tilting siphon or tipping bucket system or a weighing device.

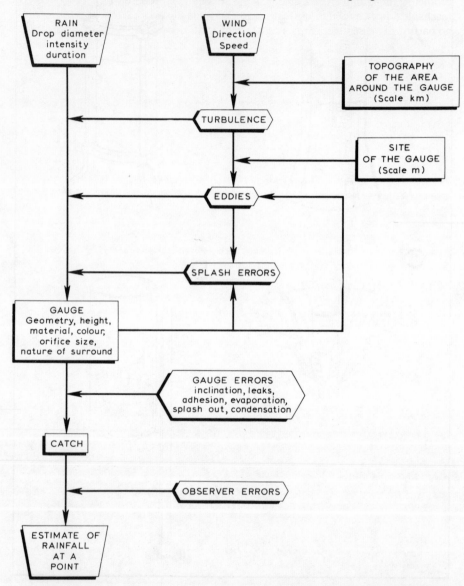

Fig. 2.11. *Diagrammatic representation of problems involved in determining rainfall with a standard gauge. (From an original diagram by J. C. Rodda in R. J. Chorley (Ed.),* Water, Earth and Man, *Methuen, London, 1969.)*

2.7 Accuracy of rainfall measurement

Despite the long history of rainfall measurement many problems and uncertainties about the collection and accuracy of rainfall data still remain and the more important of these are summarized diagrammatically in Fig. 2.11. Some of these problems, for example those involving the possibility of observer errors or mechanical failures within the instrument, are virtually unavoidable and will form no part of this discussion; others, concerning the siting and exposure of the raingauge, for example, are more relevant and considerable attention is normally paid to them during gauge installation. By far the most important sources of error, however, result from aerodynamic interactions between the falling precipitation, wind, and the gauge and its surroundings and this situation is further complicated by the fact that the standard instrumental procedures adopted by various countries differ quite considerably. In Britain, for example, the standard 127 mm gauge is exposed with its rim 305 mm above ground level; in the United States a gauge diameter of 203 mm or 305 mm is normally used; and in some European countries raingauges are installed so that the rim is either 1 or 1.5 metres above the ground, although in the Soviet Union the standard gauge is set at 2 metres and is surrounded by a shield. The W.M.O. Interim Reference Precipitation Gauge was introduced in an attempt to provide a basis for comparison [91] and expected differences between readings from this and the various national gauges range from 5 to 15 per cent [21]. Clearly then, what is recorded as 20 mm of precipitation on one side of a national frontier could be registered as something different on the other side, so that the existing global and continental precipitation maps are not as meaningful as they might be with uniform measurement conditions [91].

Turbulence and splash errors

A raingauge acts as an obstacle to the flow of the wind over the ground surface and, therefore, causes turbulence and eddying. Air flow is speeded up, as was demonstrated by Robinson and Rodda [87] (see Fig. 2.12) and by Green and Helliwell [42] and falling rain tends to be carried past, rather than fall into, the gauge. This effect will inevitably vary with the speed of the wind and with the height of the gauge but will result, at whatever intensity it occurs, in a deficiency of catch. If, on the other hand, the rim of the gauge is set too close to the ground surface, insplashing may occur, resulting in an excess catch. The precise location of the raingauge rim must, therefore, be a compromise, as is the height of 305 mm adopted by the Meteorological Office, aimed at minimizing both effects. Alternatively, the gauge may be set at a sufficient height to obviate all insplashing, in which case some device is needed to minimize the effects of increased turbulence, or turbulence effects may be completely avoided by locating the gauge rim at ground level and installing devices to prevent insplashing. Both lines of investigation have been followed up for many years, and Bleasdale [16] referred to some of the earlier contributions.

The hydrologist frequently needs to measure rainfall in highly turbulent conditions. The measurement of rainfall in upland areas, for example, normally involves siting gauges on exposed slopes where wind-speeds are high. Catch deficiencies may be avoided in these circumstances by the erection of a turf wall

33

round the gauge, having an inside diameter of 3 metres and a crest which is 150 mm wide and 300 mm high [78]. In studies of interception (see chapter 3), however, the need to measure rainfall above the vegetation canopy may necessitate the location of gauges at heights of up to 10 metres or more, and in these conditions turbulence may be reduced by fitting shields to the gauge itself.

Some of the most valuable and most rigorous work in recent years was reported by Robinson [86] and Robinson and Rodda [87] and showed that, in comparison with various types of Meteorological Office and U.S. Weather Bureau gauges, a simple funnel-shaped gauge gave substantially better results. This type of gauge was aerodynamically the most satisfactory in a series of wind tunnel trials and generated less lift and acceleration over itself than the standard gauges. This could prove an important finding since a simple funnel gauge would be much easier and cheaper to install and maintain than the ideal ground level gauge but would yield only slightly less satisfactory results.

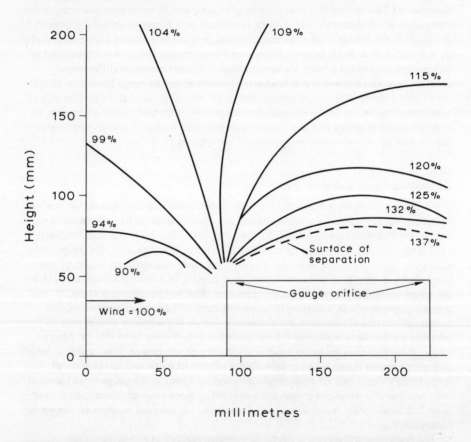

Fig. 2.12. *The structure of the windfield above a standard British raingauge. At a wind velocity of 3.5 m/s air speed over the gauge is increased by up to 37 per cent. (From an original diagram by A. C. Robinson and J. C. Rodda, Meteorological Magazine, 98, 1969. By permission of the Controller of Her Majesty's Stationery Office.)*

Ground-level gauges

Ideally, the correct level at which to measure the rain actually falling on the ground surface would be that of the ground surface itself. Although, as Rodda [92] observed, the use of ground level or pit gauges was advocated by Stevenson [96] 130 years ago, rainfall measurements with such gauges were, until recently, unsuccessful because of failures to obviate insplash despite experimentation with various types of anti-splash surface. Bleasdale [15] [16] proposed the experimental use of a nine-hole gauge by means of which a direct check could be kept on the amount of insplash (if any) from a horizontal surface of outward facing metal slats (see Fig. 2.13A). Since the central hole is farthest from the surrounding surface it should catch the smallest total of water if any splashing occurs, whereas the corner holes should catch the most. In a number of experiments similar amounts were caught in all nine holes indicating the effectiveness of this type of surround. Several contrasting shapes and sizes of metal and plastic grid have been compared, notably by Rodda in Britain [90] and by various workers in the Soviet Union [18] [98] [104], and it now seems that in Britain the design most likely to be generally adopted is that illustrated in Fig. 2.13B, i.e., a metal or plastic grid made up of open-ended cubes [91]. Extensive comparisons between this type of gauge and an adjacent standard Meteorological Office Mark II gauge showed that the ground level gauge caught 6.4 per cent more rain over a period of eight years and that the average difference between the gauges was greatest in winter and least in summer due to seasonal contrasts in wind speed and rain drop size [92]. Similar seasonal differences were found by McGuinness and Vaughan [73] between lysimeter and raingauge totals in northern Ohio.

Gauge errors

Further reference to Fig. 2.11 indicates that apart from the problems of turbulence and associated splash errors, errors in the gauge itself may be significant. Some of these, e.g., leaks, are purely mechanical, others depend upon the design of the gauge or upon its inclination.

Gauge networks

Errors in estimating areal rainfall from a given gauge network occur because of the random nature of storms and their passage between gauges [110] but conditions will vary depending on terrain and storm type. Thus, more gauges will be required in steeply sloping terrain [59] and for convectional precipitation, than in flat terrain or for cyclonic precipitation.

Generally speaking, of course, estimates of areal precipitation will increase in accuracy as the density of the gauging network increases but a dense network is difficult and expensive to maintain and would normally be used only for a short period in order to determine a smaller and more convenient network. Analysis by the U.S. Weather Bureau [106] of precipitation data for relatively flat terrain yielded the network density-area-error relationships shown in Fig. 2.14. This indicates that, for a given network density, the error increases as the size of the area is reduced.

The World Meteorological Organization [114] established guidelines for the minimum density of precipitation networks in various geographical regions as follows: small mountainous islands with irregular precipitation, 25 km^2 per gauge; temperate, mediterranean, and tropical mountainous regions, 100-250 km^2 per gauge; flat areas in temperate, mediterranean, and tropical regions, 600-900 km^2 per gauge; arid and polar regions, 1500-10 000 km^2 per gauge. In Great Britain there are about 6000 gauges [27], giving an average density of about 39 km^2 per gauge, but in the United States this figure rises to about 600 km^2 per gauge [72], and elsewhere in the world areas of 2500 km^2 per gauge are quite common.

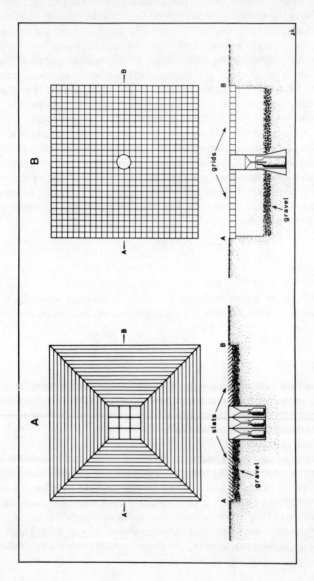

Fig. 2.13. Two types of ground-level raingauge: (A) shows plan and section of an experimental nine-hole gauge with slatted anti-splash surround and (B) shows a currently typical installation of the standard 127 mm gauge with open cuboid anti-splash surround. (A) from an original diagram by A. Bleasdale, Weather, 14: 12-18, 1959. Crown copyright by permission of the Controller of Her Majesty's Stationery Office.)

More recent work at varying scales has indicated that the density of the gauge network alone may not be all-important and that an improvement in accuracy may be effected by incorporating a selective spatial and directional component into the network. Thus Catterall [23] discussed raingauge 'domains' defined in terms of altitude, local relief, and exposure, McKay [74] referred to the index approach in network design whereby gauges are located in areas which contribute most to runoff, with one gauge in each homogeneous contributing area, and Hershfield [52], in an analysis of more than 300 storms, found that rainfall volume appeared to be estimated adequately with only a small number of 'properly located' gauges as compared with the number required for a good definition of the isohyetal pattern. Using similar correlation analysis, Caffey [22] and Hendrick and Comer [49] showed that meteorological factors such as storm size, storm direction, or direction of atmospheric moisture flux predominated over topographic features in

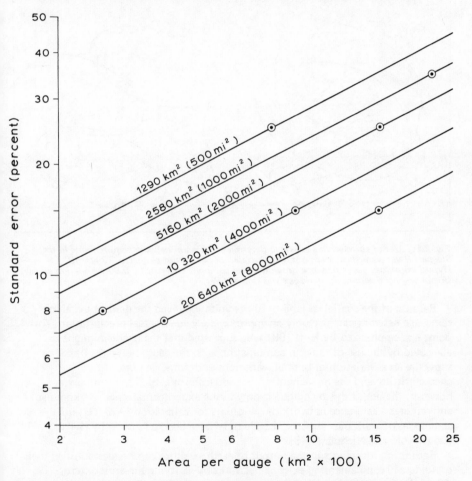

Fig. 2.14. *Raingauge network density and percentage standard error. (From an original diagram by US Weather Bureau,* Hydromet. Rept.,**5,** *1947. With acknowledgment to NOAA.)*

determining spatial association in precipitation variations and so concluded that the most effective gauging network would be one with a strong directional component. Figure 2.15A shows the elliptical shape of the correlation field about one of the 23 gauges in the Sleepers River watershed in Vermont, the direction of the long axis probably reflecting the direction of moisture flux into the watershed. Figure 2.15B illustrates that, if the accuracy requirements for precipitation measurement over the watershed can be expressed in terms of a certain minimum correlation between ungauged points and the network gauges (in this case $r = 0.90$ is assumed to provide an adequate solution), then a basic network design (in this case four gauges with minor supplementation) could be advocated. Stol [97] also used correlation analysis in the Netherlands to evaluate the relative efficiency of different network densities.

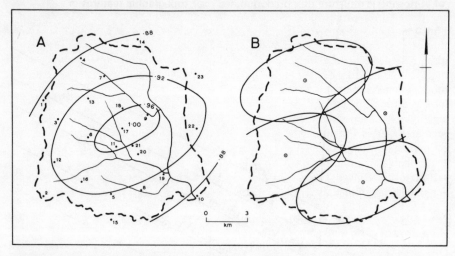

Fig. 2.15. *Use of correlation analysis to determine minimum network requirements in the Sleepers River watershed, Vermont. (A) Correlation field about gauge No. 17 and (B) Hypothetical four-gauge network providing near-complete coverage by 0.90 correlation ellipses. (From original diagrams by Hendrick and Comer [49].)*

Because of the spatial variability of precipitation, even the densest existing raingauge network can give only an approximate value of areal precipitation. This point was emphasized by Huff [58] who suggested that the problem might be alleviated by the use of radar in combination with the gauge network. Radar can show the areal variation of rainfall, variations with time, i.e., intensity characteristics, and the movement of individual storm cells [7]. Even now, however, the use of this technique is only in the experimental stage, is expensive, and requires a reference network of raingauges for calibration [110]. Useful summaries of work over the past decade or so have been published by Harrold [47] and Harrold and Nicholass [48].

Recent improvements in the output of earth satellite data have established their usefulness in supplementing existing networks by verifying the areal extent, direction of movement, and character of rain storms. A number of lines of approach have been followed in the hope of achieving a World Meteorological

Organization objective of estimating 12-hourly rainfall intensity from weather satellite data [115]. Thus use has been made of the fact that precipitating clouds may be distinguished from non-precipitating clouds through differences in emitted radiation, c.f. [70], or differences in reflection characteristics, c.f. [113]. Satellite evidence has also been used to detect previous rainfall through the relatively lower reflectivity of wetted terrain [57]. The most promising approach to date, however, would seem to be that of Barrett [4] [5] who estimated monthly and daily rainfall on the basis of statistical relationships between satellite and conventional weather data.

Problems in the analysis of precipitation data

Quite apart from problems concerned with the collection and accuracy of precipitation data, there are further difficulties associated with the use and interpretation of these data which are of direct concern to the hydrologist. For example, there is often only a short run of data available for a given station, and this information must be extrapolated by reference to long-term data from nearby stations. An even more basic requirement is to convert a number of point measurements of rainfall into an estimate of the average rainfall over an area, or to determine the pattern of a given storm from a number of point readings, from comparatively widely separated gauges. Finally, hydrologists have long been faced with the problem of estimating how large a rainfall is physically possible over a given area, i.e., what is the Probable Maximum Precipitation? Unless such problems as these are satisfactorily resolved, the value of the data from individual raingauges, however accurately determined, is very limited.

2.8 Determining average areal precipitation

One of the basic requirements in the hydrological study of, say, a catchment area, is an accurate value of the average precipitation over that area per year or per month or perhaps during the course of an individual storm. Several methods may be used to determine this.

Arithmetic mean

The simplest method is to calculate the arithmetic mean of all the raingauge totals in the area, a technique which may give adequate results if there is an even distribution of gauges, and if the area is fairly flat. In other situations, however, and particularly in mountainous areas, where raingauge sites may be rather unrepresentative, the results obtained may be substantially in error. For the hypothetical catchment illustrated in Fig. 2.16A this method yields an average precipitation of 77.5 mm.

Thiessen polygons

The use of Thiessen polygons [99] makes some allowance for the uneven distribution of gauges throughout the area, and also enables data from adjacent areas to be

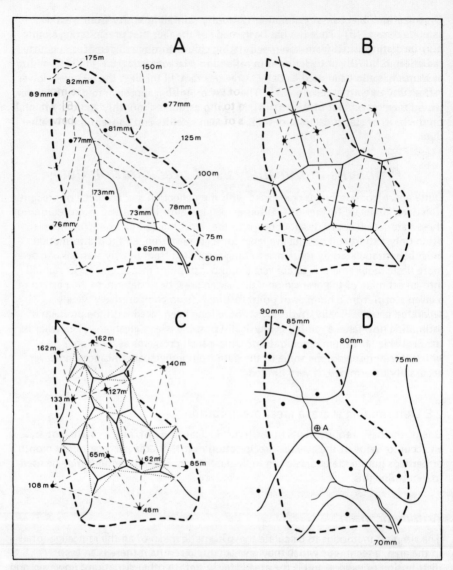

Fig. 2.16. *Methods of determining average precipitation over a catchment area showing (A) arithmetic mean of gauge totals, (B) Thiessen polygons, (C) height-balanced polygons, and (D) isohyets. (From original diagrams by Cole [28].)*

incorporated in the mean. In this method, which is illustrated in Fig. 2.16B, perpendicular bisectors are drawn through the straight lines joining adjacent gauges, leaving each gauge in the centre of a polygon which will vary in size according to the spacing of the gauges. The percentage of the total area of the catchment represented by each polygon or part of a polygon is calculated and applied to the appropriate raingauge total. These reduced totals are then added to give the average precipitation which, in the case of the catchment shown in Fig. 2.16B, is 76.2 mm.

40

Height-balanced polygons

Height-balanced polygons enable the effects of altitude, as well as raingauge distribution, to be taken into account. In this method, perpendiculars may be drawn from the straight lines joining adjacent raingauges at the mid-point in terms of altitude, rather than of distance [109]. Or, as is shown in Fig. 2.16C the mid-altitude points may be joined by straight lines (dotted) and the resulting triangles internally bisected (solid lines). In this way the solid lines delimit the bounding polygon around each gauge.

Each total is then weighted according to the relative size of its polygon, as in the Thiessen method, giving an average precipitation of 75.7 mm for the catchment shown in Fig. 2.16C.

Isohyetal method

If used with skill, the isohyetal method should provide the most accurate determination of average precipitation, since it enables a large number of factors, such as relief, aspect, and direction of storm movement to be taken into account, albeit rather subjectively. Sometimes the average precipitation in each inter-hyetal zone is taken to be the mean of the bounding isohyets. The value for each zone is then weighted according to area, totalled, and finally divided by the total area of the catchment, giving a value of 76.6 mm for the area shown in Fig. 2.16D. It will be seen from this diagram, however, that the mean rainfall for, say, the area between the 70 mm and the 75 mm isohyets, should really be closer to 75 than to 70 mm and indeed, a more accurate mean value for this inter-hyetal zone may be determined if, not only the area between the isohyets is measured, but also the lengths of the isohyets. The mean depth of rainfall between the isohyets is then given by the following simple equation:

$$r = B + \frac{i}{3} \ \frac{2a + b}{a + b} \quad (2.1)$$

where b is the length of the lower value isohyet (B), a is the length of the higher value isohyet (A), and i is the isohyetal interval $(A-B)$ [109]. When this method is applied to the catchment in Fig. 2.16D, it yields a value of 77.2 mm for the average precipitation.

Other Methods

Although the methods described briefly above are those most commonly used in hydrological analysis, other methods have been proposed from time to time. Thus, Whitmore and others [109] described a simple technique for calculating the triangular area-weighted mean, in which gauge locations on the map are connected by straight lines, thereby forming a number of triangular sub-areas. It is assumed that precipitation over the sub-area varies linearly between the three corner gauge points. In presenting this 'new procedure' more than a decade later, Akin [1] rightly suggested that it can be easily programmed for computer application. The main weaknesses of the technique, however, appear to be that, in general, the perimeter of the study area will not coincide with the perimeter of the triangular mesh, although as with the isohyetal and Thiessen methods, the use of additional gauges outside the area could obviate this problem, and secondly, because of the freedom of choice over the particular triangular mesh used no unique value for mean areal precipitation emerges [37].

Various workers, c.f. Unwin [103], Chidley and Keys [24], have discussed the use of trend surface analysis in areal rainfall estimation. Chidley and Keys [24] proposed a method whereby a polynomial surface is fitted to the rainfall data for a given period. This polynomial is integrated over the area of interest, the result being a set of weights for each gauge which, when multiplied by the observed rainfall at each gauge, give the total volume of rainfall on the area. As with the Thiessen method, once the weights have been determined for a given gauge network and area, the method is suitable for hand computation. The potential application of trend surface analysis in determining mean areal rainfall was also investigated by the Institute of Hydrology [60]. Rainfall amount was expressed as a low-order polynomial function of the point's coordinate values and computer programs were

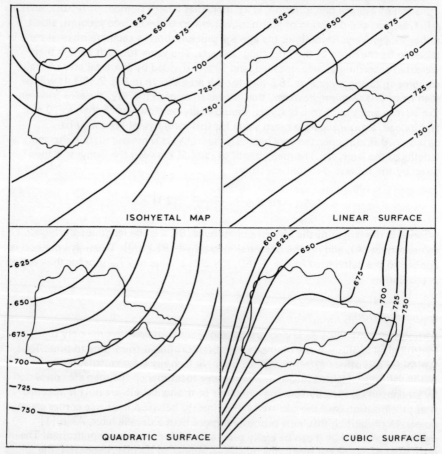

Fig. 2.17. *Distribution of annual rainfall amounts (mm) for 1967 by different methods for the River Ray catchment. (From an original diagram by Institute of Hydrology [60].)*

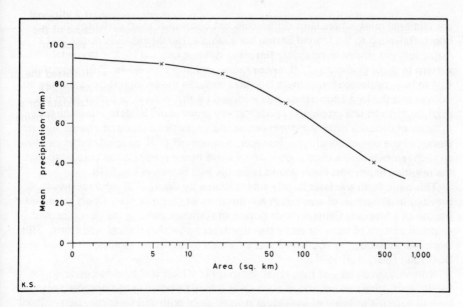

Fig. 2.18. *Depth-area curve characteristic of convectional rainstorms. (From an original diagram by D. A. Kraijenhoff,* Trans. Amer. Geophys. Union, 39: 124-127, 1958.)

developed to obtain the mean areal rainfall by integration and to give a computer plot of the fitted surfaces, examples of which are shown in Fig. 2.17.

In some circumstances still other methods may be appropriate. Where, for example, few or even no rainfall records exist, a tentative rainfall map may possibly be based upon the distribution of vegetation [109]. At the other extreme, where a wide range of data is available, it may be possible to establish multiple regressions of rainfall on dependent variates such as altitude, aspect, exposure, and continentality, thus enabling the prediction of rainfall at any desired number of additional points over a catchment area. Rodda [88] described and reviewed a number of such regression techniques.

2.9 Determining rainfall patterns

Closely related, in some respects, to the problem of determining the average precipitation over an area is the further problem of determining the pattern of storm rainfall from the individual totals recorded at a number of perhaps widely spaced raingauges. The degree to which rainfall decreases, from one or more peaks at the centre of a storm to zero rainfall at the outer margins of the storm, will obviously have a considerable bearing on its hydrological effects on a catchment area in terms of runoff, soil moisture, and groundwater changes.

Notwithstanding the caution of Collinge and Jamieson [30] that cyclonic rainfall should not be regarded as a uniform sheet of rain preceding a frontal system but rather as a series of overlapping rainfall cells which build up and die away with no apparent pattern, one can still make a general distinction between cyclonic rains where there is often little variation of daily totals over a radius of 15 km [27], and convectional rains, where large differences can occur over short distances in a few hours. Referring to the United States, for example, Hershfield [53] noted that in major summer storms in relatively flat areas it is not unusual for the isohyetal pattern to show gradients of 30 mm or more per kilometre. Boyer [19] suggested that in large cyclonic storms there is a ratio between the precipitation rate along an isohyet and the logarithm of the area enclosed by this isohyet, and that from the resulting straight-line graph, the rainfall at any point could be determined directly in terms of distance from the storm centre. As a result of a study of intense summer storms of the convectional type, however, Kraijenhoff [63] concluded that nearly all such rainfalls have a central core where extra heavy precipitation occurs, so that the resulting depth-area curve would resemble that shown in Fig. 2.18.

This basic form was later largely substantiated by Court [33] who reviewed previous investigation of area-depth relationships of storm rainfall. Court proposed the use of a bivariate Gaussian distribution of statistics yielding the now accepted elliptical pattern of isohyets and a bell-shaped cross-section through the storm. This work was subsequently confirmed and extended in numerous studies, c.f. Fogel and Duckstein [38], Huff [58].

Normally, in an area of high relief, orographic effects will tend to outweigh the variations outlined above, and in such cases it may be possible to transpose seasonal rainfall patterns to those of individual storms, since both will be largely determined by the topography [112]. This technique is most effectively used in conjunction with an isopercental map which shows the relationship between the normal seasonal pattern and that for the individual storm, and enables a fairly detailed isohyetal map to be developed from a comparatively small number of raingauges [112].

2.10 Determining the probable maximum precipitation (PMP)

Although it is easy to recognize that there must be a physical upper limit to the amount of precipitation that can fall on a specified area in a specified time, it has proved very difficult to estimate its value accurately. And yet an accurate estimate is both desirable from the academic point of view and virtually essential for most engineering design purposes. The upper limit to precipitation has become known as the Probable Maximum Precipitation (PMP) and has been concisely defined as 'that depth of precipitation which, for a given area and duration, can be reached, but not exceeded under known meteorological conditions' [110]. Miller [80] drew attention to the significant difference between the standard American Meteorological Society definition: 'The theoretically greatest depth of precipitation for a given duration that is physically possible over a particular drainage area at a certain time of year' [2] and a second definition, more operational in nature and emphasizing the application of the PMP: 'PMP is that magnitude of rainfall over a particular basin which will yield the flood flow of which there is virtually no risk of being exceeded'. In each case, however, 'probable' is intended to indicate that, with inadequate understanding of the physical processes and with imperfect precipitation and other meteorological data, it is impossible to define an absolute maximum precipitation; it is not intended to indicate a statistical probability or return period.

There are various methods of estimating PMP and Wiesner [110] discussed the more important of these in detail and reviewed the basic literature. In brief, there are two main approaches involving first the maximization and transposition of real or model storms and second the statistical analysis of extreme rainfalls. The maximization technique involves the estimation of the maximum limit on the humidity concentration in the air that flows into the space above a basin, the maximum limit to the rate at which wind may carry the humid air into the basin, and the maximum limit on the fraction of the inflowing water vapour that can be precipitated. PMP estimates in areas of limited orographic control are normally prepared by the maximization and transposition of real, observed storms whilst in areas in which there are strong orographic controls on the amount and distribution of precipitation, storm models are frequently used for the maximization procedure [80]. Sufficient work has been done on the large amount of data available in the United States to enable the production of generalized Weather Bureau maps showing PMP values for various basin sizes and storm durations.

The maximization/transposition techniques require a large amount of data, particularly volumetric rainfall data. In the absence of suitable data it may be necessary to transpose storms over very large distances despite the considerable uncertainties involved. In this case reference to published values of maximum observed point rainfalls will normally be helpful. Paulhus [85] published worldwide maximum values, while values for the United States were tabulated by Gilman [40] and for the British Isles lists of heavy falls in short periods are recorded annually in *British Rainfall.* Rodda [93] compared British and world point rainfalls and found that, in comparison with world maxima, the British falls are rather small (see Fig. 2.19). An interesting development of this approach was that by Hershfield [50] who proposed that the 24-hour PMP at a given station be estimated using a general statistical formula for the analysis of extreme value data by Chow [25]. Because of the non-universality of his empirical formula, Hershfield's method will rarely

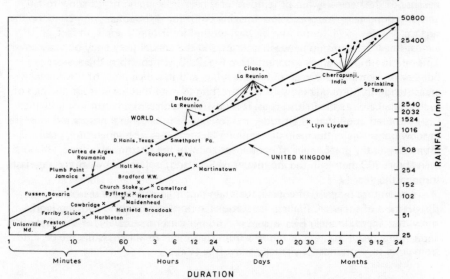

Fig. 2.19. *Magnitude-duration relationships for the world and United Kingdom extreme rainfalls. (From an original diagram by J. C. Rodda, Rainfall excesses in the United Kingdom,* Inst. Brit. Geog. Trans., **49**: *49-60, March, 1970.)*

45

yield a precise answer although it is a concise and convenient way of getting an approximate answer when the initial ignorance of the quantity to be estimated is great [82].

There is, of course, no absolute scale against which PMP estimates can be judged. If a value is too low it will probably be exceeded fairly quickly but if it is too high this may not become apparent until our knowledge has improved sufficiently to allow for more refined and accurate methods of estimating PMP [80].

Hydrological aspects of precipitation in the form of snow

Except in polar and sub-polar areas and at very high altitudes, snowfall comprises a comparatively small proportion of the annual precipitation. For example in the United States, only 100 out of the approximately 760 mm of mean annual precipitation occur in this form [111], and maps presented by Manley [75] [76] showed that over much of lowland England, snow rarely falls on more than 20 days in the year. The hydrological importance of snowfall is, however, out of all proportion to its magnitude [111], and derives essentially from the time lag between the occurrence of precipitation in the form of snow and the active participation of the snowfall, after melting, in the events of the hydrological cycle. For this reason, the hydrologist is normally less interested in *when* snow falls than in *where* it falls, in *how much* has fallen by the time that melting takes place, and in how rapidly *melting* occurs.

2.11 Distribution of snow

Inevitably, the areal distribution of snow tends to follow the pattern of relief because of the necessity for air temperatures near freezing point for snow to fall, and of ground temperatures below freezing point for it to remain, unmelted, on the surface. Manley [77] found that on high ground in Britain there is, in fact, an almost linear relationship between altitude and the annual frequency of snow cover. This relationship becomes apparent from Fig. 2.20, which shows the average number of mornings per year with snow lying, and in which most of the highland areas stand out quite markedly. It is interesting to note that, even from the map of such a small area as the British Isles, it is possible to discern a latitudinal influence superimposed upon that of altitude, and reflected in the general northward increase in snow days. Again, the main determining factor is that of temperature, resulting ultimately in the great zones of permanent snow in the circumpolar areas, although even above 900 metres in the mountains of Scotland, isolated snowdrifts rarely last through the year [77].

Temperature is again, of course, the main factor governing the seasonal distribution of snowfall. Figures for stations in northern England and Scotland show that approximately 81 per cent and 56 per cent respectively, of the days with snow lying, occur in the four months from December to March, while the period from June to September is virtually snow-free [77].

2.12 Amount of snowfall

The difficulties of measuring the amount of snowfall are even greater than those of

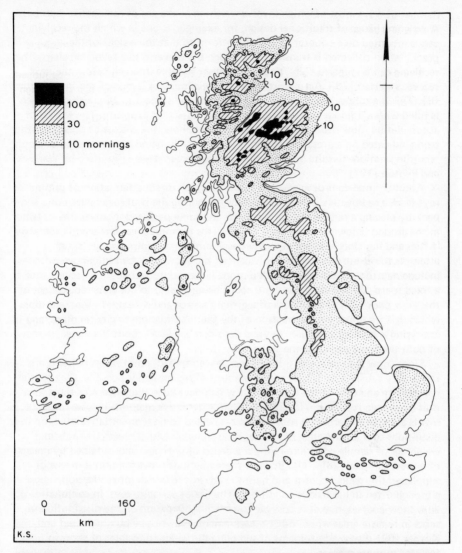

Fig. 2.20. *Average number of mornings per year with snow lying, 1912-38. (From an original map in* Climatological Atlas of the British Isles, *HMSO, 1952. By permission of the Controller of Her Majesty's Stationery Office.)*

measuring rainfall which have already been discussed. Except for slight falls, when all that is necessary is to melt the snow caught in the raingauge funnel and then to measure the resulting water in the normal way [78], the standard types of recording or non-recording gauges are of little use. The melting of the snow in the raingauge funnel, just described, emphasizes that it is not the depth but the *water equivalent* of the snow which is important, i.e., the equivalent water depth of the melted snow.

Water equivalent may be measured directly, without the need for melting, in a number of ways most of which involve weighing or the use of radioactive isotopes. A weighing gauge of traditional design, for example, is one in which the receiving can is mounted on a counter-balanced platform so that the weight of the precipitation collected is transmitted through the pen arm to a revolving chart. The receiving can is emptied daily, or at some other suitable interval, before the next record is started [78]. An important recent development is that of the snow pillow [8]. This is a rubber pillow, 3.65 m in diameter and some 5 to 10 cm thick, which is filled with a liquid which will not freeze during winter exposure. The weight of accumulating snow depresses the surface of the pillow, the change of pressure either being indicated on a pressure gauge or more normally being transmitted to a remote recording station. Results of snow pillow measurements were discussed by Warnick and Penton [107].

Another important development is one in which direct attenuation of gamma rays from a radioactive isotope is used for measuring the water equivalent of a snow pack by placing a radioactive source above the snow pack and a gamma ray detector in the ground below the snow pack [107]. This technique was initiated in the early 'fifties and has since undergone many developments and improvements. Other attempts to measure the water equivalent of the snow pack by radioactive isotopes include neutron scattering in which a probe is moved up and down an access tube at a fixed speed, the accumulated time count being related to the water equivalent of the snow pack [32], and the use of a gamma source and detector in separate tubes which utilizes the direct attenuation of the gamma emissions rather than neutron scattering and which promises to yield considerable detail about the water balance of both the snow pack and the soil profile [62].

Since snow is so prone to drifting and uneven spread, point measurements are rarely very satisfactory unless a large amount of replication is possible. Clearly, the complexity and cost of the instruments which have been discussed above are such that large-scale replication is rarely feasible. Alternative methods involve measurements along predetermined snow courses, selected to be representative of conditions over a wide area, of the depth of snowfall with a dipstick or rule and weighings of sample cores extending the depth of the snow and obtained by means of a snow sampler [26] [39]. Both of these methods provide a useful means of supplementing gauge records and have the advantage of ease of replication, although often in hazardous conditions. The use of aerial surveys, in conjunction with snow courses, makes it possible to obtain information on the distribution of snow in remote areas where other measurements may not be possible. Leaf and Kovner [69] discussed problems of sampling the water equivalent of snow in forested mountain areas.

Where only depth measurements are made, water equivalent must be estimated. This is frequently done by assuming an arbitrary ratio of 12:1 between snow depth and equivalent water depth [78]. Such an assumption is reasonable only if the density of the freshly fallen snow is at all times uniform. In fact, of course, some snow packs more closely than other snow and, therefore, has a higher water equivalent per unit of depth.

2.13 Snowmelt

Apart from its occasional ability to absorb considerable amounts of rainfall and therefore to act as a highly porous soil, snow really becomes hydrologically significant only upon melting. If melting occurs slowly and gradually, the greatest proportion of the meltwater may be absorbed by the underlying soil. If, on the other hand, melting occurs rapidly, most of the meltwater may run more or less directly into stream channels, resulting in rapid increases in stream flow. Accordingly, the melting process is of some importance.

The speed with which a given volume of snow will melt depends ultimately on the radiation balance of the snow cover, i.e., on the excess of heat absorbed over heat lost. The components of this radiation balance are numerous and their inter-relations complex. Heat is gained by direct radiation from the sun and sky and sensible heat is gained, particularly in turbulent conditions, from the overlying air. Heat is also gained in smaller quantities by conduction from the underlying ground surface and from falling rain, and by condensation on the snow surface. Because of the complexity of the snow pack energy balance, which is discussed in detail in the standard work *Snow Hydrology* [105], the relative importance of the various components of the balance is not only difficult to determine with accuracy but also varies with time both seasonally and diurnally and between open and forested sites. Dewalle and Meiman [34] suggested that late lying snow in the forest may represent optimum conditions for producing large energy gains to the snow, from net radiation and sensible heat exchange, because of the rapidly increasing radiation income, the reduced albedo of the snow pack, the warming up of the forest canopy and the presence of snow-free patches of soil. In a Colorado forest site in June they found that net radiation was the major source of melt energy (approximately 50-60 per cent) and sensible heat also accounted for large energy gains (approximately 40-50 per cent). Earlier work by Boyer [20] in an Oregon forest site late in the snow season showed that net radiation, sensible heat exchange, and latent heat exchange supplied 65, 20, and 15 per cent of melt energy respectively.

For simplification and convenience in attempts to predict the magnitude and timing of snowmelt, correlations are frequently made between aspects of air temperature and snowmelt. Some justification for this procedure may be derived from the fact that the transfer of sensible heat from the air is one of the more important factors involved in the melting process and also, from the fact that air temperature largely reflects the integration of the radiation balance of the overlying air [31] [68] [79] [84] [105] [112].

References

1. AKIN, J. E., Calculation of mean areal depth of precipitation. *J. Hydrol.*, **12**: 363-376, 1971.
2. AMERICAN METEOROLOGICAL SOCIETY, *Glossary of Meteorology*, edited by R. E. Huschke, Boston, Mass., 638 pp., 1959.
3. BALCHIN, W. G. V. and N. PYE, Local rainfall variations in Bath and the surrounding district *QJRMS*, **74**: 361-378, 1948.
4. BARRETT, E. C., The estimation of monthly rainfall from satellite data, *Mon. Wea. Rev.*, **98**: 198-205, 1970.
5. BARRETT, E. C., Forecasting daily rainfall from satellite data, *Mon. Wea. Rev.*, **101**: 215-222, 1973.
6. BARRY, R. G., Precipitation, in *Water, Earth, and Man*, Ed. R. J. Chorley, pp. 113-129, Methuen, London, 1969.

7. BATTAN, L. J., *Radar Meteorology,* Univ. of Chicago Press, Chicago, 1964. Published in revised form as *Radar Observation of the Atmosphere,* University of Chicago Press, Chicago, 324 pp., 1973.
8. BEAUMONT, R. T., Mt. Hood pressure pillow snow gage, *Proc. 33rd Annual Meeting Western Snow Conf.,* pp. 29-35, Colorado State Univ., Fort Collins, Colo., 1965.
9. BERGERON, T., Problems and methods of rainfall investigation in *Physics of Precipitation,* Geophysical Monograph 5: 5-30, Washington, 1960.
10. BERNARD, M. M., Precipitation, Chapter II in Meinzer, O. E. (Ed), *Hydrology,* McGraw-Hill, New York, 1942.
11. BILHAM, E. G., Classification of heavy falls in short periods, in *British Rainfall, 1935,* pp. 262-280, H.M.S.O., London, 1936.
12. BINNIE, A. R., On mean or average rainfall and the fluctuations to which it is subject, *Proc. ICE,* **109:** 89-172, 1892.
13. BISWAS, A. K., A short history of hydrology, *The Progress of Hydrology,* pp. 914-936, Univ. of Illinois, Urbana, 1969.
14. BISWAS, A. K., *History of Hydrology.,* North-Holland, Amsterdam, 336 pp., 1970.
15. BLEASDALE, A., A compound raingauge for assessing some possible errors in point rainfall measurements, *Meteorological Office Hydrol Memoranda* 3, 1958.
16. BLEASDALE, A., The measurement of rainfall, *Weather,* **14:** 12-18, 1959.
17. BLEASDALE, A., A. G. BOULTON, J. INESON, and F. LAW, Study and assessment of water resources, *Conservation of Water Resources,* Inst. Civ. Eng. Symposium, pp. 121-136, 1963.
18. BOGDANOVA, E. G., Relationship of readings of the Tret'yakov precipitation gauge to wind speed, *Leningrad Glav. Geofiz. Obs. Trudy,* **175** : 87-97, 1965.
19. BOYER, M. C., A correlation of the characteristics of great storms, *Trans. AGU,* **38:** 233-238, 1957.
20. BOYER, P. B., Heat exchange and melt of late season snow patches in heavy forest, *Proc. Western Snow Conf.,* **22:** 54-68, 1954.
21. BULL, G. A., Comparison of raingauges, *Nature,* **185:** 437-438, 1960.
22. CAFFEY, J. E., Inter-station correlations in annual precipitation and in annual effective precipitation, *Colorado State Univ., Hydrol. Papers* 6, Fort Collins, Colorado, 1965.
23. CATTERALL, J. W., An *a priori* model to suggest raingauge domains, *Area,* **4:** 158-163, 1972.
24. CHIDLEY, T. R. E. and K. M. KEYS, A rapid method of computing areal rainfall, *J. Hydrol.,* **12:** 15-24, 1970.
25. CHOW, V. T., A general formula for hydrologic frequency analysis, *Trans. AGU,* **32:** 231-237, 1951.
26. CHURCH, J. E., Snow and Snow Surveying : Ice, Chapter IV in Meinzer, O. E. (Ed.) *Hydrology,* McGraw-Hill, New York, 1942.
27. CLEMENTS, R. H., Rainfall: measurement, networks, data processing, Informal discussion of the Hydrological Group, *Proc. ICE,* **29:** 258-260, 1964.
28. COLE, J. A., Introduction to a colloquium on sampling density requirements for the measurement of rainfall over an area, *Water Research Assoc. Special Rept.,* **2:** 51-56, 1962.
29. COLLINGE, V. K. The frequency of heavy rainfalls in the British Isles, *Civ. Eng. and Pub. Wks. Rev.,* **56:** 341-344, 497-500, 1961.
30. COLLINGE, V. K. and D. G. JAMIESON, The spatial distribution of storm rainfall, *J. Hydrol.,* **6:** 45-57, 1968.
31. COLLINS, E. H., Relationship of degree-days above freezing to runoff, *Trans. AGU* **15:** 624-629, 1934.
32. COOPER, C. F., Sampling characteristics of neutron probe measurements in a mountain snow pack, *J. Glaciol.,* **6:** 289-298, 1966.
33. COURT, A., Area-depth rainfall formulas, *JGR,* **66:** 1823-1831, 1961.
34. DEWALLE, D. R. and J. R. MEIMAN, Energy exchange and late season snowmelt in a small opening in Colorado subalpine forest, *WRR,* **7:** 184-188, 1971.
35. DOUGLAS, C. K. M. and J. GLASSPOOLE, Meteorological conditions in heavy orographic rainfall in the British Isles, *QJRMS,* **73,** 11-42, 1947.
36. EAGLESON, P. S., *Dynamic Hydrology,* McGraw-Hill, New York, 462 pp., 1970.

37. EDWARDS, K. A., A note on the 'Calculation of mean areal depth of precipitation' by J. E. Akin, *J. Hydrol.*, **15**: 171-173, 1972.
38. FOGEL, M. M. and L. DUCKSTEIN, Point rainfall frequencies in convective storms, *WRR*, **5**: 1229-1237, 1969.
39. GARSTKA, W. U., Snow and Snow Survey, in *Handbook of Applied Hydrology*, ed. V. T. Chow, McGraw-Hill, New York, pp. 10-1-10-57, 1964.
40. GILMAN, C. S., Rainfall, in *Handbook of Applied Hydrology*, (Ed.) V. T. Chow, McGraw-Hill, New York, pp. 9-1—9-68, 1964.
41. GLASSPOOLE, J., Rainfall in relation to water supply, *QJRMS*, **81**: 268-273, 1955.
42. GREEN, M. J., and P. R. HELLIWELL, The effect of wind on the rainfall catch, *Sympos. on Distribution of Precipitation in mountainous areas, Geilo, Norway, 1972*, Vol. **2**: 27-46, WMO Publication 326, WMO, Geneva, 1972.
43. GREGORY, S., Annual rainfall areas of southern England, *QJRMS*, **80**: 610-618, 1954.
44. GREGORY, S., Some aspects of the variability of annual rainfall over the British Isles for the standard period 1901-30, *QJRMS*, **81**: 257-262, 1955.
45. GREGORY, S., Regional variations in the trend of annual rainfall over the British Isles, *Geog. J.* **122**: 346-353, 1956.
46. GREGORY, S., Annual rainfall probability maps of the British Isles, *QJRMS*, **83**: 543-549, 1957.
47. HARROLD, T. W., Estimation of rainfall using radar, *Met. Office Sci. Paper* 21, 1965.
48. HARROLD, T. W. and C. A. NICHOLASS, The accuracy of some recent radar estimates of surface precipitation, *Met. Mag.*, **101**: 193-205, 1972.
49. HENDRICK, R. L. and G. H. COMER, Space variations of precipitation and implications for raingauge network design, *J. Hydrol.*, **10**: 151-163, 1970.
50. HERSHFIELD, D. M., Estimating the probable maximum precipitation, *J. Hydraulics Div. Proc. ASCE.*, **87** (HY 5): 99-116, 1961.
51. HERSHFIELD, D. M., A note on the variability of annual precipitation, *J. Appl. Met.*, **1**: 575-578, 1962.
52. HERSHFIELD, D. M., Rainfall input for hydrologic models, in *IASH Publn. 'Geochemistry, Precipitation, Evaporation, Soilmoisture, Hydrometry'*, Publn. 78: 177-188, 1967a.
53. HERSHFIELD, D. M., Some meteorological requirements in watershed engineering research, *Proc. 3rd Annual Conf. Amer. Water Resources*, pp. 485-492, 1967b.
54. HOGG, W. H., The diurnal variation of rainfall at Rio de Janeiro, *QJRMS*, **73**: 467-470, 1947.
55. HOLLAND, D. J., Rain intensity frequency relationships in Britain, *Met. Office Hydrol. Memoranda* 33, 28 pp. 1964.
56. HOLLAND, D. J., Frequency maps of daily rainfall, *Met. Office Hydrol. Memoranda* 25, 1964.
57. HOPE, J. R., Path of heavy rainfall photographed from space, *Bull. Amer. Met. Soc.*, **47**: 371-373, 1966.
58. HUFF, F. A., Spatial distribution of rainfall rates, *WRR*, **6**: 254-260, 1970.
59. HUTCHINSON, P., A contribution to the problem of spacing raingauges in rugged terrain. *J. Hydrol.*, **12**: 1-14, 1970.
60. INSTITUTE OF HYDROLOGY, *Research 1970-71*, NERC Institute of Hydrology, 54 pp., 1971.
61. JAMES, J. W., The effect of wind on precipitation catch over a small hill, *JGR*, **69**: 2521-2524 1964.
62. JOHNSON, M. L., Research on Sleepers River at Danville, Vermont, *Paper 7184, IRI, J. Irrig. and Drainage Div., Proc. ASCE*, pp. 67-88, 1970.
63. KRAIJENHOFF VAN DE LEUR, D. A., Contribution in Discussion of 'A correlation of the characteristics of great storms' by M. C. Boyer, *Trans. AGU*, **39**: 124-127, 1958.
64. KRAUS, E. B., Secular changes in the rainfall regime of SE. Australia, *QJRMS*, **80**: 591-601, 1954.
65. KRAUS, E. B., Secular changes of east-coast rainfall regimes, *QJRMS*, **81**: 430-439, 1955.
66. KRAUS, E. B., Secular changes of tropical rainfall regimes, *QJRMS*, **81**: 198-210, 1955.
67. KUENEN, P. H., *Realms of Water*, Cleaver-Hume Press, London, 1955.

68. LEACH, H. R., H. L. COOK, and R. E. HORTON, Storm flow prediction, *Trans. AGU.,* **14**: 435-446, 1933.
69. LEAF, C. F., and J. L. KOVNER, Sampling requirements for areal water equivalent estimates in forested subalpine watersheds, *WRR,* **8**: 713-716, 1972.
70. LETHBRIDGE, M., Precipitation probability and satellite rainfall data, *Mon. Wea. Rev.,* **95**: 487-490, 1967.
71. LINSLEY, R. K., M. A. KOHLER, and J. L. PAULHUS, *Applied Hydrology,* McGraw-Hill, New York, 1949.
72. LINSLEY, R. K., M. A. KOHLER, and J. L. PAULHUS, *Hydrology for Engineers,* McGraw-Hill, New York, 1958.
73. McGUINNESS, J. L., and G. W. VAUGHAN, Seasonal variation in rain gauge catch, *WRR,* **5**: 1142-1146, 1969.
74. McKAY, G. A., Meteorological measurements for watershed research, *Research Watersheds: Proc. Hydrol. Symposium No. 4, Univ. of Guelph, Ontario,* pp. 185-209, 1965.
75. MANLEY, G., Snowfall in the British Isles, *Met. Mag.,* **75**: 41-48, 1940.
76. MANLEY, G., Snow-cover in the British Isles, *Met. Mag.* **76**: 28-36, 1947.
77. MANLEY, G., *Climate and the British Scene,* Collins, London, 1952.
78. METEOROLOGICAL OFFICE, *Handbook of Meteorological Instruments, Part I,* HMSO, London, 1956.
79. MILLER, A. A., *The Skin of the Earth,* Methuen, London, 1953.
80. MILLER, J. F., Hydrometeorological studies, *The Progress of Hydrology,* pp. 521-562, Univ. of Illinois, Urbana, 1969.
81. MILTHORPE, F. L., The income and loss of water in arid and semi-arid zones, *Plant-Water Relationships in Arid and Semi-Arid Conditions,* pp. 9-36, UNESCO, Paris, 1960.
82. MYERS, V. A., The estimation of extreme precipitation as the basis for design floods: Resume of practice in the United States, *Floods and their Computation,* Joint IASH/WMO Publn., pp. 84-104, 1969.
83. NACE, R. L., Water management, agriculture and groundwater supplies, *USGS Circ.* 415, 1960.
84. ØSTREM, G., Glacio-hydrological investigations in Norway, *J. Hydrol.,* **2**: 101-115, 1964.
85. PAULHUS, J. L. H., Indian Ocean and Taiwan rainfalls set new records, *Mon. Wea. Rev.,* **93**: 331-335, 1965.
86. ROBINSON, A. C., The aerodynamic characteristics of raingauges, *Unpublished M.Sc. thesis, University of Southampton,* 1968.
87. ROBINSON, A. C., and J. C. RODDA, Rain, wind and the aerodynamic characteristics of raingauges, *Met. Mag.,* **98**: 113-120, 1969.
88. RODDA, J. C., An objective method for the assessment of areal rainfall amounts, *Weather,* **17**: 54-59, 1962.
89. RODDA, J. C. A country-wide study of intense rainfall for the United Kingdom, *J. Hydrol.,* **5**: 58-69, 1967.
90. RODDA, J. C., The rainfall measurement problem, *Proc. Bern Assembly,—IASH Publn.* **78**: 215-231, 1967.
91. RODDA, J. C., The assessment of precipitation, in *Water, Earth, and Man,* Ed. R. J. Chorley, pp. 130-134, Methuen, London, 1969.
92. RODDA, J. C., On the questions of rainfall measurement and representativeness, *Symposium on World Water Balance, IASH Publ.* 92, **1**: 173-186, 1970a.
93. RODDA, J. C., Rainfall excesses in the United Kingdom, *Trans. IBG,* **49**: 49-60, 1970.
94. SALTER, M.deC., *The Rainfall of the British Isles,* University of London Press, London, 1921.
95. SAWYER, J. S., A study of the rainfall of two synoptic situations, *QJRMS,* **78**: 231-246, 1952.
96. STEVENSON, T., On the defects of raingauges with description of an improved form, *Edinburgh New Phil. J.,* **33**: 12-21, 1842.
97. STOL, Ph. Th., The relative efficiency of the density of raingauge networks, *J. Hydrol.,* **15**: 193-208, 1972.
98. STRUZER, L. R., I. N. NECHAYEV, and E. G. BOGDANOVA, Systematic errors in measurements of atmospheric precipitation, *Soviet Hydrol.,* pp. 500-504, 1965.

99. THIESSEN, A. H., Precipitation averages for large areas, *Mon. Wea. Rev., 39*: 1082-1084, 1911.
100. THRAN, P., and S. BROEKHUIZEN, (Eds.), *Agro-Climatic Atlas of Europe,* Elsevier, Amsterdam, 1965.
101. TODOROVIC, P., and V. YEVJEVICH, Stochastic process of precipitation, *Colorado State Univ., Hydrol. Papers* 35, 61 pp., Fort Collins, Colorado, 1969.
102. TURNER, A. K., Storm intensities versus seasonal rainfalls, *J. Hydrol., 12*: 377-386, 1971.
103. UNWIN, D. J., The areal extension of rainfall records: An alternative model, *J. Hydrol., 7*: 404-414, 1969.
104. URYVAEV, V. A., et al., Principal shortcomings of methods of observing snow cover and precipitation and proposals of the State Hydrological Institute for their improvement, *Leningrad Glav. Geofiz. Obs. Trudy, 175*: 31-58, 1965.
105. U.S. ARMY CORPS OF ENGINEERS, *Snow Hydrology,* North Pacific Div., Portland, Oregon, 437 pp., 1956.
106. U.S. WEATHER BUREAU, Thunderstorm rainfall, *U.S. Wea. Bur. Hydrometeorological Report, 5*, 1947.
107. WARNICK, C. C., and V. E. PENTON, New methods of measuring water equivalent of snow pack for automatic recording at remote mountain locations, *J. Hydrol., 13*: 201-215, 1971.
108. WHITMORE, J. S., On evidence of 'persistence' in annual rainfall in the Republic of South Africa, *J. Hydrol., 1*: 144-150, 1963.
109. WHITMORE, J. S., F. J. vanEEDEN, and K. J. HARVEY, Assessment of average annual rainfall over large catchments, *Union of South Africa, Dept. of Water Affairs, Tech. Rept.,* 14, 1960. Subsequently published in *Inter-African Conference on Hydrology, Nairobi, 1961.,* C.C.T.A. Publ. 66: 100-107, London, 1961.
110. WIESNER, C. J., *Hydrometeorology,* Chapman and Hall, London, 232 pp., 1970.
111. WILSON, W. T., Snow, Chapter 10 in Wisler, C.O. and E. F. Brater, *Hydrology,* (2nd Ed.) Wiley, New York, 1959.
112. WISLER, C. O., and E. F. BRATER, *Hydrology* (2nd Ed.), Wiley, New York, 1959.
113. WOODLEY, W. L., and B. SANCHO, A first step towards rainfall estimation from satellite cloud photographs, *Weather, 26*: 279-289, 1971.
114. WORLD METEOROLOGICAL ORGANIZATION, *Guide to Hydrometeorological Practices,* WMO, Geneva, 1965.
115. WORLD METEOROLOGICAL ORGANIZATION, The role of meteorological satellites in the World Weather Watch, *W.W.W. Planning Rept.* 18, WMO, Geneva, 38 pp., 1967.
116. YEVJEVICH, V., Fluctuations of wet and dry years, Part 1, Research data assembly and mathematical models, *Colorado State Univ., Hydrol. Papers* 1, Fort Collins, Colorado, 1963.

3. Interception

3.1 Introduction and definitions

The amount of precipitation actually reaching the ground surface is largely dependent upon the nature and the density of the vegetation cover, if this exists, or upon the existence of an artificial cover of buildings, roads, and pavements. This cover, whether natural or artificial, intercepts part of the falling precipitation and temporarily stores it on its surfaces, from where the water is either evaporated back into the atmosphere or falls to the ground.

The three main components are *interception loss,* i.e., water which is retained by plant surfaces and which is later evaporated away or absorbed by the plant: *throughfall,* i.e., water which either falls through spaces in the vegetation canopy or which drips from leaves, twigs, and stems to the ground surface: and *stemflow,* i.e., water which trickles along twigs and branches and finally down the main trunk to the ground surface. Only in the case of interception loss is the water prevented from reaching the ground surface and so taking part in the land-bound portion of the hydrological cycle. In this sense, therefore, interception loss may be regarded as a *primary* water loss, and it is evidently this component of interception which is of most concern to the hydrologist. In exceptional circumstances, however, vegetation may intercept moisture in the air which would not otherwise have fallen as precipitation, and in this case, obviously, the main hydrological interest lies in the amount of water which is transmitted to the ground as through-fall and stemflow.

As will be seen later, some authorities do not consider that interception loss as defined above is a real loss to the plant-soil-water system. Instead they propose two terms: *gross interception loss* to represent interception loss as defined previously and *net interception loss* to represent the part of the precipitation retained on and evaporated from plant surfaces that does not affect the plant's soil water use [5].

3.2 Interception in urban areas

In most urban areas interception by vegetation is comparatively unimportant. A significant percentage of precipitation may, however, be held by and evaporated from the surfaces of buildings, although reliable experimental data are virtually non-existent. In extreme cases, where water is led from a roof into a storage container and thence evaporated, the interception loss may be approximately 100 per cent, although normally such water is led into a drainage system or into the subsoil via sewers and storm drains.

3.3 Factors affecting interception loss from vegetation

The fact that following a dry spell interception loss is usually greatest at the beginning of a storm and reduces with time, reflects the interaction of the main factors which affect it. Of these, undoubtedly the most important is the *interception capacity* of the vegetation cover, i.e., the ability of the vegetation surfaces to collect and retain falling precipitation. At first, when all the leaves and twigs or stems are dry, this is high, and a very large percentage of precipitation is prevented from reaching the ground. As the leaves become wetter, the weight of water on them eventually overcomes the surface tension by which it is held and, thereafter, further additions from rainfall are almost entirely offset by the water droplets falling from the lower edges of the leaves.

Even during rainfall, however, a considerable amount of water may be lost by *evaporation* from the leaf surfaces, so that even when the initial interception capacity has been filled, there is some further fairly constant retention of water to make good this evaporation loss. Indeed, during long continued rains, the interception loss may be closely related to the rate of evaporation, so that the *meteorological factors* affecting the latter are also relevant to this discussion. While rain is actually falling, however, conditions are seldom conducive to high rates of evaporation, and in these circumstances *windspeed* tends to be the only meteorological factor of real significance. Other conditions remaining constant, evaporation tends to increase with increasing windspeed, so that during prolonged periods of rainfall the interception loss is greater in windy than in calm conditions. This observation may not, however, be applicable to rainfalls of short duration during which the effect of high windspeeds in reducing interception capacity, by prematurely dislodging water collected on vegetation surfaces, outweighs the greater evaporative losses.

The *duration of rainfall* is, thus, a secondary factor, in that it influences interception by determining the balance between the reduced storage of water on the vegetation surfaces on the one hand, and increased evaporative losses on the other. Data collected during classic work by Horton [22], and in numerous subsequent investigations, showed that interception loss increases with the duration of rainfall, but only gradually, so that the relative importance of interception decreases with time. Since *rainfall amount* and rainfall duration are closely related, many investigators have been tempted to make fallacious correlations between interception losses and rainfall amount. Evidently, however, provided that the initial interception capacity is filled, and provided that the subsequent rate of rainfall at least equals the rate of evaporation (the normal situation), rainfall amount can have no influence upon the magnitude of interception losses. Nevertheless, the relative importance of interception losses will tend to decrease as the amount of rainfall increases.

Since the greatest interception loss occurs at the beginning of a storm, when the vegetation surfaces are dry and the interception capacity is large, it will be apparent that *rainfall frequency* is of considerably greater significance than either the duration or the amount of rainfall.

Interception loss will also be affected by the *type of precipitation,* and particularly by the contrast between rain and snow, which will be discussed more fully at a later stage. Another important factor, which also merits a separate discussion below, is the variation of interception loss with the *type and morphology of the vegetation cover.*

3.4 Calculation of interception

Most of the simple equations which have been devised for calculating interception are based upon the principal factors affecting it which have been discussed above, and so take the general form

$$I = C + V.E.t \quad (3.1)$$

where I is the interception loss for a given storm, C is the interception capacity of the vegetation cover, V is the ratio of the vegetation surface area to its projected area, E is the average rate of evaporation during the storm, and t is the time duration of the storm. This type of equation was proposed by Horton [22], and has subsequently been used in slightly modified form by most authorities.

The usefulness of the equation is limited, however, since it can be applied only to individual rainstorms and not to longer-term, e.g., daily, totals and furthermore, it is applicable only to falls of rain which exceed the interception capacity (C). In the case of smaller falls, the calculated interception loss will tend to be greater than the measured precipitation above the vegetation cover. In addition, the amount of rainfall is not considered in the equation.

Meriam [39] adopted a suggestion by Linsley and others [35] that the equation should be modified to include an exponential term to consider rainfall quantities and proposed the following equations

$$I = C(1 - e^{-P/C}) + V.E.t \quad (3.2)$$

$$I = C(1 - e^{-P/C}) + K.P. \quad (3.3)$$

where P is the amount of precipitation, e is the base of natural logarithms, and K is $V.E.t/P$ and is assumed constant.

As Wigham [73] observed, these equations describe the process of interception reasonably well, allowing for the observed exponential increase in interception capacity with increased precipitation. Since the terms V and $E.t$ are difficult to evaluate separately, eq (3.3) is often favoured and although holding K constant, implying a constant relationship between E and P, may not be theoretically sound, in practical terms this may be acceptable in many cases.

3.5 Measurement of interception

In view of what has been said about the calculation of interception, direct measurement would seem to be the most accurate way of determining interception losses but, here again, numerous difficulties combine to detract from the validity of

the result. The normal method of measuring interception loss is to measure precipitation above the vegetation layer (R), ground precipitation below the vegetation
canopy (Rg) and, in many cases, stemflow (S). Then interception loss may be
simply calculated from

$$I = R - (Rg + S) \quad (3.4)$$

Problems inherent in measuring rainfall at some distance above the ground
surface have been discussed in chapter 2. Since turbulence and the raingauge errors
associated with it increase with height above the ground surface, the measurement
of R over a wooded area, for example, is likely to present more problems than the
measurement of this factor over grass or over many agricultural crops, especially
since the roughness of the vegetation surface will add further to the turbulent
effects caused by height. Experiments by Law [30], using shielded gauges within
and above a forest canopy in northern England, indicated that the gauges should
definitely be located above the canopy, in order to avoid excessive turbulence
effects among the tree-tops. Similar conclusions were reached by Laine [28], who
advocated siting the gauge some 4.5 to 6 metres above the highest tree in the
vicinity of the gauge.

Although errors associated with the measurement of R at canopy level in a
forested area may be avoided by measuring rainfall at ground level instead, either in
clearings within the forest or outside the forested area, further uncertainties are
likely to be introduced. Geiger [11], for example, reported experiments in which
the raingauge catch varied with the size of the clearings in a mixed stand of pine
and beech some 26 metres high; in a clearing 12 metres in diameter the rainfall
represented only 87 per cent of that in the surrounding open area, because rain
falling at an angle was caught by the crowns of surrounding trees, while in clearings
38 metres and 87 metres in diameter, the rainfall was 105 per cent and 102 per
cent, respectively, of that in the open, because in these cases, although the trees
were sufficiently far away to cause no direct physical obstruction, turbulence was
markedly reduced.

Ideally the clearing in the forest should subtend at the gauge an angle of 30° or
less [28], although it has been shown [49] that in a smaller clearing, giving a
clearance angle of 45°, a gauge will still catch about 98 per cent of the precipitation
falling in the open. The 45° clearing has been accepted as standard practice [32]
although even this requires considerable tree removal, especially on sloping terrain,
and vegetation regrowth immediately begins to adversely affect the gauge site.

Again, measurements on open ground outside a forested plot may be unrepresentative because rainfall tends to be reduced at the upwind edge of the plot as
a result of updraughts, and increased at the downwind edge as a result of down-
draughts [46].

The main problems associated with the measurement of throughfall or ground
rainfall beneath a forest vegetation canopy are concerned with the possibility of
sampling errors. A gauge situated immediately beneath a pronounced drip-point will
register an excessively large catch, while an adjacent gauge may catch nothing at all
during the same storm. Rutter [57] discussed the existence of drip points close to
tree trunks and avoided the problem of high gauge totals by locating the gauges at
least 15 cm from the trunks. Geiger [11] suggested that throughfall beneath a tree

conformed to a specific pattern, irrespective of the amount or intensity of rainfall, whereby it is least close to the trunk, and increases towards the periphery of the tree. This relationship has been confirmed and in some cases quantified by later workers, c.f. [54]. However, the great variability of throughfall distribution from one place to another necessitates a large number of gauges for accurate assessment [14] [28] and/or the periodic random relocation of the gauges, c.f. [30] which reduces the standard error if observations under different locations are grouped together [23]. Alternatively, accepting the approximately concentric distribution of throughfall beneath an individual tree, some authorities have discussed the use of trough gauges several metres in length [14] [28], although these pose both practical problems and problems of data comparability, or the use of regression analysis to calculate total throughfall from the relationship between a number of point measurements at measured distances from tree crown centres [23] [54].

Again, in the case of stemflow measurement, the main source of error tends to arise from inadequate sampling. Law [30], for example, found variations of 100 per cent between the stemflow down trees in the same forested plot although, at the same time, he pointed out that the total amount of water reaching the ground in this way was small, representing only seven per cent of the rain falling on the vegetation surface. Geiger [11] similarly estimated that, even during a cloudburst, stemflow amounted to less than five per cent of the rainfall. However, the amount of stemflow in forested areas depends largely upon the roughness of the bark [36]; Rowe [55] found that in the case of some smooth-barked trees, like beech, this could amount to 15 per cent of the rainfall, while Kittredge [25] found much lower values of 2 per cent to 3 per cent for rough-barked pines.

A convenient method for collecting stemflow from trees is to construct a spiral gutter around the trunk and seal the joint between the gutter and the tree surface. Water from the gutter is then led into a collecting container or measuring device. Stemflow measurements may be made for individual species or for a particular plot, depending on the type of information required. Helvey and Patric [14] recommended the latter, in which case stemflow from all the trees in the plot can be led into the same collecting container.

In many experiments, stemflow has been ignored and bearing in mind the possible errors involved in its measurement, as well as in the measurement of R and Rg, this omission may not significantly affect the validity of the results.

Although instrumental difficulties are normally greater, the same basic principle of measurement may be used to determine the interception loss from lower vegetation, such as grasses or agricultural crops. In this case, rainfall may be measured above the crop, or in an adjacent standard raingauge, to determine the gross precipitation, and in gauges sunk to ground level within the crop in order to determine the throughfall. Alternatively, the vegetation from a given area may be cut and placed on, say, a wire screen, and then sprinkled with measured amounts of water in order to determine interception losses in a variety of conditions [3], [64]. One of the most satisfactory methods of measuring grass interception losses was described by Corbett and Crouse [7]. They drove 25 cm diameter metal collars into the soil so that about 2.5 cm of the collar remained above the soil surface. When the grass height reached 5 cm, the soil surface inside the collar was sealed by first sifting on a fine layer of sand, to provide a uniform sloping surface for drainage, and then applying a latex emulsion which did not affect grass growth. A drain in the side of each collar carried combined throughfall and stemflow to a collecting container or measuring device.

An entirely different approach to the measurement of interception losses involves the use of small instrumented watersheds. Swank and Miner [65], for example, reported that the effects of converting mature hardwoods on two experimental watersheds in the Southern Appalachians to eastern white pine was to reduce streamflow after ten years by almost 10 cm. Since most of the water yield reduction occurred during the dormant season, it was attributed mainly to greater interception loss from white pine than from hardwoods. Increases in water yield, also attributable largely to interception effects, were reported by Pillsbury and others [50] and by Hibbert [19], after the conversion of chapparal-covered watersheds to grass and forbs. Rigorously applied, this method could yield valuable large-scale estimates of interception loss. Because of numerous experimental difficulties [70], however, the method is unlikely to provide more than corroborative evidence.

3.6 Interception losses from different types of vegetation

Mainly because of the fact that measurements of the relevant physical factors are relatively easier in woodlands than in other types of vegetation, much of the experimental data on interception concern forested areas. Zinke [75] provided an excellent review of North American work on forest interception effects. However, sufficient evidence exists to enable brief comments to be made upon interception from other types of vegetation as well, with a view to assessing the relative importance of each in the water balance of a catchment area.

Interception loss from woodlands

Considerable attention has been paid, in many of the studies of interception loss in woodland, to possible differences in the magnitude of loss from coniferous and deciduous forests. Despite the fact that, in most cases, the leaf density is greater in deciduous than in coniferous forest, the bulk of experimental evidence shows that interception losses are greater from the latter. Reviewing a broad range of Russian, European, and American data, Rakhmanov [52] suggested that coniferous forests, together with sparse woods and inhibited stands on peat bogs and other marshy terrain, intercept an average of 25-35 per cent of the annual precipitation compared with 15-25 per cent by broad-leaved forests. Figure 3.1 shows data from F. E. Eidmann and E. Hoppe, quoted by Penman [46] and Geiger [11], respectively, in which this contrast is clearly illustrated. Both types of woodland show the reduction in relative importance of interception loss as rainfall amounts increase but, over the complete range of rainfall totals, interception loss is markedly greater from the spruce forests. Geiger [11] suggested that one of the reasons for this somewhat surprising contrast was that, while water droplets remain clinging to separate spruce needles, they tend to run together on the beech leaves, and so drop off or flow on to twigs and branches. It is also likely that the open texture of coniferous leaves allows the freer circulation of air, and consequently more rapid evaporation of the retained moisture.

With regard to seasonal contrasts, winter and summer losses appear to be about

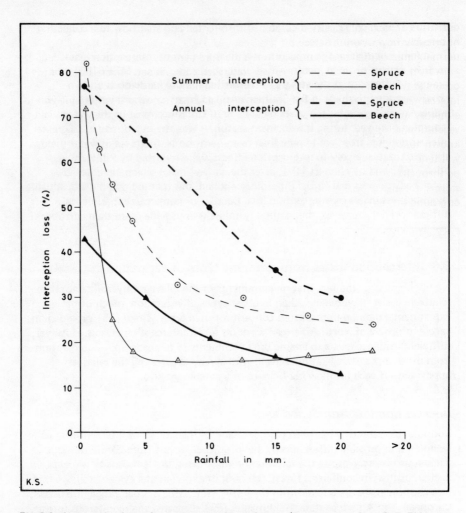

Fig. 3.1. *Interception losses from spruce and beech forests. (Based on data from F. E. Eidmann quoted by Penman [46] and from E. Hoppe quoted by R. Geiger [11].)*

the same from coniferous trees [24], [74]. Indeed, Fig. 3.2 shows that, in some circumstances, winter losses from a spruce plantation in northern England may be slightly higher than summer losses, despite the fact that, theoretically, evaporative conditions are considerably more favourable during the latter period [30]. As would be expected, however, interception losses from deciduous trees are greatest during the period of full leaf. Lull [36] quoted figures showing that, in northern hardwood and aspen-birch forests, interception losses with the trees in leaf were 15 per cent and 10 per cent of the precipitation respectively, whereas with leafless trees the losses were 7 per cent and 4 per cent. Similarly, Wisler and Brater [74] suggested that summer losses for deciduous trees are something like two or three times greater than winter losses although, in certain conditions, a considerable

60

quantity of freezing rain may be stored on bare twigs and branches, and as this does not readily flow down the trunk, it is more susceptible to evaporation than is normal liquid rainfall at this time of year [6].

A further important aspect of interception loss in wooded areas is that this often occurs at two, or even more, levels within the vegetation cover. Precipitation is first intercepted by the tree crowns; some of the throughfall is then intercepted again, either by undergrowth, or by a layer of ground litter, which acts like living vegetation in holding a film of water that is subject to evaporation [6]. Little is known about the relative importance of this secondary interception, although it will tend to increase with the amount of rainfall, because during light rains little or no throughfall from the crown canopy will occur, whereas during long, heavy storms, throughfall will probably fill the interception capacity of the undergrowth or ground litter.

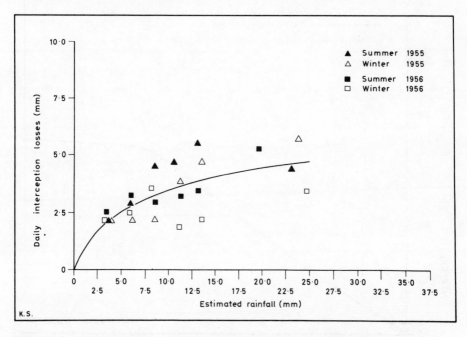

Fig. 3.2. *Seasonal interception losses from sitka spruce. (From an original diagram by Law [30].)*

Interception loss from grasses and shrubs

The total leaf area of a continuous cover of mature grass or shrub closely resembles that of a closed canopy forest, so that interception losses from the former are similar in magnitude to those from trees during the season of maximum development. Since their season is short, however, total annual interception loss from grasses is considerably less than from, say, deciduous woodland. Furthermore, in areas where grass is either cut for hay or silage, or heavily grazed in the field, interception losses will tend to be small, approaching zero in extreme cases [31].

Kittredge [25] found that in California undisturbed grass of *avena, stipa, lolium,* and *bromus* species intercepted 26 per cent of an 826 mm seasonal rainfall, while in Missouri, bluegrass intercepted 17 per cent of the rainfall in the month before harvesting [42]. In neither case was a correction made for stemflow, which was shown by Beard [4] to account for between 35 and 45 per cent of total rainfall during a period when grass interception loss was about 13 per cent of the total rainfall. Work by Corbett and Crouse [7], in which throughfall and stemflow were measured, showed that annual interception losses from *bromus* in southern California averaged 7.9 per cent. Interception experiments with cut vegetation or with artificial sprinkling have given widely divergent results.

As Leyton and others [34] suggested, there is a paucity of data on interception by herbaceous and shrubby covers typical of the heaths and moorlands of Europe.

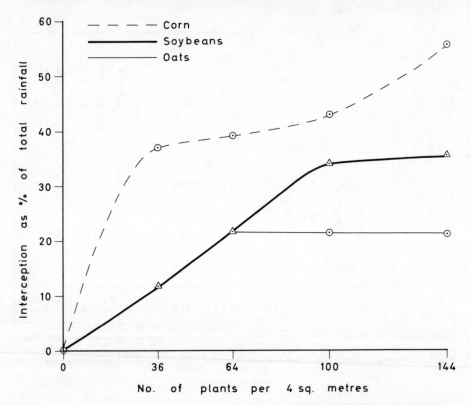

Fig. 3.3. *Interception by three agricultural crops. (Based on data from Wollny quoted by Baver [2].)*

Measurements on heather in Scotland showed that between 35 and 66 per cent of precipitation penetrated the canopy, the amount increasing as precipitation increased, although no measurements were made of stemflow [1]. Interception losses from mature chaparral cover in southern California averaged 12.8 per cent of the annual precipitation [7].

Interception loss from agricultural crops

Slightly more information is available on the interception loss from agricultural crops, and from cereals in particular, although again, the data are sparse in relation to those from forested areas. Figure 3.3 is based upon early data of Wollny reported by Baver [2], and shows interception by corn, soybeans, and oats. Interception increases initially with increasing crop density. After a certain coverage has been attained, however, the subsequent increase of interception is slight, as shown by the latter portions of the curves, and in general it can be said that average interception by fully developed oats, soybeans, and corn is 23, 35, and 40 to 50 per cent respectively. Since no measurements appear to have been made of stemflow, these figures would have to be reduced by an appropriate amount to be representative of the real interception loss. Lull [36] reported observations by Haynes for the same three crops during the growing season, which showed that interception losses from oats, soybeans, and corn respectively were about 7, 15, and 16 per cent of the total rainfall. Recent Russian experiments [26] showed that interception by spring wheat during the growing season was about the same as, or a little less than, that by forests in leaf for the same period, amounting to between 11 and 19 per cent of the total precipitation.

3.7 Interception of snow

Experimental evidence concerning the interception of snow has frequently been

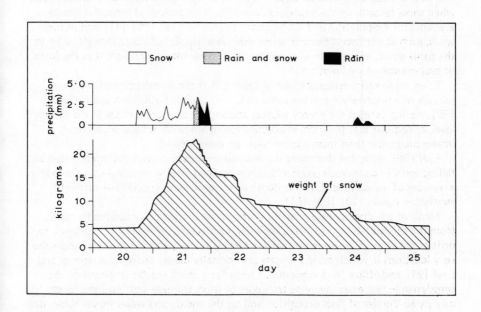

Fig. 3.4. *Interceptograph trace illustrating different aspects of the snow-interception water balance. See text for explanation. (From an original diagram by D. R. Satterlund and H. F. Haupt,* Water Resources Res., **6,** *p. 650, Fig. 1, © American Geophysical Union.)*

unsatisfactory and confusing and is largely restricted to woodland vegetation. Geiger [11] reported experiments by J. Schubert and H. Hesselman which indicated that most of the snow falling on a forest reaches the forest floor. Schubert found that the ratio of snow outside to that inside forested areas was 100 to 90, while Hesselman found the same depth of snow in a pine forest as in cuttings. Later work by West and Knoerr [71] in the Sierra Nevada suggested that interception of snow in dense coniferous stands might amount to about 8 per cent of the fall during the winter period.

There are both mechanical and thermodynamic factors which support this view of the relative unimportance of snow interception losses. For example, snow accumulating on vegetation surfaces is prone to large-scale mass release by rainwash and sliding resulting from its own weight, frequently aided by wind-induced movement of the vegetation, and also the smaller-scale release of snow particles and meltwater drip. Such release mechanisms mean that in many areas snow remains on trees for only a few days before falling to the ground, so that opportunities for evaporation are small. A typical sequence of events in Idaho is illustrated in the interceptograph trace in Fig. 3.4 which shows the continuous weight record of snow load on a tree [60]. On days 20 and 21 snow began and accumulated on the tree but as the storm turned to rain part of the intercepted snow was washed to the ground. On day 22 warm temperatures resulted in minor mass release, indicated by the irregular decline, and large mass release indicated by the vertical trace. There was a slight decline on day 23 resulting from the evaporation of intercepted snow and then, on day 24, rain washed the remaining snow from the tree leaving it bare and wet. On day 25 the liquid water evaporated leaving the tree bare and dry. Even when snow remains on the vegetation cover for long periods of time, the energy available for evaporation and sublimation is minimal [21] [40] [41] and in some areas, such as northwest Europe, when there *is* a transfer of water it tends to be to the snow cover, in the form of condensation, rather than away from it in the form of evaporation or sublimation [47].

Even so, some investigators have suggested that the interception of snow and rain are quantitatively much the same [24] [56], while Satterlund and Eschner [59], having defined the energy sources and vapour pressure deficits at the canopy level, concluded that these could account for significantly greater losses from intercepted snow than from a snow cover on open ground.

Lull [36] suggested that there is a marked contrast between the interception of falling snow in coniferous and deciduous woodland, a view seemingly supported by a number of measurements of the depth of snow accumulation under different vegetation covers [10] [37] [61].

Many of the discrepancies evident in the above discussion undoubtedly result from the difficulties of measuring snowfall, particularly in view of its tendency to drift at the edge of pronounced barriers such as forests, and in forest clearings—the very locations in which measurements are normally made. There is, as Hoover and Leaf [21] and Miller [41] suggested, a need for a much greater emphasis on the aerodynamic processes involving transport of snow through and from the forest canopy to the site of final deposition and on the energy and water fluxes above the snow cover. Preliminary consideration of these groups of factors, however, would seem to confirm Geiger's view that evaporative losses from snow interception are likely to be small.

3.8 Horizontal interception

In conditions of high atmospheric humidity, it is known that some types of vegetation, particularly trees, are able to extract moisture from the air which would not otherwise have reached the ground as precipitation. Water droplets are formed on leaves, twigs, and branches by direct condensation, often in such quantities that they fall to the ground like rainfall. This phenomenon is commonly experienced under trees during fog, particularly when windspeeds are relatively high, and may be utilized for agricultural purposes. In parts of Japan, for example, forested belts are maintained to filter out a large percentage of the water from fogs moving in off the sea [8]. This process is, however, also observed in agricultural crops and has been held responsible for the lodging of grain crops which often occurs in calm, misty conditions [48].

In the sense that this *horizontal interception* or *fog drip* represents measurable precipitation beneath the vegetation canopy where none is recorded in the open, Kittredge [25] suggested that it could be regarded as negative interception. Reference to the three main components of interception as defined on page 54, however, indicates that although interception *loss* in these conditions is negative, throughfall and stemflow are positive, and it is these two components of interception which assume the greatest hydrological significance.

Experimental evidence has frequently emphasized that horizontal interception is essentially an edge effect, and that its importance decreases markedly away from the borders of, say, a forested plot or area of relatively taller vegetation [11] [44].

The border nature of the phenomenon is also apparent from results of various experiments with fog gauges. Nagel [43], for example, used a fog gauge in which a wire gauze cylinder was placed above the funnel of a normal raingauge, on the basis that the gauze would intercept horizontally blown cloud and fog droplets in just the same way as would natural vegetation surfaces, such as leaves and twigs and stems. In measurements on Table Mountain, it was found that, during a period of twelve months, fog drip from the 'table cloth' or orographic cloud was 3294 mm, in comparison with a measured rainfall of 1940 mm. Since edge effects are so large with the type of instrument used, it is not possible to extrapolate the results in order to show what the natural vegetation on Table Mountain is capable of intercepting from cloud moving over it. Even a significant reduction of the 1354 mm excess fog drip, however, would still leave an impressive contribution to the income side of the water balance of the mountain [46]. Similar difficulties arise when attempting to relate measurements of fog drip, by Twomey [67], to natural horizontal interception on Mount Wellington, Tasmania. In this case, the fog gauge, which consisted of a 203 mm raingauge surmounted by a 183 X 92 cm wire mesh screen, caught 1117 mm in a 10-day period during February, in comparison with the 104 mm catch of an adjacent normal raingauge.

It is, thus, virtually impossible to determine a valid figure for the magnitude of natural horizontal interception or fog drip. From what has been said above, however, it may be deduced that at the windward edge of areas of upstanding vegetation, and on high crests and ridges, in foggy localities, normally measured rainfall may be exceeded by a factor of two or three [25]. Certainly, it has been suggested that, in those parts of Tasmania where soil and vegetation types are typical of a considerably higher rainfall than that actually measured, the anomalies can probably be attributed to the interception of cloud water by vegetation [67].

In North America, too, such additions to the normal precipitation are important in determining plant distributions along the west central coast [36] and, as Reynolds [53] observed, a real gain in precipitation has also been demonstrated in Israel [69], Hawaii [9], and Chile [27].

3.9 Interception and the water balance

Despite the long history of experimental investigation of interception, there remain conflicting views about its hydrological significance and particularly about its quantitative significance in the water balance of a drainage basin. There are three basic views which may, in part, be complementary.

First, some authorities have emphasized that interception losses are essentially evaporative and that, since there is only a certain amount of energy available at any time, this will tend to be used either to evaporate water from within the leaf, i.e., in transpiration, or to evaporate water from the surface of the leaf, i.e., interception loss. In their experiments on grasses, Burgy and Pomeroy [5] and McMillan and Burgy [38] concluded that wet foliage evaporation was, in fact, equally balanced by a reduction in transpiration loss, so that with this type of vegetation the net interception loss was zero. Similar conclusions were reached by Rakhmanov [51] in his discussion of forest interception losses. In such cases, interception loss is clearly an alternative and not an addition to evapotranspiration and would therefore have little, if any, effect upon the water balance of a catchment area.

A second view, increasingly supported by additional experimental evidence, was that put forward at an early stage by Hirata [20] and Kittredge [25] who regarded interception unequivocally as a loss of precipitation that would otherwise have been available at the ground surface for direct evaporation, for infiltration through the surface, or for overland flow. This view seems to be implicit in such comments as that of Whitmore [72] that, in South Africa, interception probably amounts to about 5 to 15 per cent of the annual rainfall and it was certainly the view of Law [29], whose experiments in a small spruce plantation in the Pennines showed that water losses from a forested area exceeded those from adjacent grassland by some 280 mm in a year.

Undoubtedly, evidence is accumulating, mostly from forested areas, to support the conclusion that intercepted water evaporates much faster than transpired water, and therefore much of the interception loss represents an additional loss in the catchment water balance [17]. Thorud [66], for example, applied water artificially to the foliage of potted ponderosa pine and found that 91 per cent of the applied water was a net loss to soil moisture. More realistically, field experiments by Rutter [57], [58], Patric [45], Helvey [13], and Leyton and others [34], indicated that during the winter period the loss of intercepted water considerably exceeded the transpiration rate in the same environmental conditions and, of course, results from small watershed studies have shown that substantial increases in water yield result from the removal of the forest vegetation [16] [18] and that decreased yields result from the conversion of hardwood forests to pine forests [65], largely as a result of interception effects.

An additional energy supply must be postulated in order to explain the greater rate of evaporation of intercepted water, which in Rutter's experiments exceeded the equivalent intensity of net radiation by a considerable margin. Rutter's view that the extra energy comes from the air was supported by his observations that when intercepted water was being evaporated the foliage was measurably cooler than the air and that the resulting temperature gradient was sufficient to yield a heat flux to supply the energy deficiency [58]. In some other conditions, for example winter conditions in cold climates, it may be possible that transpiration is limited more by the availability of water than of energy [12] and that, by increasing the amount of available water, interception inevitably increases the total loss of water from a catchment area. The evaporation of water intercepted by dead, and therefore non-transpiring, vegetation and by a litter layer would certainly represent a net interception loss [5] [38], the only factor involved in this case being the interception storage capacity and its depletion by evaporation. It was the storage aspect of interception loss which Zinke [75] considered might play the greatest part in affecting the catchment water balance.

Third, in certain circumstances, it can be argued that the interaction of water loss and gain in vertical and horizontal interception respectively, may result in a *net gain* of water in a catchment area. In view of what was said earlier about horizontal interception, this is evidently most likely to be true in forested areas of high relief, where fogs or low cloud are prevalent.

On a more local scale, interception affects catchment hydrology because of its effects on the areal distribution of precipitation actually reaching the ground. Thus, throughfall and dripping meltwater are concentrated at the edges of tree crowns, whilst concentrated drip close to the trunk and stemflow itself often result in high values of infiltration and soil moisture recharge [62] [68] and even the initiation of minor rills and channels in the surface [33].

3.10 Surface storage

Surface storage, or *depression storage,* as it is commonly known, is an aspect of water losses which is closely allied to interception. It is, in fact, another example of precipitated moisture, most of which is returned to the atmosphere by evaporation before taking an active part in the land-bound portion of the hydrological cycle. Surface storage comprises the water retained in hollows and depressions in the ground surface during and after rainfall. This water is then either evaporated directly, or is used by vegetation, or else infiltrates into the soil, so that none of it appears as surface runoff [74].

Although surface storage is commonly thought of as a small-scale phenomenon related to minor depressions and puddles, it may also assume a considerable importance on a larger scale where topographical conditions are particularly favourable. On the Canadian prairies for example, glaciation has created a great deal of surface storage in most of the catchment areas [63].

References

1. ARANDA, J. M., and J. R. H. COUTTS, Micrometeorological observations in an afforested area in Aberdeenshire: rainfall characteristics, *J. Soil Sci.,* 14: 124-133, 1963.
2. BAVER, L. D., *Soil Physics,* (3rd Ed.), Wiley, New York, 1956.
3. BEARD, J. S., Results of the Mountain Home rainfall interception and infiltration project on Black Wattle, 1953-1954, *J. S. Africa Forestry Assoc.,* 27: 72-85, 1956.

4. BEARD, J. S., Rainfall interception by grass, *J. South African Forestry,* **42**: 12-15, 1962.
5. BURGY, R. H., and C. R. POMEROY, Interception losses in grassy vegetation, *Trans. AGU,* **39**: 1095-1100, 1958.
6. COLMAN, E. A., *Vegetation and Watershed Management,* Ronald, New York, 1953.
7. CORBETT, E. S., and R. P. CROUSE, Rainfall interception by annual grass and chaparral ... losses compared, *USFS Research Paper* PSW-48, 12 pp., Berkeley, 1968.
8. DAUBENMIRE, R. F., *Plants and Environment,* (2nd Ed.), Wiley, New York, 1959.
9. EKERN, P. C., Direct interception of cloud water on Lanaihale, Hawaii, *Proc. SSSA,* **28**: 419-421, 1964.
10. ESCHNER, A. R., and D. R. SATTERLUND, Snow deposition and melt under different vegetative covers in Central New York, *USFS Research Note* NE-13, 6 pp., 1963.
11. GEIGER, R., *The Climate Near the Ground,* (Translation), Harvard Univ. Press, Massachusetts, 1957.
12. GOODELL, B. C., A reappraisal of precipitation interception by plants and attendant water loss, *J. Soil and Water Cons.* **18**: 231-234, 1963.
13. HELVEY, J. D., Interception by eastern white pine, *WRR,* **3**: 723-729, 1967.
14. HELVEY, J. D., and J. H. PATRIC, Design criteria for interception studies, *Symposium on Design of Hydrological Networks, IASH Publn.* **67**: 131-137, 1965(a).
15. HELVEY, J. D., and J. H. PATRIC, Canopy and litter interception of rainfall by hardwoods of eastern United States, *WRR,* **1**: 193-206, 1965.
16. HEWLETT, J. D., and A. R. HIBBERT, Increase in water yield after several types of forest cutting, *Bull. IASH,* **6**: 5-17, 1961.
17. HEWLETT, J. D., and W. L. NUTTER, *An outline of Forest Hydrology,* Univ. of Georgia Press, Athens, 1969.
18. HIBBERT, A. R., Forest treatment effects on water yield in Sopper, W. E. and H. W. Lull (Eds.), *Forest Hydrology,* pp. 527-543, Pergamon, Oxford, 1967.
19. HIBBERT, A. R., Increases in streamflow after converting chaparral to grass, *WRR,* **7**: 71-80, 1971.
20. HIRATA, T., *Contributions to the problem of the relation between forest and water in Japan,* Imp. Forest Expt. Sta. Meguro, Tokyo, 41 pp., 1929.
21. HOOVER, M. D., and C. F. LEAF, Process and significance of interception in Colorado subalpine forest, in Sopper, W. E. and H. W. Lull (eds.) *Forest Hydrology,* pp. 213-224, Pergamon, Oxford, 1967.
22. HORTON, R. E., Rainfall interception, *Mon. Wea. Rev.,* **47**: 603-623, 1919.
23. JACKSON, I. J., Problems of throughfall and interception assessment under tropical forest, *J. Hydrol.,* **12**: 234-254, 1971.
24. JOHNSON, W. M., The interception of rain and snow by a forest of young Ponderosa-pine, *Trans. AGU,* **23**: 566-570, 1942.
25. KITTREDGE, J., *Forest Influences,* McGraw-Hill, New York, 1948.
26. KONTORSHCHIKOV, A. S., and K. A. EREMINA, Interception of precipitation by spring wheat during the growing season, *Soviet Hydrol.,* **2**: 400-409, 1963.
27. KUMMEROW, J., Quantitative measurements of fog in the Fray Jorge National Park, *For. Abstr.,* **24**: 4576, 1962.
28. LAINE, R. J. de, Measuring rainfall on forest catchments, *J. Hydrol.,* **9**: 103-112, 1969.
29. LAW, F., The effect of afforestation upon the yield of water catchment areas, *J. Brit. Waterworks Assoc.,* **38**: 489-494, 1956.
30. LAW, F., Measurement of rainfall, interception and evaporation losses in a plantation of Sitka spruce trees, *IASH Gen. Ass. of Toronto, Proc.,* **2**: 397-411, 1958.
31. LEE, C. H., Transpiration and total evaporation, Chapter 8 in Meinzer, O. E. (Ed.), *Hydrology,* McGraw-Hill, New York, 1942.
32. LEONARD, R. E., and K. G. REINHART, Some observations of precipitation measurement on forested experimental watersheds, *USFS Research Note* NE-6, 4 pp., 1963.
33. LEOPOLD, L. B., M. G. WOLMAN, and J. P. MILLER, *Fluvial Processes in Geomorphology,* Freeman, San Francisco, 1964.
34. LEYTON, L., E. R. C. REYNOLDS, and F. B. THOMPSON, Rainfall interception in forest and moorland, in Sopper, W. E. and H. W. Lull (Eds.), *Forest Hydrology,* pp. 163-178, Pergamon, Oxford, 1967.

35. LINSLEY, R. K., M. A. KOHLER, and J. L. H. PAULHUS, *Applied Hydrology,* McGraw-Hill, New York, 1949.
36. LULL, H. W., Ecological and silvicultural aspects, Section 6 in Chow, V. T. (Ed.), *Handbook of Applied Hydrology,* McGraw-Hill, New York, 1964.
37. LULL, H. W., and F. M. RUSHMORE, Snow accumulation and melt under certain forest conditions in the Adirondacks, *USDA, N.E. For. Exp. Sta. Paper* No. 138, 16 pp., 1960.
38. McMILLAN, W. D., and R. H. BURGY, Interception loss from grass, *JGR,* **65**: 2389-2394, 1960.
39. MERRIAM, R. A., A note on the interception loss equation, *JGR,* **65**: 3850-3851, 1960.
40. MILLER, D. H., The heat and water budget of the earth's surface, *Advances in Geophys.,* **11**: 175-302, 1965.
41. MILLER, D. H., Sources of energy for thermodynamically-caused transport of intercepted snow from forest crowns, in Sopper, W. E. and H. W. Lull (Eds.), *Forest Hydrology,* pp. 201-211, Pergamon, Oxford, 1967.
42. MUSGRAVE, G. W., Field research offers significant new findings, *Soil Cons.,* 3: 210-214, 1938.
43. NAGEL, J. F., Fog precipitation on Table Mountain, *QJRMS,* **82**: 452-460, 1956.
44. OBERLANDER, G. T., Summer fog precipitation on the San Francisco peninsula, *Ecology,* **37**: 851-852, 1956.
45. PATRIC, J. H., Rainfall interception by mature coniferous forests of southeast Alaska, *J. Soil and Water Cons.,* 21: 229-231, 1966.
46. PENMAN, H. L., *Vegetation and Hydrology,* C.A.B., Farnham Royal, 1963.
47. PENMAN, H. L., in discussion of Hoover, M. D. and C. F. Leaf, Process and significance of interception in Colorado subalpine forest, in Sopper, W. E. and H. W. Lull (Eds.), *Forest Hydrology,* pp. 213-224, Pergamon, Oxford, 1967.
48. PENMAN, H. L., in discussion of Delfs, J., Interception and stemflow in stands of Norway spruce and beech in West Germany, in Sopper, W. E. and H. W. Lull (Eds.), *Forest Hydrology,* pp. 179-185, Pergamon, Oxford, 1967.
49. PEREIRA, H. C., J. S. G. McCULLOCH, M. DAGG, and others, Assessment of the main components of the hydrological cycle, *East African Agr. and Forestry J.,* **27**: 8-15, 1962.
50. PILLSBURY, A. F., R. E. PELISHEK, J. F. OSBORN, and T. E. SZUSZKIEWICZ, Effects of vegetation manipulation on the disposition of precipitation on chaparral-covered watersheds, *JGR,* **67**: 695-702, 1962.
51. RAKHMANOV, V. V., Are the precipitations intercepted by the tree crowns a loss to the forest?, *Botanicheskii Zhurnal,* 43: 1630-1633, 1958. Trans PST Cat. No. 293, Office Tech. Serv. US Dept. Commerce, Washington, DC.
52. RAKHMANOV, V. V., *Role of Forests in Water Conservation,* Goslesbumizdat, Moscow, 1962; Trans. and edited by A. Gourevitch and L. M. Hughes, Israel Program for Scientific Translations Ltd, Jerusalem, 192 pp., 1966.
53. REYNOLDS, E. R. C., The hydrological cycle as affected by vegetation differences, *J. Inst. Water Eng.,* 21: 322-330, 1967.
54. REYNOLDS, E. R. C., and L. LEYTON, Measurement and significance of throughfall in forest stands, in Rutter, A. J. and F. H. Whitehead (Eds.), *The Water Relations of Plants,* pp. 127-141, Blackwell Scientific Publications, Oxford, 1963.
55. ROWE, P. B., Some factors of the hydrology of the Sierra Nevada foothills, *Trans. AGU,* **22**: 90-100, 1941.
56. ROWE, P. B., and T. M. HENDRIX, Interception of rain and snow by second growth of Ponderosa pine, *Trans. AGU,* **32**: 903-908, 1951.
57. RUTTER, A. J., Studies in the water relations of *pinus sylvestris* in plantation conditions, *J. Ecol.,* **51**: 191-203, 1963.
58. RUTTER, A. J., An analysis of evaporation from a stand of Scots pine, in Sopper, W. E. and H. W. Lull (Eds.), *Forest Hydrology,* pp. 403-417, Pergamon, Oxford, 1967.
59. SATTERLUND, D. R., and A. R. ESCHNER, The surface geometry of a closed conifer forest in relation to losses of intercepted snow, *USFS Res. Paper* NE-34, 16 pp., 1965.
60. SATTERLUND, D. R., and H. F. HAUPT, The disposition of snow caught by conifer crowns, *WRR,* 6: 649-652, 1970.
61. SCHOMAKER, C. E., Comparison of snow interception by a hardwood and a conifer forest, *Res. in the Life Sciences,* **16**: 35-43, 1968.

62. SPECHT, R. L., Dark Island Heath (Ninety-Mile Plain, South Australia). IV. Soil moisture patterns produced by rainfall, interception and stem-flow, *Australian J. Bot.,* **5**: 137-150, 1957.

63. STITCHLING, W., and S. R. BLACKWELL, Drainage area as a hydrologic factor on the glaciated Canadian prairies, *IASH Gen. Ass. of Toronto, Proc.,* **3**: 365-376, 1957.

64. STOLTENBERG, N. L., and T. V. WILSON, Interception storage of rainfall by corn plants, *Trans. AGU,* **31**: 443-448, 1950.

65. SWANK, W. T., and N. H. MINER, Conversion of hardwood-covered watersheds to white pine reduces water yield, *WRR,* **4**: 947-954, 1968.

66. THORUD, D. B., The effect of applied interception on transpiration rates of potted Ponderosa pine, *WRR,* **3**: 443-450, 1967.

67. TWOMEY, S., Precipitation by direct interception of cloud-water, *Weather,* **12**: 120-122, 1957.

68. VOIGT, G. K., Distribution of rainfall under forest stands, *Forest Sci.,* **6**: 2-10, 1960.

69. WAISEL, Y., Fog precipitation by trees, *La-Yaaran,* **9**: 29, 1960.

70. WARD, R. C., *Small Watershed Experiments,* Occasional Papers in Geography, No. 18, Univ. of Hull, 254 pp., 1971.

71. WEST, A. J., and K. R. KNOERR, Water losses in the Sierra Nevada, *J. Amer. Wat. Wks. Assoc.,* **51**: 481-488, 1959.

72. WHITMORE, J. S., Agrohydrology, *South African Dept. of Water Affairs Tech. Reprt.,* **22**, 1961.

73. WIGHAM, J. M., Interception, Section IV in Gray, D. M. (Ed.), *Handbook on the Principles of Hydrology,* National Research Council of Canada, Ottawa, 1970.

74. WISLER, C. O., and E. F. BRATER, *Hydrology* (2nd Ed.), Wiley, New York, 1959.

75. ZINKE, P. J., Forest interception studies in the United States, in Sopper, W. E. and H. W. Lull (Eds.), *Forest Hydrology,* pp. 137-161, Pergamon, Oxford, 1967.

4. Evaporation

4.1 Introduction and definitions

Evaporation, briefly defined, is the process by which a liquid or a solid is changed into a gas. In the context of the hydrological cycle, this involves the conversion to water vapour and the return to the atmosphere of the solid or liquid precipitation which reaches the earth's surface. Indirectly, the most important form of evaporation is probably that which takes place from the seas and oceans, since this forms the main source of all the water on the land areas, and is one of the principal factors in the large-scale transfer of water and water vapour between the oceans and the continents [4]. Other important forms include the evaporation of intercepted moisture from vegetation surfaces which was discussed in chapter 3, and the evaporation from bare soil surfaces, which is normally small per unit area in relation to that which takes place from streams, rivers, and lakes. Direct evaporation from falling precipitation can be ignored by the hydrologist, who is concerned only with the amount of precipitation which, in fact, reaches the earth's surface.

4.2 The process of evaporation

The molecules comprising a given mass of water, no matter whether this is a large lake or a very thin layer covering an individual soil grain only a fraction of a millimetre in diameter, are in constant motion.

Adding heat to the water causes the molecules to become increasingly energized and to move more rapidly, the result being an increase in distance between liquid molecules and an associated weakening of the forces between them. At high temperatures, therefore, more of the molecules near the water surface will tend to fly off into the lower layers of the overlying air. At the same time, of course, water vapour molecules in these lower air layers are also in continual motion, and some of these will penetrate into the underlying mass of water. The rate of evaporation at any given time will, therefore, depend on the number of molecules leaving the water surface, less the number of returning molecules. If this is a negative quantity, i.e., if more molecules are returning to the water surface than are leaving it, condensation is said to be taking place.

Generally speaking, the evaporation of water from a given surface is greatest in warm, dry conditions, and least in cold, calm conditions, because when the air is warm, the saturation vapour pressure (E) of water is high, and when the air is dry, the actual vapour pressure (e) of the water in the air is low. In other words, in warm, dry conditions, the saturation deficit $(E - e)$ is large and, conversely, in cold moist conditions it is small. There is, thus, a basic proportionality between the size of the saturation deficit and the rate of evaporation.

Since the process of evaporation involves the net movement of water vapour molecules from a water surface into the overlying air, it will be apparent that a continuation of this process will eventually lead to the saturation of the lowest layers of the overlying air, and the consequent cessation of evaporation when $(E - e)$ is zero. In absolutely calm conditions this situation would soon occur. Normally, however, air movement in the form of turbulence or convection mixes the lowest layers with the overlying air, thereby effectively reducing the water vapour content and permitting further evaporation to take place. Clearly, the stronger the wind, the more vigorous and the more effective will be the turbulent action in the air; and the greater the difference between surface and overlying air temperatures, the greater will be the effect of convection.

The basic factors of the evaporation process were first expressed in quantitative terms by Dalton [9], who suggested that if other factors remain constant, evaporation is proportional to the windspeed and the vapour pressure deficit, i.e., the difference between saturation vapour pressure at the temperature of the water surface, and the actual vapour pressure of the overlying air. Dalton's Law, although never expressed by the author in mathematical terms [47], has formed the starting point of much of the subsequent work on evaporation.

Factors affecting evaporation from free water surfaces

A number of factors, both meteorological and physical, affect the rate of free water evaporation although it is often difficult to assess the relative importance of each of them.

4.3 The meteorological factors

Radiation

It was reported [24] that Aristotle, pondering whether the sun or the wind was the most important factor determining evaporation, favoured the wind because it removes the water vapour once it has been produced. To the extent, however, that the change in state of water from a liquid to a gas involves the expenditure of approximately 590 calories per gram of water at ordinary field temperatures, it is evident that solar radiation will be a factor of considerable importance, and that it will set the broad limits, and will govern the main variations, in the rate of evaporation. Indeed, solar radiation is now generally regarded as by far the most important single factor involved, and it has been suggested that the term 'solar evaporation' [22] is basically applicable to losses from a free water surface. This is emphasized in Fig. 4.1, where there is a very close relationship between the curves for solar energy and evaporation.

Temperature

Since air and water temperatures are largely dependent upon solar radiation, it is to

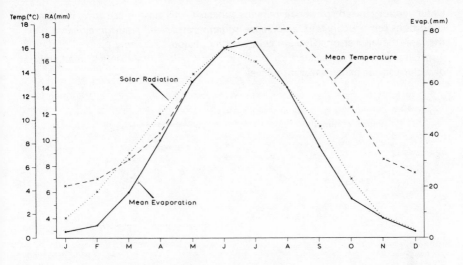

Fig. 4.1. *Graphs of mean monthly totals of evaporation, temperature, and solar radiation (mm evaporation equivalent). (Evaporation and temperature data for Camden Square, London, from Miller [27]; solar radiation data from tables.)*

be expected that there will be a fairly close correlation between them and the rate of evaporation. Figure 4.1 shows, however, that largely because air temperature shows a delayed response to solar energy the air temperature curve does not 'fit' as closely as the curve for solar energy. The temperature of the water surface is important in that it governs the rate at which water molecules leave the surface and enter the overlying air; a change in water surface temperature may, therefore, have a profound short-term effect upon the rate of evaporation.

Humidity

The direct relationship between the actual humidity of the air and the rate of evaporation has already been discussed, i.e., the rate of evaporation is proportional to the difference between the actual humidity and the saturated humidity at given temperatures. The actual vapour pressure varies only slightly from time to time throughout the day (see Fig. 4.2), except with a change of air mass—a frequent occurrence in the British Isles. Because of its different water vapour content, each air mass affecting the British Isles will tend to have a different evaporating capacity, other conditions, e.g., temperature, being equal.

In contrast, relative humidity is a much more variable characteristic (see Fig. 4.2). As the relative humidity of the air over the evaporating surface rises, proportionately fewer of the water vapour molecules leaving the evaporating surface can be retained in the air, and so the rate of evaporation, i.e., the total number of molecules leaving, minus the total number of molecules returning, is gradually reduced, although even at 100 per cent relative humidity some evaporation normally takes place.

Since relative humidity increases as the air temperature falls, even though the water vapour content of the air remains constant, it is easy to see why, if other conditions remain constant, a decrease in temperature will result in a decrease in the rate of evaporation. Thus, in cold weather, evaporation may be smaller than in warm weather simply because the overlying air is able to hold only a small amount of water vapour below saturation level.

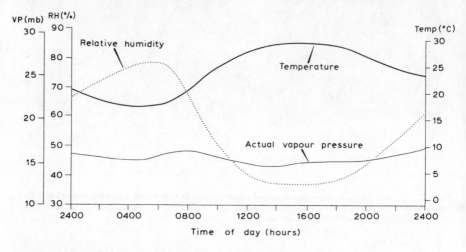

Fig. 4.2. *An example of the diurnal variation of vapour pressure, relative humidity, and air temperature. (From an original diagram in American Society of Civil Engineers,* Hydrology Handbook, *New York, 1949.)*

Wind

Evaporation from a free water surface into a completely still layer of air will be continually slowed down as the air layer approaches saturation point, and as more of the water vapour molecules in the air re-enter the water surface. Some sort of air movement is, therefore, necessary to stir up the air and to remove the lowest moist layers in contact with the water surface and to mix them with the upper, drier layers. In fact, absolutely calm air, even above a water surface, is very rarely experienced in natural conditions, so that the rate of evaporation is almost always influenced to some extent by air movement.

In this respect, turbulent movement is more important than the strength of the wind although in fact the degree of turbulence is closely related to wind velocity and to surface roughness—the latter factor being more important in relation to evaporation from a land surface.

However, the relationship between windspeed and evaporation holds good only to a certain point: beyond a certain critical value any further increase in windspeed leads to no further increase in evaporation. This critical windspeed will obviously vary as other conditions influencing evaporation change, but the associated turbulence will evidently remove water vapour molecules from the air in contact with the water surface sufficiently rapidly to enable evaporation to proceed at the maximum rate governed by the existing radiation, temperature, and humidity conditions. Even if the vapour-laden air were removed more rapidly, the rate of evaporation would not be increased. In this sense, wind does not actually *cause* evaporation but, by 'clearing the air', permits a given rate of evaporation to be maintained.

4.4 The geographical factors

Water quality

The rate of evaporation from water surfaces exposed to identical climatic conditions may vary according to the quality of the water. For example, evaporation decreases by about 1 per cent for every 1 per cent increase in salinity, so that evaporation from sea water with an average salinity of about 3.5 per cent is some 2 to 3 per cent less than evaporation from fresh water. This reduction is brought about by the reduced vapour pressure of the saline water. However, except when salinities are very high, e.g., in sea water, this is normally small enough to be discounted when comparing evaporation rates from different water bodies.

The turbidity of the water—another aspect of its quality—probably has little direct effect upon evaporation although, by affecting the albedo (or reflectivity) of the water, and consequently its heat budget, and temperature, it may have an indirect effect.

Depth of the water body

The effect of water depth upon the rate of evaporation may be quite considerable. The seasonal temperature regime of a shallow water body, e.g., a small lake, will normally approximate closely to the seasonal air temperature regime, so that maximum water temperatures are reached in the mid- to late summer months and minimum water temperatures in the mid- to late winter months. This means that maximum rates of evaporation from a shallow water body will be experienced during the summer, and minimum rates during the winter, as illustrated by curve A in Fig. 4.3.

Large, deep lakes, however, not only have a much higher capacity for heat storage than small water bodies but in middle and higher latitudes they normally experience a marked thermal stratification which also affects evaporation from their surfaces. Thus, during the spring and summer, heat entering the water surface is mixed downward for only a limited distance by wind action. This results in the formation of an upper layer of water, the epilimnion, in which temperatures are relatively uniform and higher than those in the rest of the water body, and beneath which there is the thermocline, a water layer in which temperature decreases, and density increases, rapidly with depth. With cooling of the surface waters during autumn and their associated increase in density, the cooled surface waters settle away from the surface and are replaced by warmer water from below. In ice free conditions this turnover may continue throughout the winter as a result of wind action [57]. In any case the turnover results in approximately uniform temperatures and densities throughout the depth of water so that turbulence resulting from wind action can easily activate deep mechanical mixing. There is, therefore, a slow release by the deep water body of stored heat during the autumn and winter months which means that a supply of heat energy in excess of that received directly from the sun is made available for evaporation at that time of the year. The net

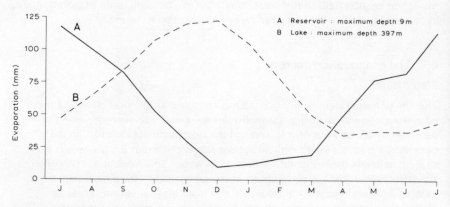

Fig. 4.3. *Graphs showing the trend of evaporation through the year from (A) Kempton Park Reservoir, near London and (B) Lake Superior. (Based on data from Lapworth [19] and S. T. Harding in Hickman [15].)*

result of this heat storage on relative air and water temperatures is that water temperatures are lower than air temperatures during the summer, and higher than air temperatures during the winter. Since, according to Dalton's law, evaporation is proportional to the difference between saturation vapour pressure at the temperature of the water surface, and the actual vapour pressure of the overlying air, it will be apparent that the highest rates of evaporation from deep water bodies should occur during the winter, as shown by curve B in Fig. 4.3. Furthermore, at that season, water-vapour-laden air will be rapidly lifted away from the underlying water surface as a result of convectional activity, encouraged by the temperature gradient, whereas during the summer, the colder water will tend to cool and stabilize the air immediately above it, and so inhibit the removal of vapour-laden air.

Despite the seasonal discrepancies which have been noted above, it is unlikely that the annual evaporation from deep and shallow water will differ very markedly [13].

Size of water surface

The relationship between the size of a water surface and the rate of evaporation from it has long been a subject of investigation. Figure 4.4 shows qualitatively how, with constant windspeeds, the rate of evaporation is related to the size of the evaporating surface and to the relative humidity. The reasons for this phenomenon are not difficult to understand. Thus, air moving across a large lake (see Fig. 4.5) may have a low water vapour content at the upwind edge; evaporation from the lake surface will gradually increase the water vapour content but, at the same time, there will be a reciprocal decrease in the rate of evaporation as the vapour 'blanket' increases in thickness. It is evident that the larger the lake, the greater will be the total reduction in the *depth* of water evaporated, although of course the total *volume* of water evaporated will almost inevitably increase with the size of water surface.

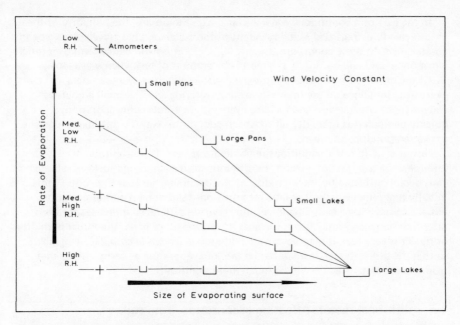

Fig. 4.4. *Diagram to indicate the relationship between the rate of evaporation and the size of the evaporating surface and the relative humidity. (From an original diagram by C. W. Thornthwaite and J. R. Mather,* Publ. in Climat., **8***: 1-86, 1955.)*

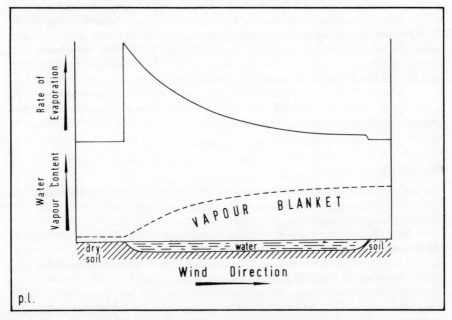

Fig. 4.5. *A simple illustration of the increase in humidity and decrease in evaporation as air moves across a large water surface.*

In the case of a continuous water surface, e.g., the oceans, the humidity of the air can clearly be regarded as independent of the distance it has travelled, except in coastal areas. In these conditions, therefore, evaporation will be uniform over much larger areas, and will be closely related to the amount of heat energy available.

At the other extreme, small evaporating surfaces such as evaporimeters and pans exert little influence on the humidity of the overlying air. The small amount of water vapour which leaves the surface, even with high rates of evaporation, is quickly dissipated as more dry air moves in and, in this way, a continuous high rate of evaporation is maintained.

However, with other conditions remaining constant, the magnitude of differences in evaporation rates from different-sized evaporating surfaces will be considerably affected by the humidity of the 'incoming' air (see Fig. 4.4). If this is initially high, then clearly it will be modified only slightly, even after a lengthy passage across a large lake. There will thus be little difference in the evaporation rates from large and small water surfaces. If, on the other hand, the initial humidity of the air is very low, it may be considerably increased while crossing a large lake, so that the proportional decrease in the rate of evaporation as compared to that from a small water surface will be much more significant.

4.5 Additional factors affecting evaporation from soils

The rate of evaporation from a soil surface will be governed by the same meteorological factors that govern the evaporation loss from a free water surface since soil evaporation is merely the evaporation of the films of water surrounding the soil grains and filling the spaces between them. However, in the case of free water evaporation, the supply of moisture is always, by definition, so plentiful that it exerts no limiting influence on the rate of water loss; in other words, what Horton called the 'evaporation opportunity' is always 100 per cent. On the other hand, evaporation from soils is often less than evaporation from a free water surface, not because the climatic conditions are different, but because there is not a sufficient supply of water in the soil to be evaporated; in other words, the 'evaporation opportunity' is less than 100 per cent. Thus, the most important additional factors affecting evaporation from soils are those which influence the evaporation opportunity.

Soil moisture content

The actual moisture content of the soil surface obviously exerts the most direct influence on the evaporation opportunity. Experiments have shown that, with saturated soils, the evaporation opportunity is often greater than 100 per cent [14], [47], because the countless minute irregularities of the soil cover comprise a larger evaporating surface than that of an 'identical' area of water. The finer the soil texture and the smoother the surface, the closer will the evaporation opportunity approximate to 100 per cent at saturation. Early investigators found that the relationship between soil moisture content and evaporation was quite close [10], [14], [54], [56], [57]; evaporation decreases rapidly as the surface moisture content falls until, with a dry soil surface, it is zero. This pattern has since been confirmed by many experiments: Veihmeyer and Brooks [52], for example, reported the data illustrated in Fig. 4.6, which shows the relationship between open water and soil evaporation during a dry period after irrigation. The soil was at field capacity on 9 July and, although for the first few days the rate of evaporation was high, the soil curve soon flattened out and, by the end of the period illustrated, total evaporation from the water surface had exceeded total evaporation from the soil surface by 372 mm.

It should be emphasized that it is the moisture content of the few centimetres of surface soil which plays the decisive role; the subsoil may be saturated but, because of the slow movement of soil moisture, it may have a negligible effect on the rate of evaporation from the surface. There may, of course, be some movement of water vapour from the subsoil to the surface, but this will normally represent only a minute fraction of the total evaporation loss. The way in which rainfall replenishes the soil moisture from above, therefore, assumes considerable importance. If other conditions remain constant, evaporation will be greater from a soil surface frequently wetted by intermittent showers than from a soil surface thoroughly soaked by the same quantity of rain falling in one storm.

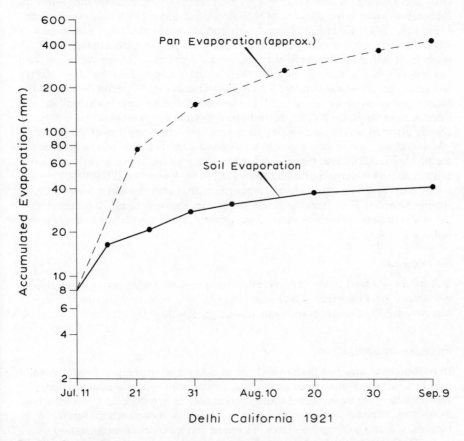

Fig. 4.6. *Evaporation from open water and from soil during a dry period after irrigation. (Based on data presented by Veihmeyer and Brooks [52].)*

Soil capillary characteristics

In areas where rainfall is not a frequent source of soil moisture replenishment, the evaporation opportunity tends frequently to vary in relation to those soil factors which affect the capillary rise of moisture stored in the soil layers immediately beneath the surface. The upward movement of soil moisture by capillarity is conditioned partly by the size and arrangement of the soil particles. In fine-textured soil, capillary movement is effective over larger vertical distances, e.g., a metre or so. In coarse-textured soil, it may be only a few centimetres. However, the speed of movement tends to vary inversely with the height of capillary lift so that, in neither case, except where special conditions prevail, does the supply of water to the soil surface by capillary activity significantly increase the total amount of evaporation.

Water table depth

One special condition which may modify the part played by capillary movement in evaporation losses is the presence of shallow ground water. Numerous experiments [16], [30], [44], [51], [52], have shown that soil evaporation is at a maximum when the water table is at the surface, and decreases quite rapidly at first, as the water table retreats downwards; but that, after a depth of about one metre has been reached, any further decrease in water table height is accompanied by only a slight change in the rate of evaporation. This is illustrated in Fig. 4.7, which is based on figures, presented by Veihmeyer [51], of evaporation from waterlogged soils at Davis, California. Naturally, the critical depth beyond which evaporation is only slightly affected, will depend largely on the capillary characteristics of the soil, and will, therefore, tend to vary with soil texture and particle size. It will be greatest in the fine textured silts and loams, and least in coarse sands. Classic experiments on this phenomenon were reported by Keen [16], who found that capillary lift contributed insignificantly to total evaporation losses after the water table had receded to about 35 centimetres below the surface in coarse sand: 70 centimetres below the surface in fine sand; and 85 centimetres below the surface in heavy loam soil (see Fig. 4.8).

Soil colour

Soil colour will tend to affect evaporation because darker soils, having a low albedo, will absorb more heat than lighter soils. The resulting increase in surface temperature may modify the evaporation rate quite significantly.

Presence of vegetation

In relation to an area like the British Isles, discussion of evaporation from bare soil may be largely academic, since extensive areas completely devoid of vegetation, ploughed land, for example, are usually found only during the colder months when, in any case, climatic conditions favour only a low rate of evaporation. Again, most natural and cultivated surfaces in summer carry a continuous vegetation cover. During the spring and early summer months, however, climatic conditions may favour considerable evaporation losses from soil partially covered by immature,

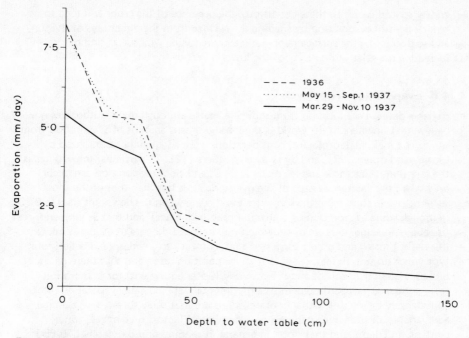

Fig. 4.7. *Graphs showing the relationship between soil evaporation and water table depth. (Based on data presented by Veihmeyer [51].)*

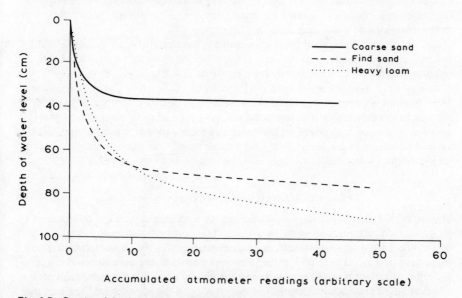

Fig. 4.8. *Graphs of declining water table level plotted against cumulative evaporation. The flattening out of each curve indicates the level at which capillary movement of water to the soil surface becomes insignificant. (From an original diagram by B. A. Keen,* Proc. and Papers 1st Internat. Cong. Soil Sci., **1***: 504-511, 1927.)*

widely spaced crops. In these conditions, the presence of the crops will tend to reduce the soil evaporation by shading the surface from the direct rays of the sun, and so decreasing the surface temperature; by reducing windspeed, and by increasing the relative humidity of the lower layers of the air.

4.6 Evaporation from snow

Reliable data on evaporation from snow are sparse although many authorities now believe that nowhere in the world is there a significant amount of evaporation from snow. The main meteorological considerations were adequately summarized by Linsley and others [22], and by Gray and others [12]. Of particular importance is the fact that, since snow and ice melts at 0° C, and since evaporation can occur only when the vapour pressure of the overlying air is less than that of the snow surface, evaporation from the snow itself will cease when the dewpoint rises to 0° C (although some evaporation of meltwater may take place), and that as temperatures rise above freezing, the rate of snowmelt must exceed the rate of evaporation. On this basis, Linsley and others suggested that, contrary to a widespread belief, fohn type winds do not, in fact, result in high evaporation rates, and that there is probably an upper limit of about 5 mm/day (water equivalent) for evaporation from a snow surface, although most experimental data indicate evaporation rates considerably below this. Other considerations are that snow surfaces have a high and variable albedo which depends on a number of factors such as age, water content, and depth and that since, in general, it requires approximately 675-680 cal/g to sublimate snow whereas the heat of fusion is only about 80 cal/g the mass of water melted is substantially greater than the amount evaporated [12]. The limited energy normally available for snow evaporation has already been referred to in the discussion of snow interception (see section 3.7).

Unfortunately, for several practical reasons it has been difficult to substantiate theory by measurement. Nevertheless, interesting and valuable results have been obtained from time to time, particularly by Croft [7], Clyde [6], Baker [2], Kittredge [17], Garstka and others [11] and by the U.S. Army Corps of Engineers [49]. Kittredge's work in the western Sierra Nevada was very comprehensive and led him to the conclusion that the outstanding feature of snow evaporation in this area was 'the small magnitude' of the measured losses. Indeed, the data reported by various investigators are fairly consistent on this point, and indicate that, during the spring months, evaporation averages about 25 mm/month or less [21].

The estimation of evaporation

The important role played by evaporation in the hydrological cycle is reflected in a long history of attempts to improve the accuracy of its estimation and measurement. Apart from empirical work, associated mainly with the Daltonian expression, there have developed two main theoretical approaches to the estimation of evaporation: the turbulent transfer approach and the energy balance approach—each of which is based upon a consideration of the physics of the atmosphere. The fact that these differ between themselves, and from the simple Dalton approach, does not imply that either of these approaches is incorrect, but merely that in each a different weight is given to the various factors involved.

4.7 Turbulent transfer approach

Air, like any other fluid, can have a *laminar* or a *turbulent* motion. In the first case, it moves in straight lines or along smooth, regular curves in one direction. In the second, air particles follow irregular, tortuous, fluctuating paths, evidenced in cross-currents and gusts of wind with intervening brief lulls. True laminar flow is never observed in the lower layers of the atmosphere [42]; instead, friction between the air and the ground surface induces eddies and whirls, and other irregular air movements. This is the turbulent layer, i.e., the zone above the earth's surface in which the frictional effects of the ground are felt [47]. The depth and strength of the turbulence in this layer is largely dependent upon the roughness of the ground surface and the strength of the wind, and increases as these two factors increase. It is also greatly influenced by the stability of the air, which varies diurnally. Normally, during the daytime, the lapse rate increases, reaching a maximum in the early afternoon. The associated convectional activity, and the increased buoyancy of the air, together reinforce the degree of turbulence and deepen the turbulent layer. At night, on the other hand, the ground surface cools, and lapse conditions are normally replaced by an inversion of temperature. This decreases the buoyancy of the air, damps down convectional activity, and effectively suppresses the turbulent motion of the air until, on calm nights, an almost laminar flow is experienced. As would be expected, this diurnal variation of turbulent activity is greatest with clear skies, and scarcely apparent in completely overcast conditions [42].

With regard to evaporation, it is the mixing and diffusion processes associated with turbulence that are of the greatest importance. The greater the intensity of turbulence, the more effectively are water vapour molecules dispersed or diffused upwards into the atmosphere. When, however, the level of turbulence is low, as in the evening, the consequent reduction in the mixing and diffusion of water vapour through the lower layers of the atmosphere is evidenced in the greatly increased humidity of a thin layer of air immediately above the ground surface, and the probable formation of a shallow layer of ground fog.

The irregular eddying motion of air in the turbulent layer will quite quickly tend to establish a uniform moisture content at all heights, provided that no water vapour is added or removed. If water vapour *is* added to the bottom of the turbulent layer by evaporation from a soil or water surface, it will be dispersed and diffused upwards; but as long as the evaporation continues, the moisture content of the air will be highest at the ground surface, and lowest at the top of the turbulent layer. This moisture gradient will persist until evaporation ceases. A reverse gradient will develop if water vapour is removed from the bottom of the turbulent layer by condensation in the form of dew or frost. It follows from this that, with a constant intensity of mixing, an increase in the rate of evaporation will be reflected in an increased moisture gradient in the turbulent layer. Alternatively, with a constant rate of evaporation, variation of the moisture gradient will reflect changes in the intensity of mixing. Accordingly, it should be possible to determine the rate of evaporation from any surface by reference to the moisture gradient and the intensity of turbulent mixing. Measurements are needed of the moisture content at a minimum of two known heights within the turbulent layer, and of the windspeed at two or more levels.

Takhar [43] pointed out that within the turbulent transfer approach there have been three main lines of development based on the work of Prandtl [35], Taylor [45] [46], and Richardson [37] [38]. Associated with these different approaches, numerous evaporation formulae have been developed. Since, in each case, the rate of evaporation is determined on the basis of small humidity and windspeed differences over a narrow height range within the turbulent layer, the frequency and the accuracy of the instrumental observations must be very high. Certainly, for most hydrological problems where replication is desirable, and where the use of existing standard climatic data would enable the derivation of long-term evaporation estimates, the turbulent transfer approach makes unacceptable demands upon instrumental and data accuracy.

4.8 Energy balance approach

The energy balance approach constitutes a different and much simpler view of evaporation. Instead of concentrating attention upon the upward diffusion of water vapour from the evaporating surface, this approach emphasizes that, in order for water to change in state from a liquid to a gas, a certain amount of energy (approximately 590 cal/g at normal air temperatures) is needed. If, then, the total amount of radiation received by a given area of soil or water could be determined, and if all this radiation were used in evaporation, it would be a simple matter to calculate the depth of water evaporated.

In fact, of course, only a part of the incoming radiation is used for evaporation (H_E). Some is immediately reflected back into the atmosphere (R_E). Of the remainder which is absorbed, some is returned as longwave radiation from the soil or water surface (R_A); some is used to heat the overlying air (H_A); and some to heat the soil or water body (H_B). Finally, a negligible part is used in plant growth to produce carbohydrate [31] (H_C). An energy balance equation can, therefore, be written in the following terms where (R) represents the total incoming radiation:

$$R - R_E - R_A \quad = \quad H_E + H_A + H_B + H_C \quad (4.1)$$

Net radiation income   Radiation expenditure

The energy balance approach thus involves two main problems: first, to determine the total radiation income and, second, to apportion its outgo under the various headings outlined above in order that the amount used in evaporation may be accurately determined.

The use of radiometers to determine the income expression in eq. (4.1) is increasing. Measurements may be made either of the total incoming radiation, i.e., R in eq. (4.1), in which case R_E and R_A must be separately determined, or preferably of net radiation, i.e., $R - R_E - R_A$ in eq. (4.1). The net radiometer (see Fig. 4.9) is an essentially simple device whereby outgoing radiation from the ground, and incoming radiation from the sun and sky are measured by thermocouples between blackened plates. Two transparent hemispheres form a windshield round the sensitive element, and are kept free of internal condensation by means of dry nitrogen gas or desiccated air. The two radiation values are then integrated, and the resulting *net* radiation is recorded either on a register or on a continuous chart.

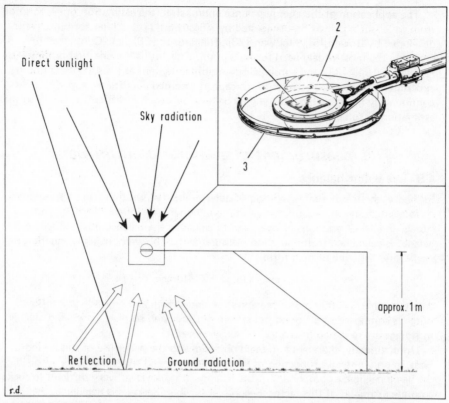

Fig. 4.9. *General principles of the net radiometer shown diagrammatically. Detailed drawing of the sensitive head (inset) shows: (1) thermocouples between blackened plates, (2) transparent hemispheres, (3) anti-condensation heating ring which cannot be 'seen' by the sensitive elements.*

When direct measurements are not available, it is possible to make a less reliable estimate of net radiation from bright sunshine data using empirical relationships between radiation and the duration of sunshine [3], and the reflection coefficient of the soil or water surface.

Various estimates have been made of the way in which the incoming radiation is apportioned. In average English summer conditions it has been suggested [32] that, of the total incoming radiation over a land surface, 20 per cent is reflected (R_E), 34 per cent is radiated back into the atmosphere (R_A), 4 per cent is used to heat the overlying air, 2 per cent to heat the soil, and 1 per cent in plant growth, leaving 39 per cent for evaporation. These figures can obviously serve only as a general guide. The proportion of reflected radiation (R_E), for example, will depend upon the albedo of the soil surface, which will be much higher for a light-coloured soil than for, say, a black, peaty soil. Again, as the soil dries out, less radiation will be used in evaporation, and consequently more will be available to raise the surface temperature. In the case of an open water surface, whose albedo is of the order of 0.05 compared with 0.17 [28] for bare soil, more of the net radiation will be available for evaporation than is indicated by the figures given above.

85

The application of the energy balance approach to the estimation of evaporation from water surfaces was first suggested by Ångström [1] and later formulae were developed by Bowen [5], McEwan [23], Richardson [36], and Cummings [8]. Probably the most widely used formula in this category is the one first proposed by Penman in 1948 [33] and subsequently modified by him as a result of continuing experimental work (see chapter 5). Except in specified conditions, e.g., a continuously moist surface, the energy balance approach is difficult to apply to the estimation of soil evaporation.

The measurement of evaporation from water

4.9 The water balance

In theory, the most reliable method of determining evaporation from a free water surface should be to measure the various components of inflow, outflow, and storage in a lake, reservoir, or river, and to calculate the balance, just as the energy balance is calculated from radiation measurements. The water balance equation would thus take the normal form.

$$\text{Income} = \text{Outgo} \pm \Delta \text{Storage} \qquad (4.2)$$

Where income is composed of precipitation and stream and groundwater inflow; outgo is composed of evaporation, stream outflow, and seepage losses; and changes in storage are reflected in changes of surface level.

Unfortunately, there are two main objections to the wider use of this method. First, in only a small proportion of lakes, reservoirs, and rivers is seepage through the bed a negligible component of the balance; it is, however, very difficult to make accurate estimates of this factor. Second, even where seepage is negligible, the cumulative error involved in the separate determination of the other components, i.e., precipitation (particularly in winter months when snow is frequent), stream flow, and water levels is usually so large as to render the final evaporation estimate no more reliable than that derived from the better formulae.

4.10 Evaporation pans

The simplest and most widespread method of obtaining a direct measurement of the evaporation from a free water surface is to measure the loss from an evaporation pan. For a number of reasons, however, the loss so recorded will differ from the evaporation from a large body of water, such as a lake or reservoir, which is exposed to the same meteorological conditions. The magnitude of this difference may depend to a large extent on the installation and exposure of the pan. If, for example, the pan is installed on supports above the ground surface (c.f. US Weather Bureau Class A pan), the evaporation rate will be modified by the extra radiation which falls on the sides of the container and is transmitted through to the water; by the heat exchanges between the pan and the atmosphere, through the bottom as well as through the sides of the container; and by the 'oasis' effect caused by the advection of warm dry air over the limited water surface, which maintains a continuous high rate of evaporation during the day.

Although some of these sources of discrepancy may be obviated by burying the pan in the ground (c.f. British Standard (MO) pan), others result from this method of installation. It has been found, for example, that there is an appreciable exchange of heat between sunken evaporation pans and the surrounding soil [29]. Roberts [39] suggested that this could be partly overcome by keeping the soil surrounding the pan in a moist condition throughout the year. Again, because of its screening effect, the height of the vegetation round the instrument is more important with this type of installation. Since the effect of both these factors will tend to diminish as the size of the pan is increased, it has been found that large sunken pans give better results than small ones.

A third type of evaporation pan actually floats in the lake or reservoir and, since it is surrounded by an extensive water surface, it should give the most accurate results. This type is, however, fast losing what popularity it had, mainly because of the difficulties and expense involved in routine observations and maintenance, and because the effect of wave splash in rough weather makes the data unreliable.

Certain problems are common to all three types of installation. In each case, for example, air flow across the water surface in the pan is affected either by an upstanding rim or by the fact that the water level is lower than the surface surrounding the instrument. Evaporation, therefore, differs from that which would be experienced from a large body of water. Again, the rate of evaporation from a pan is known to be influenced by the level of the water surface below the rim [29], and discrepancies may result if the water level is allowed to fall much below the recommended standard for a given type of pan.

The combined effect of the various sources of discrepancy is normally that pan evaporation is greater than the evaporation from a large water surface exposed to the same meteorological conditions. It is, thus, necessary to apply a correction factor or pan coefficient in order to derive the 'true' evaporation figure. These coefficients, which are fairly constant for each type of pan, have been empirically determined from experimental evidence.

The measurement of evaporation from soils

The measurement of evaporation from soils is complicated by the fact that only rarely is the evaporation opportunity 100 per cent (see page 78). In most cases the depletion of soil moisture reserves reduces the total evaporation from a soil surface compared with that from a water surface exposed to the same meteorological conditions. This means that soil evaporation can only be measured indirectly by calculating changes in the weight, volume, or equivalent depth of water in a given volume or depth of soil, or directly by measuring the water vapour flux above the evaporating soil surface.

4.11 Lysimeters and drain gauges

Lysimeters have been used in evaporation studies since the late seventeenth century [18]. For much of this time, however, they have been used to measure evapotranspiration rather than soil evaporation as such and will, therefore, be considered in the following chapter. At this stage discussion will be limited to examples of a particular type of lysimeter, i.e., the drainage or percolation gauge first used in Great Britain by Dalton [9].

Drain gauges or percolation gauges are exposed to the prevailing meteorological conditions and measurement is made of the quantity of water percolating through the topsoil layers exposed to natural conditions, and then evaporation is equal to rainfall minus percolation. Since there is no free water surface present in the gauge, the soil surface will dry out during rainless periods, so that evaporation is negligible. In dry years, therefore, the measured evaporation will tend to be lower than in wet years, and considerably lower than the values of free water evaporation obtained from evaporation pans. A major problem in measuring soil evaporation is to ensure that soil conditions in the gauge are truly representative of natural undisturbed conditions; otherwise, the amount of moisture moving through the profile to the evaporating surface may be unduly large or small [20] [34].

4.12 Other methods

Most of the remaining methods of measuring soil evaporation involve measurement of the moisture profile, either within the soil or within the shallow air layer immediately above the evaporating surface and will be further discussed in sections 6.7 and 5.21 respectively.

Seasonal and areal variations of evaporation

4.13 Free water surfaces

It has already been shown that evaporation from free water surfaces is largely determined by the amount of solar energy available. Accordingly, both seasonal and areal variations of evaporation are closely related to similar variations in the duration and intensity of solar radiation. A few selected examples will suffice to illustrate this point.

Smith [41] analysed evaporation pan data obtained between 1935 and 1961 at Harrogate (54°N), and the mean monthly totals for this twenty-six year period are plotted in Fig. 4.10. Curve A shows that the highest rates are experienced in June and July, and the lowest in December and January, and that there is a fairly uniform gradation between these two extremes. It should be noted that 299 mm (61.4 per cent) of evaporation occurs in the four 'summer' months from May to August, while only 44 mm (9.1 per cent) occur in the four 'winter' months from November to February. Long-period (1885-1938) monthly evaporation totals from the evaporation pan at Camden Square, London, are also plotted for comparison, and it will be seen that curve B is almost identical in shape with curve A, although in this case, summer (May-August) evaporation represents 68.6 per cent (270 mm) of the mean annual total, while winter (November-February) evaporation accounts for only 4.4 per cent (18 mm).

In few parts of the world is the network of evaporation pans sufficiently close or sufficiently long-established to enable accurate maps of the areal variation of evaporation to be made. However, a tentative evaporation map has been constructed for the British Isles (Fig. 4.11A), and shows a rational pattern with relatively high evaporative losses in the south and at various points along the east coast, suggesting that a simple latitudinal response to solar radiation is modified by maritime influences, and particularly by the advection of moist oceanic air from the

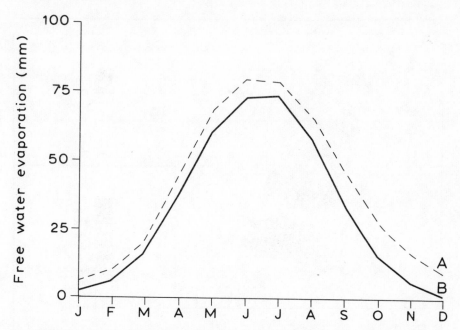

Fig. 4.10. *Mean monthly totals of free water evaporation at (A) Harrogate, 1935-61 and (B) Camden Square, London, 1885-1938. (Based on data from Smith [41] and Meteorological Office [25].)*

south-west, which considerably depress evaporation totals over much of Wales and south-west England. Although there are over 500 Class A evaporation pan installations in the United States [40], evaporation from lakes yields a more reliable distribution (Fig. 4.11B). As would be expected, the resulting pattern reflects the dominant control of energy availability as affected by latitude and continentality.

Again, the close relationship between free water evaporation and potential evapotranspiration, which was emphasized by Penman [33] (see later discussion in chapter 5), enables a schematic pattern of annual evaporation over Europe to be derived from Fig. 4.12. Particularly noteworthy is the gradual decrease in evaporative losses from south to north (from less than 600 mm in Norway to more than 1200 mm in southern Greece), and the fact that the latitudinal evaporation gradient, as evidenced by the spacing of the isolines, becomes steeper towards the south. The over-riding control of solar radiation is more apparent in this map than in Fig. 4.11B and indeed the comparatively regular pattern of the isolines is broken only by the mountains of central and southern Europe where relatively lower evaporation rates prevail.

4.14 Soil surfaces

Reference to the discussion of additional factors affecting the evaporation from soil surfaces (see pages 78-82) will indicate that both seasonal and areal variations of this parameter are far less rhythmic and predictable than the variations of free water evaporation.

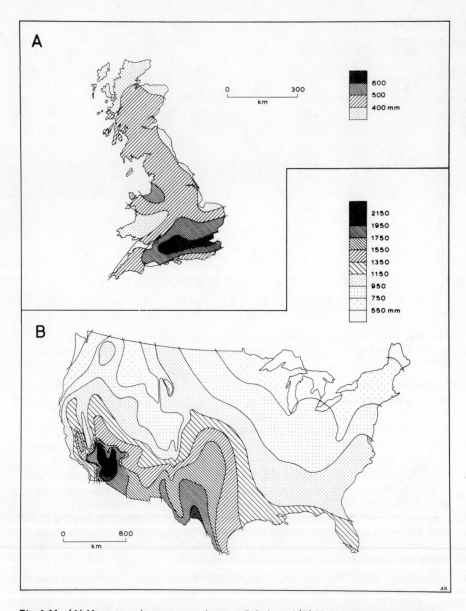

Fig. 4.11. *(A) Mean annual pan evaporation over Britain and (B) Mean annual lake evaporation over the conterminous USA. ((A) based on data in* British Rainfall, 1961 *[26] and on an original diagram by J. Wadsworth.* Weather, **3***: 322-324, 1948 and (B) on an original diagram by J. J. Geraghty, D. W. Miller, F. van der Leeden, and F. L. Troise,* Water Atlas of the United States, *Water Information Center Inc., New York, 1973.)*

A detailed analysis of conditions typical in the British Isles was made by Penman and Schofield [34] in a study of long-period data from the Rothamsted bare soil drain gauges. They concluded that although annual evaporation is nearly constant,

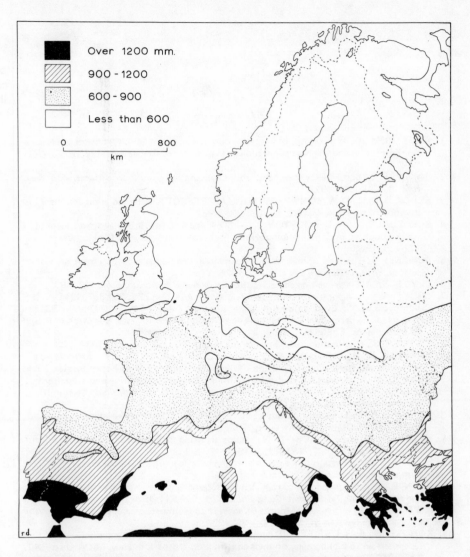

Fig. 4.12. *The distribution of annual potential evapotranspiration over Europe. (From an original map by J. C. J. Mohrmann and J. Kessler, Water deficiencies in European agriculture, Internat. Inst. Land Reclamation and Improvement, Publ., 5, 1959.)*

it tends to be higher in wet years and lower in dry years. This difference is largely to be accounted for by differences in the totals of summer evaporation in view of the fact that, during the summer months, soil evaporation is considerably less than the evaporation from a free water surface but that, the more frequently it is re-wetted, the closer it will approximate to this latter quantity.

Similarly, the areal pattern of annual soil evaporation will vary considerably from year to year, depending on the interaction between meteorological factors on the one hand and the wetness of the surface layers of the soil on the other, basically

91

References

1. ÅNGSTRÖM, A., Applications of heat radiation measurement to the problems of evaporation from lakes and the heat convection at their surfaces, *Geografisca Ann.,* **2**: 237-252, 1920.
2. BAKER, F. S., Some field experiments on the evaporation from snow surfaces, *Mon. Wea. Rev.,* **45**: 363-366, 1917.
3. BLACK, J. N. C., C. W. BONYTHON, and J. A. PRESCOTT, Solar radiation and the duration of sunshine, *QJRMS,* **80**: 231-235, 1954.
4. BLEASDALE, A., A. G. BOULTON, J. INESON, and F. LAW, Study and assessment of water resources, in *Conservation of Water Resources,* Institution of Civil Engineers, pp. 121-136, 1963.
5. BOWEN, I. S., The ratio of heat losses by conduction and by evaporation from any water surface, *Physical Review,* **27**: 779-787, 1926.
6. CLYDE, G. D., Snow-melting characteristics, *Utah Agric. Expt. Sta. Bull.,* 231, 1931.
7. CROFT, A. R., Evaporation from snow, *Bull. Amer. Met. Soc.,* **25**: 334-337, 1944.
8. CUMMINGS, N. W., Evaporation from water surfaces, *Trans. AGU,* **17**: 507-509, 1936.
9. DALTON, J., Experimental essays on the constitution of mixed gases, *Manchester Lit. and Phil. Soc. Mem.,* **5**: 535-602, 1802.
10. FORTIER, S., Evaporation losses in irrigation, *Eng. News,* **58**: 304-307, 1907.
11. GARSTKA, W. U., L. D. LOVE, B. C. GOODELL, and S. A. BERTLE, *Factors affecting snowmelt and streamflow,* U.S. Dept. of the Interior and U.S.D.A. Forest Service, 1958.
12. GRAY, D. M., G. A. McKAY, and J. M. WIGHAM, Energy, evaporation and evapotranspiration, in Gray, D. M. (Ed.), *Handbook on the Principles of Hydrology,* Section III, National Research Council of Canada, Ottawa, 1970.
13. HARDING, S. T., Evaporation from free water surfaces, Chapter 3 in Meinzer, O. E. (Ed.), *Hydrology,* McGraw-Hill, New York, 1942.
14. HARRIS, F. S., and J. S. ROBINSON, Factors affecting the evaporation of moisture from the soil, *J. Agric. Res.,* **7**: 439-461, 1916.
15. HICKMAN, H. C., Evaporation experiments, in Hydrology of the Great Lakes, A Symposium, *Trans. ASCE,* **105**: 807-849, 1940.
16. KEEN, B. A., The limited role of capillarity in supplying water to the plant roots, *Proc. and Papers of the 1st Internat. Cong. Soil Sci.,* **1**: 504-511, 1927.
17. KITTREDGE, J., Influences of forests on snow in the Ponderosa-subar pine-fir zone of the central Sierra Nevada, *Hilgardia,* **22**: 1-96, 1953.
18. KOHNKE, H., F. R. DREIBELBIS, and J. M. DAVIDSON, A survey and discussion of lysimeters and a bibliography on their construction and performance, *USDA Misc. Publ.,* 372, 1940.
19. LAPWORTH, C. F., Evaporation from the water surface of a reservoir, Cyclostyled copy of a paper given before the Internat. Union of Geodesy and Geophysics, 1956.
20. LAPWORTH, C. F., J. GLASSPOOLE, and D. LLOYD, Report on standard methods of measurement of evaporation, *J. Inst. Water Eng.,* **2**: 257-266, 1948.
21. LINSLEY, R. K., M. A. KOHLER, and J. L. H. PAULHUS, *Applied Hydrology,* McGraw-Hill, New York, 1949.
22. LINSLEY, R. K., M. A. KOHLER, and J. L. H. PAULHUS, *Hydrology for Engineers,* McGraw-Hill, New York, 1958.
23. McEWEN, G. F., Results of evaporation studies, *Scripps Inst. Oceanog. Tech. Ser.,* **2**: 401-415, 1930.
24. MENGEL, O., Critique des instruments et des methodes et observation, *IASH Assembly of Edinburgh, Publn.* **22**: 99-113, 1936.
25. METEOROLOGICAL OFFICE, *British Rainfall,* p. 145, 1938.

26. METEOROLOGICAL OFFICE, *British Rainfall, 1961,* H.M.S.O., 1967.
27. MILLER, A. A., *The Skin of the Earth,* Methuen, London, 1953.
28. MONTEITH, J. L., and G. SZEICZ, The radiation balance of bare soil and vegetation, *QJRMS,* **87**: 159-170, 1961.
29. NORDENSON, T. J., and D. R. BAKER, Comparative evaluation of evaporation instruments, *JGR,* 67: 671-679, 1962.
30. PARSHALL, R. L., Experiments to determine rate of evaporation from saturated soils and river bed sands, *Trans. ASCE,* **94**: 961-999, 1930.
31. PEARL, R. T., R. H. MATHEWS, L. P. SMITH, H. L. PENMAN, E. R. HOARE, and E. E. SKILLMAN, The calculation of irrigation need, *Min. Agric. and Fisheries Tech. Bull.* 4, 1954.
32. PENMAN, H. L., Theory of porometers used in the study of stomatal movements in leaves, *Proc. Roy. Soc., Ser. B.,* **130**: 416-434, 1942.
33. PENMAN, H. L., Natural evaporation from open water, bare soil and grass, *Proc. Roy. Soc., Ser. A.,* **193**: 120-145, 1948.
34. PENMAN, H. L., and R. K. SCHOFIELD, Drainage and evaporation from fallow soil at Rothamsted, *J. Agric. Sci.,* **31**: 74-109, 1941.
35. PRANDTL, L., Bericht über Untersuchungen zur Turbulenz, *Z. Angew. Mech. Math.,* 5: 136-139, 1925.
36. RICHARDSON, B., Evaporation as a function of insolation, *Trans. ASCE,* **95**: 996-1019, 1931.
37. RICHARDSON, L. F., Atmospheric diffusion shown on a distance-neighbour graph, *Proc. Roy. Soc. Ser. A.,* **110**: 709-737, 1926.
38. RICHARDSON, L. F., and H. STOMMEL, Note on the eddy diffusion in the sea, *J. Met.,* **5**: 238-240, 1948.
39. ROBERTS, D. F., A comparison of evaporation loss from Symons and Class A pans in South Africa, *Union of South Africa, Dept. of Water Affairs, Tech. Rept.,* 13, 1960.
40. ROBERTS, W. J., Significance of evaporation in hydrologic education. *The Progress of Hydrology,* University of Illinois, Urbana, Illinois, 1: 666-693, 1969.
41. SMITH, K., A long-term assessment of the Penman and Thornthwaite potential evapotranspiration formulae, *J. Hydrol.,* 2: 277-290, 1965.
42. SUTTON, O. G., *The Challenge of the Atmosphere,* Hutchinson, London, 1962.
43. TAKHAR, H. S., *Evaporation Theory and Practice,* Univ. of Manchester, Simon Engineering Lab., 49 pp., 1971 (cyclostyled).
44. TANNER, C.B., Factors affecting evaporation from plants and soils, *J. Soil and Water Cons.,* 12: 221-227, 1957.
45. TAYLOR, G. I., Diffusion by continuous movements, *Proc. London Math. Soc.,* **20**: 196-212, 1922.
46. TAYLOR, G. I., Statistical theory of turbulence, *Proc. Roy. Soc. Ser. A.* 151: 494-512, 1935.
47. THORNTHWAITE, C. W., and B. HOLZMAN, The determination of evaporation from land and water surfaces, *Mon. Wea. Rev.,* **67**: 4-11, 1939.
48. THORNTHWAITE, C. W., and J. R. MATHER, The water balance, *Publ. in Climat.,* 8: 1-86, 1955.
49. U.S. ARMY CORPS OF ENGINEERS, *Snow Hydrology,* Summary report of snow investigations, U.S. Army Corps of Engineers, N.Pacific Div., Portland, Ore., 1956.
50. VARIOUS CONTRIBUTORS, Symposium on evaporation from water surfaces, *Trans. ASCE,* **99**: 673-747, 1934.
51. VEIHMEYER, F. J., Evaporation from soils.and transpiration, *Trans. AGU,* **19**: 612-615, 1938.
52. VEIHMEYER, F. J., and F. A. BROOKS, Measurements of cumulative evaporation from bare soil, *Trans. AGU,* **35**: 601-607, 1954.
53. WADSWORTH, J., Evaporation from tanks in the British Isles, *Weather,* 3: 322-324, 1948.
54. WHITNEY, M., and F. K. CAMERON, Investigations in soil fertility, *USDA Bur. Soils Bull.,* **23**: 6-21, 1904.
55. WIDTSOE, J. A., Irrigation investigations; factors influencing evaporation and transpiration, *Utah Aa. Exp. Sta. Bull,* **105**, 64 pp., 1909.

56. WIDTSOE, J. A., *The Principles of Irrigation Practice,* 496 pp., New York, London, 1914.
57. ZUMBERGE, J. H., and J. C. AYERS, Hydrology of lakes and swamps, in Chow, V. T. (Ed.), *Handbook of Applied Hydrology,* Section 23, McGraw-Hill, New York, 1964.

5. Evapotranspiration

5.1 Introduction and definitions

Evaporation was defined and discussed in the previous chapter. Transpiration has been defined as '. . . the process by which water vapour escapes from the living plant, principally the leaves, and enters the atmosphere' [88]. The present chapter will be concerned with the combined effects of evaporation and transpiration, normally referred to as *evapotranspiration,* or sometimes as *consumptive use,* thereby emphasizing that which is a necessary part of the food-fibre production mechanism is not really a 'loss' but a 'use' of water [60], or *total evaporation.* Since evapotranspiration comprises' . . . the sum of the volumes of water used by the vegetative growth of a given area in transpiration or building of plant tissue and that evaporated from adjacent soil, snow, or intercepted precipitation on the area in any specified time' [202], it obviously represents the most important aspect of water loss in the hydrological cycle, accounting for the disposal of nearly 100 per cent of the annual precipitation in arid regions and about 75 per cent in humid regions [60]. It therefore merits rather more detailed consideration than was given individually to either evaporation or interception in chapters 3 and 4.

Only over those areas of the earth's surface where no vegetation is present, e.g., ice and snow fields, bare rock slopes, some desert areas, and water surfaces, will purely evaporative water losses occur. Elsewhere, whether in semi-desert scrublands or in the dense evergreen equatorial forests, transpiration will be a more or less dominant factor in the total losses of water from the land surface.

5.2 Transpiration as a botanical process

The plant transpiration system which is illustrated in Fig. 5.1 consists first of the *stomata,* the leaf orifices through which water vapour escapes to the atmosphere. Lenticels are occasional holes in bark tissue which perform a similar function on a much smaller scale and in addition, some transpiration takes place through the cuticle of leaves [63]. Secondly, there are the *spongy leaf cells,* from the surface of which vaporization takes place, the thin film of water covering the cell surfaces being supplied from inside the cells. These supplies of water are in turn replaced by the translocation of water through the *xylem,* which is a low resistance hydraulic conductor, from the roots. Finally, the *root system* obtains water by absorption at those surfaces which are in contact with soil moisture.

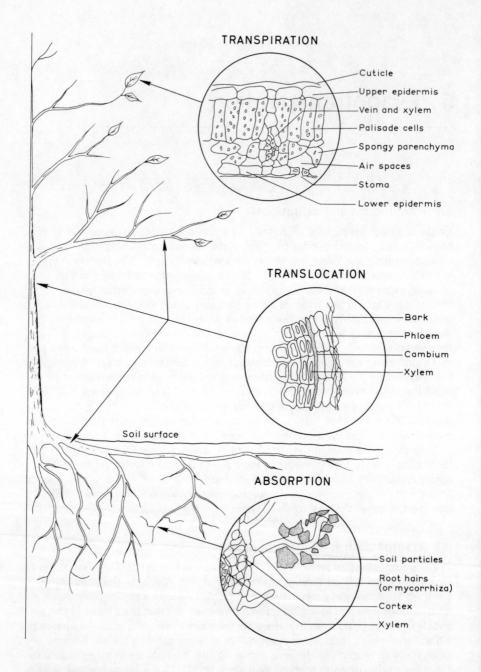

TRANSPIRATION

— Cuticle
— Upper epidermis
— Vein and xylem
— Palisade cells
— Spongy parenchyma
— Air spaces
— Stoma
— Lower epidermis

TRANSLOCATION

— Bark
— Phloem
— Cambium
— Xylem

Soil surface

ABSORPTION

— Soil particles
— Root hairs (or mycorrhiza)
— Cortex
— Xylem

Fig. 5.1. *A simplified diagram of the movement of water into and through the plant system. (From an original diagram by J. D. Hewlett and W. L. Nutter,* An Outline of Forest Hydrology, *Univ. of Georgia Press, 1969.)*

The transpiration process depends, of course, upon a supply of energy. A simplified view, outlined by Hewlett and Nutter [63] and expressed in Fig. 5.1, is that radiant and sensible heat energy supplied to the leaves and stems of plants causes evaporation within the leaf and that, when the resulting water vapour has diffused through the stoma and has been carried away by turbulent mixing, the resulting water loss produces a water deficit within the leaf cells. This water deficit represents a suction force or potential which is transmitted from cell to cell right through the system from the leaf to the root and is capable of drawing up moisture through even the highest trees against the force of gravity.

The water potential, negative in a transpiring plant cell, may be considered as

$$\psi = O + P + Z \qquad (5.1)$$

where ψ is total potential, O is osmotic potential, P is cell wall pressure potential, and Z is the gravity potential. In an actively transpiring plant, the osmotic potential of the cell sap in root and leaf will be strongly negative, cell wall pressure will approximate zero, and the gravity potential will be positive downwards. Some workers consider that the osmotic potential, developed at the root surfaces because of the concentration difference between the soil solution and the sap solution, is the primary energy source for the transpiration process [38], capable of developing a pressure difference of the order of 10 atmospheres [61]. Certainly, the rate of supply of water to the leaves may exceed the rate at which moisture is removed from the leaves by transpiration, in which case the stomatal cavities fill with moisture and liquid may fall from the leaves. This process, *guttation,* is common with some types of plant such as willow [38].

Transpiration is thus a flow process through a conducting medium of varying resistance, the soil-plant-atmosphere continuum (SPAC) [133], involving both chemical and phase changes. A simplified representation of the process as a series of driving forces and resistances was attributed by Hewlett and Nutter [63] to Van den Honert who, in 1948. suggested that

$$V = \frac{\psi_{soil} - \psi_{root}}{R_{cortex}} = \frac{\psi_{root} - \psi_{leaf}}{R_{xylem}} = \frac{\psi_{leaf} - \psi_{air}}{R_{stoma}} \qquad (5.2)$$

where V is the flow of transpiration water towards the leaf in cm/s; ψ is the total potential in bars at various levels, and R is the resistance offered at each level. This equation applies only if transpiration is steady through time which, of course, it seldom is.

Most physicists studying the transpiration process ascribe primary 'control' to the stomatal apparatus because the resistance is largest in the vapour pathway from the parenchyma cell through the stoma [63]. Other evidence, however, suggests that in some conditions it is the root resistance which is the most important link in the transpiration flow [28].

5.3 Potential and actual evapotranspiration

For a number of reasons which will become apparent in the following pages, considerable attention has been devoted in recent years to the concept of *potential evapotranspiration* (PE), and we must now distinguish between this parameter and (actual) evapotranspiration as it was defined in section 5.1. An essential distinction lies in the fact that the concept of PE involves a supply of moisture, either from the soil or from the atmosphere as precipitation, which is at all times sufficient to meet the demands of the transpiring vegetation cover. It was defined, for example, by Thornthwaite [175] as the 'water loss which will occur if at no time there is a deficiency of water in the soil for use of vegetation'.

Subsequent modifications to this concept have limited its application to the evapotranspiration which takes place from a continuous and unbroken green vegetation cover. Thus, PE was more explicitly defined by Penman and others [132] as the '. . . evaporation from an extended surface of short green crop, actively growing, completely shading the ground, of uniform height and not short of water'. If it is assumed that advective effects are excluded from consideration by the term 'extended', i.e., of sufficient extent to minimize advectional influences, that the effects of vegetation height are excluded from consideration by the term 'short', the effects of shape and roughness by 'uniform', and the role of soil moisture movement by the term 'never short of water', then clearly, according to this definition, PE is a climatic parameter which will not be affected by the movement of water through the soil and which will be affected by plant behaviour and plant type only in so far as these affect colour (and therefore albedo) and stomatal closure [196].

Equally clearly, a concept so restrictingly defined is likely to be of only limited practical significance. Thus, provided that all the restrictive conditions are fulfilled, the Penman concept suggests that PE represents the maximum possible evaporative loss from a vegetation-covered surface. However, these conditions are likely to be fulfilled, if at all, only for a very large surface of close-mown grass in a humid environment. In all other conditions, theoretical argument and experimental evidence indicate that the shape and height of the vegetation cover and the supply of large-scale advective energy affect the transpiration rate in such a way that, even in the humid conditions of Britain or the Netherlands, the actual rate of evapotranspiration under conditions of optimum water supply can exceed PE.

Because small departures from the ideal situation defined by Penman and others [132] may result in large differences in PE values, the concept of PE has been redefined, and renamed, on a number of occasions. Thus, a more realistic definition of the maximum evaporative loss from a vegetation-covered surface was proposed by van Wijk and de Vries [199], in terms of *wet surface evaporation,* defined as the maximum evaporation rate from a wet surface of similar shape, colour, and dimensions as the crop under consideration. The evaporation rate so defined is clearly a more useful concept, since it gives the maximum value under any climatological conditions [145]. Pruitt [134] developed the notion of *potential maximum evapotranspiration* to allow for the (normal) situation in which advected energy is an important factor in determining evapotranspiration rates. Finally, van Bavel [3] in a major contribution to the study of potential evapotranspiration, noted that PE can be defined for any vegetation cover in terms of the appropriate meteorological variables and the radiational and aerodynamic properties of that cover, suggesting that: 'When the surface is wet and imposes no restriction upon the flow of water vapor, the potential value is reached'.

Despite its weaknesses, some of which will be discussed in more detail in the following pages, the definition of PE by Penman and others [132] has been a veritable cornerstone in the development of evapotranspiration investigations in recent years and is implicit in subsequent references to PE in this book.

Factors affecting evapotranspiration

Except for essentially physiological reactions such as stomatal opening, the factors controlling the rate of transpiration are very similar to those controlling the rate of evaporation, since transpiration is, in effect, the evaporation of water from the leaf cells. It is, however, at this point that the distinction between potential and actual evapotranspiration becomes important. Potential evapotranspiration is controlled essentially by meteorological factors, whereas actual evapotranspiration, using the term in the sense of *sub*-potential evapotranspiration, is also considerably affected by plant and soil factors. For this reason, it is probably convenient to discuss first the factors affecting PE, and then to discuss the additional factors which come into play when water is a limiting factor.

5.4 PE and meteorological influences

Important meteorological influences on PE include solar radiation and temperature, and atmospheric moisture and wind; in general terms, PE tends to increase as the temperature, sunshine, and windspeed increase and as the humidity decreases. With regard to solar radiation and temperature, all plant processes, including photosynthesis and the circulation of water through the root-stem-leaf system, are speeded up by an increase in the amount and intensity of solar radiation. Rates of transpiration thus tend to show both a diurnal and a seasonal fluctuation, transpiration being greatest during the day and during the summer months, and least at night and during the winter months.

Evidence on the influence of windspeed tends to be conflicting. It might be expected that increased turbulence, associated with higher windspeeds, would result in increased diffusion and hence higher transpiration, although Penman and Long [130] suggested that a crop surface tends to seal itself as windspeed increases, so that turbulent mixing is restricted in depth. It has certainly been found that wind enhances photosynthesis by increasing the concentration of carbon dioxide at the leaf surface [92], and that wind-induced movement in vegetation exposes more leaves to full radiation, so that photoefficiency increases [62]. On theoretical grounds, however, van Wijk and de Vries [199] suggested that the influence of windspeed on transpiration may not be all that important. For example, on a bright sunny day in middle latitude areas, they estimated that well over 60 per cent of the available energy is used to provide the latent heat of evaporation when the wind velocity is low, so that PE cannot increase very much even with a marked increase in wind velocity. Again, de Vries and van Duin [194] found that, compared with the effect of solar radiation, the influence of wind speed on PE was only a secondary one, but suggested that, in given humidity conditions, it will tend to increase as the temperature falls.

Advection

One meteorological influence of considerable geographical significance is that of advection—the exchange of energy or moisture due to horizontal heterogeneity in conditions at the earth's surface [142]. It has already been pointed out that, although the use of the term 'extended surface' in the classical Penman definition of PE is clearly intended to obviate the advection effect, it fails to do so even in humid, mid-latitude conditions, so that in England and the Netherlands observed evapotranspiration rates can exceed PE by a factor of between 1.0 and 1.4 [144]. Clearly, in arid conditions, large-scale advection must occur for if PE requires, by definition, an upwind moist evaporating surface so large that advection is obviated then the climate is no longer arid [22]. Rijtema [144], in fact, showed that in arid climatological conditions maximum evapotranspiration may exceed PE by a factor as high as 1.9. Advective transfers of heat are common where different kinds of surface lie side by side in a typically variegated landscape which occurs in humid lands but which is especially sharply pronounced where irrigated fields inter-mingle with desert [105].

Knoerr [81] observed that advection was also important within a vegetation cover where the exterior leaves of the vegetation canopy act as a sink for solar radiation from which sensible heat may be transferred to the shaded leaves of the canopy interior.

Aerodynamic roughness

The effect of aerodynamic roughness on the turbulent transfer of moisture and energy from the vegetation surface is an important factor in evapotranspiration, as was emphasized by van Bavel [3]. This is true even if the conditions of complete canopy cover and uniform height are maintained for, in general, evapotranspiration increases with vegetation height due to increased roughness and zero plane displace-ment àt a given windspeed [22]. The coefficient of turbulent exchange, in fact, increases by a factor of 2 with a change in vegetation height from 10 cm to 90 cm, and by a factor of more than 5 with a change from a short cut green surface at about 2 or 3 cm (as implied in the classical Penman definition of PE) to a vegetation height of 90 cm [148]. This large variation in the turbulent exchange coefficient is quite sufficient to explain why PE, as defined by Penman, is frequently exceeded by the evapotranspiration rate from, say, typical agricultural crops [147]. Relations between aerodynamic characteristics and PE for some types of tropical rainforest were discussed by Brunig [14] while Penman [127], generalizing about conditions over forest and grass, suggested that the greater roughness of the trees will always be a dominant factor in evapotranspiration.

5.5 PE and plant influences: vegetation type

Considerable controversy has been aroused in the past by suggestions that the plant itself can influence the rate of PE. A partial answer was provided by Penman and others [132], who restated the earlier Penman view that, with certain restrictions, PE appears to be largely independent of the type of crop. The opposing argument, however, has also been supported by experimental evidence [13] [141].

One aspect of vegetation type which must inevitably affect the rate of PE is the

100

reflectivity or albedo of the vegetation surface which materially influences the energy balance at the evaporating surface. A light-coloured surface will obviously reflect away more (and therefore absorb less) solar radiation than a dark-coloured surface. But most plant surfaces fall within a fairly narrow colour range, and this is particularly true during the summer months when transpiration is greatest. In fact the albedo of most green agricultural crops is very nearly constant at about 0.25, irrespective of the shade of green [109]. For forest vegetation there is great uncertainty arising from the lack of adequate measurements, although in the case of some coniferous forests, the figure may be as low as 0.15 [126] or even 0.10 [179]. Uncertainty also exists about the effect of vegetation wetness on the albedo, although Penman [128] suggested that, if there is any difference between wet and dry leaves, it would take the form of a reduced reflection coefficient for the wet leaves.

5.6 PE and plant influences: vegetational 'control'

A number of ways in which the plant might physically control the rate of PE have been suggested and argued from time to time and it seems likely that, at the present time, controversy is more marked than ever before as to whether or not a well-watered plant cover may be considered as a passive evaporating 'wick'. Some aspects of possible vegetational 'control' of PE will now be briefly considered.

A defence mechanism?

It has been suggested that plants transpire as a means of self-defence [151] in order to reduce the temperature of their leaves. Certainly, the absorption of solar energy by the leaves will raise their temperature above that of the surrounding air and, in some cases, the differences may be as much as $3°-5°$ C [6]. Linacre [93] found that the published evidence concerning thin-leaved plants suggested a tendency for equality of leaf and air temperatures at a temperature of approximately $33°$ C. Below this temperature of equality, leaves tend to be warmer than the air. By increasing the vapour pressure gradient, this temperature difference encourages the transpiration of water from the leaves and, therefore, it is not surprising that the rate of transpiration follows a diurnal cycle which is directly related to the diurnal variation of temperature and the intensity of solar radiation. This is in no sense, however, a self-defence reaction on the part of the plant, but a purely physical response to an increase in vapour pressure gradient and, although the transpiration process undoubtedly contributes to the control of leaf temperature, other processes, such as conduction and the reradiation of heat, are usually more effective than transpiration in keeping the leaves cool [6].

The effect of vegetation growth

Another possible vegetational 'control' on PE, which has been advanced at various times, is the rate of vegetation growth. As Penman [126] explained, this was a point of view put forward by Warington at the turn of the century and encouraged in the early part of this century by the publication of transpiration ratios, mainly by Briggs and Shantz, on the basis of experiments with plants in pots. Little of this

101

information is relevant or meaningful and it is seldom possible to determine whether the ratios were derived for crops grown only in conditions of abundant water supply and completely covering the ground. More recently, however, van Hylckama [72] reported experiments in Arizona with saltcedar which showed that, even though water seemed to be freely available, water losses diminished in parallel with a decrease in growth and development of the vegetation.

The opposing view has been vigorously argued [55] [151], and Bernstein [6] and Penman [126] suggested that the transpiration-growth relationship must be treated with considerable caution. In brief, it was shown in section 5.2 that transpiration is a flow process responding to potential drops within the soil-plant-atmosphere continuum. Certainly the circulation of water is also part of the plant's life function without which it could not survive. Water, for example, is essential for the transport of materials in solution through the plant and is one of the raw materials in photosynthesis, although the amount of water so used is very small in comparison with the amount used in transpiration. It is unlikely, therefore, that photosynthesis is ever affected by variations in *potential* evapotranspiration or *vice versa*, although it has been found that levels of cytokinin and abscisic acid are affected by water stress in a manner detrimental to plant growth [29] [74] [201].

An environmental response?

Penman has always rejected the notion of vegetational 'control' over transpiration. The gist of his argument is contained in the following statement:

> Rejecting the suggestion that transpiration is a vital process controlled by the plant, a more simple-minded approach regards it as something that cannot be avoided because it is a response of the plant to its physical environment. If purpose is to come in at all, it is best introduced as by Kramer (1949) who suggests that the stomatal structure of leaves is designed to permit the uptake of carbon dioxide in the process of assimilation, and as the stomata is neither a valve nor a semi-permeable membrane, it cannot prevent outward movement of water vapour while it is open: transpiration is a leakage process [126].

In short, provided there is an adequate supply of water in the leaves to be transpired—provided, in other words, that water can be transmitted through the soil-root-stem-leaf system at a rate equal to the rate of transpiration—the rate of transpiration will depend almost entirely upon the speed with which liquid water in the leaves can be transformed into water vapour and 'exhaled' through the stomata, and this, in turn, will depend upon the amount of energy which is available at the leaf surface.

This view has been generally supported by a number of workers, c.f. Idso [73] and van Bavel [4], but firmly rejected by others, c.f. Lee [89] [90] [91], Gardner [43], and Shepherd [156]. Lee [90] summarized his argument effectively when, in pleading for the acceptance of the active roles of both environment *and* plant in water losses, he stated:

> If plants are indeed 'wicks', the plant scientist must argue that they are wicks of a unique kind . . . with varying hydraulic conductivities . . . [and] . . . coated with an epidermis that changes its permeability to gaseous exchange diurnally and seasonally and shows characteristic variations with plant species and type. . . (Reprinted with permission from R. Lee, *Water Resources Research,* 4: 669, 1968, © American Geophysical Union).

Much of the confusion over this problem has been related to the part played by stomatal aperture in controlling the rate of transpiration. In this respect three related problems have been first, the range of stomatal aperture within which stomatal opening can exert a control on transpiration, second, the timing of stomatal opening and closing, and third, the factors affecting stomatal aperture. Each of these problems will now be briefly discussed.

It seems likely that when the stomata are wide open the transpiration rate depends upon the factors governing evaporation, and that these factors continue to be dominant until the stomata are at least half-closed. From then on stomatal opening apparently determines water loss quite accurately, whatever the meteorological conditions [26].

With regard to the timing of stomatal opening and closure, Chang [22] drew attention to the distinction between conventional and non-conventional plants. In the case of the former, the stomata are open during the daytime and closed at night, with the result that the opportunity for evapotranspiration is decreased considerably during the hours of darkness, and this is reflected both in the estimates of the relationship between daytime and night-time evapotranspiration, and in observations of evapotranspiration phenomena. Many authorities consider that evapotranspiration during the night hours is negligible, never exceeding five to ten per cent of the daytime value [108], [174], [194], and a more direct indication of this is given by continuous traces of shallow water table fluctuations during the summer months when the common trend of the graphs is as shown in Fig. 5.5 (page 120). During the daytime, the removal of water by evapotranspiration considerably exceeds the inflow of groundwater and, as a result, the water table level falls. At night, however, when the rate of evapotranspiration is reduced, inflow exceeds withdrawal, and the groundwater level recovers.

Even with conventional plants, however, complications may arise. For example, in high evapotranspiration conditions internal plant-stresses can develop so that, even with optimal soil moisture availability, the stomata may close, thereby reducing the transpiration rate [147]. Thus, a number of authorities have reported daytime stomatal closing in wet tropical rainforest, under the influence of a high heat load, while data for irrigated spring wheat and potatoes were reported by Rijtema [147] and Endrödi and Rijtema [40] respectively. However, in now classical experiments, van Bavel and others [5] and Fritschen and van Bavel [42] found that unsurpassed daily rates of evapotranspiration were maintained from a patch of sudangrass with no visible evidence of moisture stress and no mid-day depression of transpiration which would have been convincing evidence of stomatal influence.

Finally, as Stigter [168] pointed out, there is still considerable controversy over the exact mechanisms which are responsible for the influence which the stomata are able to exercise. As intimated in the earlier quotation from Penman [126], the carbon dioxide content of the substomatal cavities is often held responsible in many species for the opening or closing of the stomata in conditions of adequate moisture [84] [137] [158]. Interesting studies of the efficiency of water use by various species of plant have been carried out by comparing the total resistance to water vapour loss *from* leaves with the total resistance to the flow of carbon dioxide *into* these leaves and its subsequent fixation in the primary production process of photosynthesis [30]. Clearly, a high resistance to water vapour loss, together with a low resistance to carbon dioxide entry, would indicate a relatively high efficiency of water use. Experiments in Australian irrigated areas have demonstrated large differences between species [30].

The strong influence of light on the carbon dioxide content of the sub-stomatal cavities has drawn attention to the relationship between stomatal opening and illumination [39], [77], [86], [167], [185], while the special influence of blue light on the opening mechanism has been reported by Raschke [138] and Meidner and Mansfield [102] although, as Stigter [168] observed, its influence within the canopy is not clear.

Zelitch [204], however, quoted extensive experimental evidence rejecting the role of carbon dioxide concentration in the normal opening and closing of stomata, and other evidence which substantiated the role of light-induced biochemical processes in influencing the osmotic potential of the stomatal guard cells relative to the adjacent epidermal cells. These views were later supported by the experimental work of Hopmans [65] [168].

Apart from the possible influence of all these factors and others, such as temperature, whose effects remain difficult to evaluate [12] [102] and plant internal moisture stress, known endogenous rhythms and different kinds of short period fluctuations [64] [65] [139] mean that the interpretation and measurement of stomatal behaviour and its influence on transpiration remain extremely difficult [168].

5.7 PE and soil moisture

It now remains to consider the interaction between potential evapotranspiration and soil moisture. It is frequently stated that evapotranspiration will take place at the potential rate only when the vegetation cover is not short of water [132], but it has proved difficult to determine at what stage of soil moisture depletion evapotranspiration begins to fall below the potential rate. Veihmeyer [187] and Veihmeyer and Hendrickson [189], [190] reported 'pot' experiments with young prune trees which demonstrated that water losses were the same when the soil moisture content had been reduced almost to the wilting point as when the soil was at field capacity. The wilting point is the lowest soil moisture content at which the plant can extract water from the soil, and the field capacity is the normal quantity of water held in the soil pore spaces against the force of gravity. In other words, according to Veihmeyer and Hendrickson, throughout the period of depletion of the soil moisture storage available to the plant between the wilting point and field capacity, evapotranspiration does take place at the potential rate, and this view was re-affirmed by Halkias and others [56]. Linsley and others [94] suggested that the total amount of the stored soil moisture so available varies with soil type, the variation ranging from about 4 cm for each 100 cm of depth in a sand, to 17 cm or more for each 100 cm of depth in a clay loam. Similarly, Gardner and Ehlig [44] reported experiments with birdsfoot trefoil, cotton, and pepper, which showed that there was little variation in transpiration rates as the soil moisture content decreased, until the plants began to wilt; thereafter, there was a virtually linear relationship between water content and transpiration rate.

The opposing view is that evapotranspiration decreases as the soil moisture decreases. This was argued, for example, by Kramer [85] and Lassen and others

[87]. Later, Thornthwaite [178] discussed the rate of soil moisture depletion in a sandy loam, and noted that, although at first evapotranspiration occurred at approximately the potential rate, '... within a week when three centimetres of water have been lost from the soil, the depressing effect on evaporation has become significant. Within three weeks the soil moisture has been reduced to a point where the evapotranspiration is only about 25 per cent of the potential rate'. Makkink and van Heemst [100] found that the actual transpiration from grass-covered lysimeters fell below PE soon after the soil moisture content fell below field capacity and that the rate of this reduction in transpiration was dependent on the moisture tension of the soil and on the rate of PE. Actual evapotranspiration is able to keep pace with PE only up to a certain intensity, which will vary according to the wetness of the soil. Similar conclusions were reached by Bierhuizen and others [8] as a result of experiments with tomato plants in pots. Visser [191], [192] extended the reasoning of Makkink and van Heemst, and suggested that the relationships between average daily rates of PE and the total water content of the soil are less important than the actual number of hours during each day when maximum rates of PE cannot be satisfied. This means that the shape of the diurnal curve of PE is more important than daily totals of PE (see Fig. 5.2).

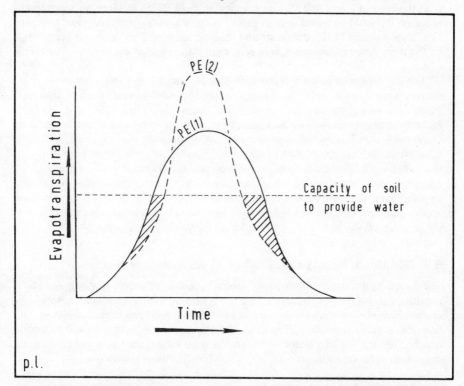

Fig. 5.2. *Diurnal variations of PE throughout a 24-hour period. Total PE represented by each curve is identical but more evapotranspiration in fact occurs at the potential rate in the case of curve (1) than in the case of curve (2), the difference being represented by the shaded area of the graph. (From an original diagram by Visser [192].)*

Penman [129] argued that although this second view sounds plausible it cannot be supported by physical theory whereas the Veihmeyer hypothesis can, especially where roots ramify thoroughly throughout the soil. The fact, however, that the Veihmeyer hypothesis so frequently breaks down in field conditions led Penman to the formulation of the 'root constant' concept [118] [122] which suggests that because moisture, on the whole, moves relatively slowly through the soil, the readily available moisture is effectively restricted to that water in the immediate proximity of the root system [118], and it will, therefore, be limited in amount partly by the depth of the soil available, and partly by the soil type. The root constant was the term given to the available moisture so defined by Penman [122], and may be described as the maximum soil moisture deficit that can be built up without checking transpiration [118]. As the term itself implies, the root constant is primarily a plant characteristic, although it may be modified by other factors such as soil type [118], and its value for grass has been estimated at between 7 and 12 cm. This means, in effect, that some 7 to 12 cm of soil moisture can be removed by grass before transpiration falls below the potential rate; after which it is assumed that transpiration falls to about one-tenth of the potential rate. Obviously, since the main factor affecting the root constant is the depth of rooting, its value will vary with the type of plant, and may be as much as 25 to 30 cm or more for some trees, although it should be noted that all these figures are merely estimates, and that, in fact, later research [115] demonstrated that the value of 7 to 12 cm for grass may be too high. Any re-assessment, however, must depend upon the results of further research.

The assumption that the transpiration rate decreases as the soil moisture decreases has formed one of the basic premises of several successful water balance studies carried out by workers in the United States [18], [19], [20], [70], [71], and the apparent accuracy of the calculations, as evidenced by correlations between measured and estimated values of soil moisture and runoff, may be considered to substantiate this approach. At the same time, however, the root constant principle has been used with equal success in irrigation practice. Penman [126] remarked that these views could not be simultaneously correct, but suggested that the errors involved are probably a lot less important than other sources of uncertainty in their application, and that because field measurements, in any case, are always so crude, either assumption could adequately describe the soil drying process.

5.8 Additional factors affecting actual evapotranspiration

The foregoing discussion of the relationships between soil moisture status and PE provides a convenient introduction to a consideration of the additional factors which affect evapotranspiration rates in conditions other than those permitting potential evapotranspiration. The most important of these conditions will be that in which water is a limiting factor, although there are others, such as a non-continuous vegetation cover or a vegetation cover which is not actively growing.

Inadequate moisture supply

When the vegetation surface is dry and when the soil moisture content falls below optimum values the actual evapotranspiration rate will be limited, by the resistances to water movement imposed by the soil pores and by the plant, to a value lower than the potential rate. Visser [192] [193] described the flow of soil moisture to plant roots and evolved a formula for this flow. It was concluded that the extraction of soil moisture by the roots tends to cause a dry cylinder around each root and that the conductivity of this relatively dry cylinder will considerably increase the resistance for liquid flow.

One implication of this is that the extent and efficiency of the root system will help to determine the total amount of moisture available to the plant cover and the rapidity with which, in drying conditions, the rate of actual evapotranspiration falls below the rate of PE. Thus, the root-surface ratio, the ratio of root surface to soil volume, is an important factor. This ratio, which normally decreases sharply with depth, is influenced by the ability of the root system to extend into moist soil and to dry out additional cylinders around the new roots. In addition, fungal growths on the roots of many trees serve to increase the surface area of absorbing roots and thus increase the availability of soil water [63].

Numerous experiments have demonstrated the effect of root range on actual evapotranspiration and some of these were reviewed by Penman [126] and Douglass [32]. Generally speaking, it is found that a forest will transpire more than a pasture because trees, on the whole, are more deeply rooted than grass.

Of the various resistances to moisture flow within the plant itself, it is generally considered that the stomatal resistance exerts the greatest influence on transpiration rates (see section 5.2). With regard to the relationships between soil moisture stress within the plant and stomatal aperture (and therefore stomatal resistance) it has conventionally been considered that the stomata open if the moisture stress is low and close gradually as the moisture stress increases. As Visser [193] pointed out, it is often assumed that they are closed completely at a tension of $pF=4.2$ although there is increasing evidence that, in fact, total closure can only be expected at much higher tensions. Indeed, Australian experiments [29] on a number of species indicated that, in every case, resistance to the flow of liquid water through the plant *decreased* with an increase in rate of flow as the atmospheric water stress was increased. In this way the plants minimized the accompanying fall of leaf water potential, which might have been expected, in order to provide the increased driving force for the increased rate of water movement, with the important consequence that leaf water potential never fell to such a low level that stomatal closure set in.

Other factors

Other plant factors may result in a marked departure of the actual evapotranspiration rate from the potential rate. One such factor is the density of the vegetation cover, which may be expressed either as a leaf-surface ratio, i.e., the ratio of leaf surface to ground area, which is commonly of the order of 5 to 10 [63], or as plants per areal unit. Douglass [32] reviewed some of the experimental work relating stand density and evapotranspiration from forested areas and concluded that, in general, a reduction in stand density results in reduced evapotranspiration, largely as a result of changes in root patterns and interception characteristics. In some cases, however, evapotranspiration losses may be higher from a non-continuous than from a continuous vegetation cover. For example, Hanks and others [57], studying the effects of advection in grain sorghum planted in rows, found that a substantial amount of energy originally used to heat the soil between the rows subsequently became an additional source of energy for transpiration from the crop itself.

Actual evapotranspiration losses will also differ according to the main stages of plant development. Thus, in the seedling stages, the type of advectional effect investigated by Hanks and others [57] will tend to increase evapotranspiration, whilst at later stages, e.g., after the ripening of grain crops or during the autumn when chlorophyll is disappearing from deciduous forest leaves, evapotranspiration may be reduced considerably below the potential rate, c.f. [45].

5.9 The components of evapotranspiration

That part of the net energy available at the earth's surface for evaporation must itself be divided between the evaporation of intercepted water (see chapter 3), the transpiration of water from vegetation, and the separate evaporation of water from the soil in which the vegetation is growing. Until comparatively recently, few

Table 5.1. *Water use in different types of forest (From tables in Penman [126])*

Aspen (Tellerman Forest, Voronezh)

Age years	Evapn. from soil (mm/growing season)	Transpiration (mm/growing season)	Ratio E:T
8	74	203	1:2.7
25	53	230	1:4.3
36	69	201	1:2.9
63	90	185	1:2.1

Aspen (mm) (per year ?)

Age years	Evapn. from soil	Transpiration + Interception	Ratio Evapn.: Transpiration + Interception
20	78	314	1:4.0
40	84	282	1:3.4
60	97	224	1:2.3

Scots Pine (mm) (per year ?)

Age years	Evapn. from soil	Transpiration + Interception	Ratio Evapn.: Transpiration + Interception
20	48	363	1:7.6
40	67	400	1:6.0
60	87	340	1:3.9
80	100	320	1:3.2
100	100	290	1:2.9
120	100	263	1:2.6
140	103	246	1:2.4
160	105	222	1:2.1

serious attempts were made to separate these components of evapotranspiration or to assess their respective contributions to total water losses.

In terms of the simple comparison between soil evaporation and the other components of evapotranspiration, experimental evidence suggests that, in general, soil evaporation is the least important component. For example, where the vegetation cover is continuous, or almost so, it must severely limit opportunities for direct evaporation from the ground surface. This was partially substantiated by Russian experiments [107], [160], reported by Penman [126], on the use of water by various forest stands. A summary of the relevant results is given in Table 5.1, where it can be seen that in all cases transpiration or transpiration plus interception is estimated as being at least twice as great as evaporation from the soil, and that in many cases the ratio is in excess of 1:3.

Or again, in areas of shallow groundwater, it is probable that transpiration is considerably more important than soil evaporation. Veihmeyer [188], for example, carried out experiments which showed that when the water table was some 120 cm below the ground surface, losses of water by transpiration were approximately 100 times greater than the surface evaporation from the soil.

Where the vegetation cover is not continuous, however, or where it has a relatively small plant mass, energy balance considerations suggest that the relationship between the components of evapotranspiration may be rather different from that outlined above. This was indicated by Ritchie [149] and in work by Baumgartner [2] in which comparisons between forest, meadow grass, and cultivated crops showed soil evaporation accounting for 10, 25, and 45 per cent respectively of the total evapotranspiration losses. Similarly, Thornthwaite and Hare [179] suggested that, from the standpoint of energetics, soil evaporation in an open row crop will contribute considerably to the total water loss. Thus King [80] reported that, in corn crops in the mid-west of the U.S.A., soil evaporation was about 30 per cent of total evaporation in dry years and as much as 50 per cent in wet years, whilst Tanner and others [174] found that net radiation at the soil surface within a maize crop was 40 per cent of that above the crop, a substantial part of the net energy available at the soil surface being infrared emission from the heated crop.

With regard to the relative importance of transpiration and the evaporation of water intercepted by the vegetation surface, it has comparatively recently become clear that this may be of considerable importance in the search for ways of reducing evapotranspiration by modifying the vegetation cover. Rutter [152] made the plea that, in the case of water losses from forest vegetation, these two components should be clearly separated on the grounds that even though the initial observations of unexpectedly high rates of evaporation of interception were accepted with great reservation they have since been fully substantiated and a theoretical explanation now exists. Detailed discussion of this problem occurred in chapter 3 (section 3.9) and in earlier sections of the present chapter and it will be sufficient at this stage to summarize the argument thus: For evapotranspiration to occur both aerodynamic and stomatal resistances must be overcome; the aerodynamic resistance, which is inversely proportional to surface roughness, is smaller for forest than for low vegetation while the stomatal resistance is much higher. When leaf surfaces are wet with intercepted precipitation stomatal resistance has no effect, so that the evaporation of intercepted precipitation should be greater than transpiration from the forest or from low vegetation under the same conditions.

Measuring evapotranspiration

5.10 Introduction

The measurement of evapotranspiration has attracted the attention of scientists of many disciplines since classical times and even today has not been entirely satisfactorily resolved. Some of the difficulties involved have already been touched upon, and not least among these is the problem of determining the extent to which the plant itself influences water losses. This is a particularly important problem because, if it is accepted that transpiration is normally the principal factor involved in evapotranspiration, it follows that attempts to estimate evapotranspiration by means of formulae should theoretically place more emphasis on the factors which influence transpiration than on those which influence evaporation. Again, it has been shown (chapter 4) that there are problems associated with the physics of evapotranspiration and still other uncertainties and problems associated with the measurement of the relevant physical quantities. For these reasons, in particular, no completely successful technique for measuring or estimating evapotranspiration has been devised.

During recent years there have been numerous literature reviews and publications of experimental evidence concerning comparative assessments of measured and calculated evapotranspiration, c.f. Rosenberg and others [150], Harrold [60], Federer [41], Ward [196], and McGuinness and Bordne [96]. Apart from the publications referred to here, and some excellent theoretical discussions by, for example, Rijtema [143], [145], [146] and Chang [22], much of this comparative work is unsatisfactory because it is based on too short a run of data. The use of a 26-year period of observations by Smith [161] was exceptional in this respect. In addition, the discrepancies between the results of different methods are often large in comparison with the magnitude of other hydrological variables such as precipitation or streamflow, and frequently fall clearly outside an acceptable margin of experimental error. Finally, although such discrepancies indicate that some, if not all of the methods for determining evapotranspiration are in error, there is no absolute standard against which results from given formulae or instruments may be assessed.

Potential evapotranspiration

Three main approaches to the measurement of PE have been evolved. These are first, the transformation of evaporation measurements made from non-vegetated surfaces, e.g., evaporimeters and evaporation pans, secondly, the direct measurement of water losses from moist vegetated surfaces, e.g., evapotranspirometers, and third, the use of more or less empirical formulae.

5.11 Evaporimeters and evaporation pans

This approach to the measurement of PE relies on the transformation of measurements from evaporimeters and evaporation pans, (see section 4.10). At best evaporimeters provide a measure of the drying power of the air, although it is doubtful if they do even that consistently and accurately. Very rarely are they able to provide a value which closely approximates potential water loss.

Although one would not expect a water surface to behave like a normal vegetated one, Holland [103] drew attention to the fact that, in some circumstances, the British pan gives results which may approximate PE (i.e., a pan coefficient of unity) and it is of interest to note that, under normal climatological conditions in the Netherlands, maximum evapotranspiration from a grass cover of 10-15 cm height closely approximates the evaporation rate from sunken pans. In addition, Rijtema [144] discussed other Dutch data which indicated a close agreement between measured water loss from sunken pans and the E_o values calculated with the Penman equation (see section 5.16), whilst Chang [22], in a review of evaporation pan and other measured data, concluded that the evaporation pan is as accurate as any formula or field instrument for estimating PE in a humid climate and, when correctly sited, the potential maximum evapotranspiration in an arid climate.

5.12 Evapotranspirometers

The second approach to the measurement of PE involves the use of irrigated evapotranspirometers, in which the soil moisture content is maintained at a level which permits water losses to occur at the potential rate and for which, therefore, a simple balancing of precipitation and irrigation inputs against drainage output suffices to permit the calculation of PE. These simple instruments consist basically of three or more watertight tanks. At least two of the tanks are filled with soil supporting a continuous vegetation cover (normally grass), and are connected by piping to collecting containers which are housed in a central tank. This type of installation is shown in Fig. 5.3 and it will be seen from the diagram that water can enter the soil tanks only from the atmosphere, either as natural or artificial precipitation, and that it can leave only via the bottom outlets as percolation. This being so, the difference between these two measured quantities, i.e., the amount of water which enters and the amount of water which leaves the tanks, will represent the amount of water lost by evapotranspiration, if allowance is made for changes of moisture storage within the soil tanks.

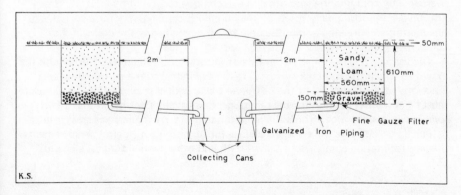

Fig. 5.3. *Diagram of an evapotranspirometer showing two soil tanks; a third soil tank ensures a greater degree of reliability.*

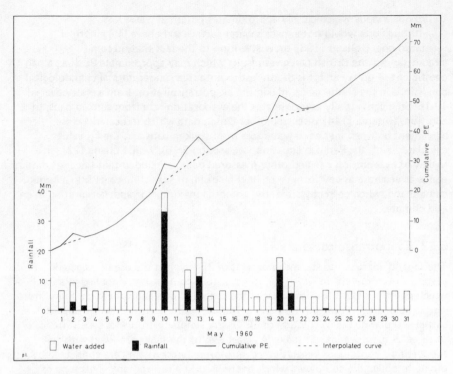

Fig. 5.4. *Diagram showing the cumulative plotting of evapotranspirometer data and the interpolation of a smooth curve.*

Such variations in soil moisture storage are of no great significance if long-term, e.g., monthly or seasonal, totals of PE are being considered. Many hydrological problems, however, demand an accurate assessment of PE over much shorter periods, e.g., ten days, one week, or even one day, and in these cases storage changes could result in considerable discrepancies. This problem can be largely overcome by applying the 'smooth curve' technique suggested by F. H. W. Green. This involves plotting daily totals of 'measured' PE cumulatively for a period, e.g., one month, and drawing a smooth curve which passes below the peaks in the cumulative graph caused by excessive precipitation (see Fig. 5.4).

The validity and usefulness of evapotranspirometer results depend largely on the extent to which the soil tank surfaces are representative of the surrounding area. Since one of the functions of the instrument is to collect precipitation, its exposure must be at least as good as for the raingauge. It is usually considered that a fairly level, open, grassed plot affords the most suitable site, particularly as grass is internationally accepted as the standard surface for other meteorological measurements, thereby facilitating comparisons between PE and other meteorological elements [51].

5.13 Evapotranspiration formulae

Numerous formulae have been proposed for estimating PE. Some of these are based largely upon the turbulent transfer approach, others on the energy budget approach while most of the better-known examples represent more or less empirical adap-tations of one or both approaches. Four well-known examples will be discussed to illustrate the scope and diversity of the more popular methods.

5.14 The Thornthwaite method

Thornthwaite's formula for the estimation of PE was presented and modified in a number of papers between 1944 and 1954 [175], [176], [178]. He began by estimating a monthly heat index i by means of the equation

$$i = (t/5)^{1.514} \qquad (5.3)$$

where t is the mean temperature of the month under consideration expressed in °C. Twelve monthly indices are then added together to give an annual heat index I. Monthly PE is then calculated by means of an equation presented by Thornthwaite in 1948:

$$e = 1.6\, b(10t/I)^a \qquad (5.4)$$

where e is monthly PE in cm; t is the mean monthly temperature in °C; a is a cubic function of I; and b is a factor to correct for unequal day length between months.

It will be seen that the only factors taken into account are mean air temperature and hours of daylight. The method, therefore, appears to be extremely empirical and somewhat complex, and has inevitably been criticized on both grounds. Serra [155] also questioned the pseudo-precision of, for example, the three decimals of the exponent 1.514, arguing that, since an approximation of the order of one per cent would be quite acceptable for any empirical expression, the equation for i could be written

$$i = (t/5)^{3/2} \text{ or } i = 0.09\, t3/2 \qquad (5.5)$$

He also suggested a considerable simplification in the coefficient a when the formula is applied to temperate regions.

In fact, as Hare [58] pointed out, Thornthwaite's choice of an empirical formula was quite deliberate as also was his choice of mean air temperature as the main parameter influencing potential evapotranspiration. Thornthwaite [178] justified the selection of mean air temperature on the grounds that there is a fixed relationship between that part of the net radiation which is used for heating, and that part which is used for evaporation when conditions are suited to evapotrans-piration at the potential rate, i.e., when the soil is continuously moist. This means, in effect, that although the Thornthwaite method is empirical, it nevertheless estimates PE by an indirect reference to the radiation balance at the earth's surface.

The Thornthwaite approach to the estimation of PE has been more widely adopted than any other. Paradoxically, it is probably because the formula expresses PE simply as a function of mean air temperature and day length, two quantities which are independent of the rate of evaporation, that it is applicable over a wide range of climatic conditions [157]. Certainly, experimental and documentary evidence has justified its application to the estimation of PE in almost every climatic area in the world [17], [18], [19], [20], [24], [66], [67], [68], [70], [71], [101], [136], [153], [154], [177]. This is not to say that, in all of these cases, the Thornthwaite formula has permitted a continuously precise and accurate estimation of monthly or short-term PE. From what has already been said, it would obviously be unrealistic to expect this to be so. It is likely, however, that in each case the Thornthwaite approach has provided a convenient and easily applicable method whose proven weaknesses lend weight to its superiority over other less well-tried techniques. Only over very short periods of time when mean temperature is not a suitable measure of incoming radiation [150] and in areas like the British Isles, where the advection effects resulting from the continual interchange of polar and tropical, continental and maritime air masses result in frequent rapid changes in mean air temperature and humidity, is it likely that the Thornthwaite approach will be relatively unsuccessful. Even so, during the summer months, when PE is of greatest significance, more stable airmass conditions tend to prevail, thereby strengthening the relationship between temperature, the radiation balance, and evaporation.

5.15 The Blaney-Criddle method

Blaney and Criddle [10] developed a simplified formula for estimating consumptive use when water supply to the plant cover is not a limiting factor. As in the Thornthwaite method, it is assumed that the heat budget is shared in fixed proportion between heating the air and evaporation so that mean air temperature can be used as an important index of water loss. An earlier equation derived by Blaney and Morin [11], and based on correlations between pan evaporation, monthly temperature, hours of daylight, and relative humidity, was used by Blaney and Criddle as the basis of the method in which the mean monthly temperature (t) is multiplied by the monthly percentage of daytime hours of the year (p) to obtain a monthly consumptive use (= PE) factor (f) thus:

$$f = (t \times p)/100 \text{ where } t \text{ is in } °F. \qquad (5.6)$$

Then it is assumed that when water supply is non-limiting, consumptive use is directly proportional to f. A simple equation expressing this relationship and intended for use mainly during the growing season is:

$$u = kf \qquad (5.7)$$

where u is monthly consumptive use in inches and k is a crop coefficient.

Although the formula looks attractively simple, and uses readily available data, it has several disadvantages, not the least of these being its empirical nature. The coefficient k, for example, was empirically determined for a range of crops, its main function being to correct for the length of growing season and for climatic variations. Despite this, however, Blaney [9] found it necessary to present a range of values (see Table 5.2) for different geographical and climatic areas.

Again, the factor f has been criticized [20] on the grounds that the f values may be applied differently according to the crop, and also differently for each crop

Table 5.2 *Crop coefficient k for selected crops for use in Blaney's equation (From a table in Penman [126]*

Crop	k coastal areas	k arid areas
Orchard, citrus	0.50	0.65
Beans	0.60	0.70
Cotton	0.60	0.65
Orchard, deciduous	0.60	0.70
Potatoes	0.65	0.75
Sugar beet	0.65	0.75
Grain, sorghums	0.70	
Orchard, walnuts	0.70	
Tomatoes	0.70	
Corn	0.75	0.85
Grains, small	0.75	0.85
Pasture, grass	0.75	
Lucerne	0.80	0.85
Flax	0.80	
Pasture, Ladino Clover	0.80	0.85
Rice	1.00	1.20

according to its geographical location, which means that the estimated PE will not always be the same in areas where identical climatic conditions occur.

Criddle [27] presented nomograms for the rapid calculation of u and the U.S. Soil Conservation Service published a modification of the method designed to extend its use to short periods of time, e.g., from 5 to 30 days [186].

5.16 The Penman method

Penman [121] devised a formula for potential evapotranspiration which combined the turbulent transfer and the energy balance approaches. This formula was later restated by Penman in slightly modified forms [123], [124], [125], [126]. Basically, there are three equations—the first of which is a measure of the drying power of the air. This increases with a large saturation deficit, indicating that the air is dry, and with high windspeeds. The first equation is, therefore, derived from the basic pattern of the turbulent transfer approach, and takes the form:

$$Ea = 0.35(e_a - e_d)\ (1 + u/100) \text{ mm/day} \qquad (5.8)$$

where e_a is the saturation vapour pressure of water at the mean air temperature in millimetres of mercury; e_d is the saturation vapour pressure of water at the dew-point temperature, or the actual vapour pressure at the mean air temperature; and u is the windspeed in miles per day at a height of two metres above the ground surface.

The second equation provides an estimate of the net radiation available for evaporation and heating at the earth's surface and takes the form:

$$H = A - B \text{ mm/day} \qquad (5.9)$$

where A is the short-wave incoming radiation and B is the long-wave out-going radiation as estimated in the following expressions:

$$A = (1 - r)Ra(0.18 + 0.55n/N) \text{ mm/day} \qquad (5.10)$$

$$B = \sigma Ta^4 (0.56 - 0.09 \sqrt{e_d}) (0.10 + 0.90n/N) \text{ mm/day} \quad (5.11)$$

where Ra is the theoretical radiation intensity at the ground surface in the absence of an atmosphere, expressed in evaporation units; r is the reflection coefficient of the evaporating surface; n/N is the ratio of actual/possible hours of bright sunshine; σTa^4 is the theoretical back radiation which would leave the area in the absence of an atmosphere, Ta being the mean air temperature, and σ being Stefan's constant; and e_d is as in eq. (5.8).

Penman assumed that the heat flux into and out of the soil, which usually represents about two per cent of the total incoming energy [118], is small enough to be conveniently ignored, so he simply divided the net radiation between heating the air and evaporation, and the proportion of it used in evaporation is then estimated by combining eqs. (5.8) and (5.9) to give:

$$E = (\Delta/\gamma H + E_a)/(\Delta/\gamma + 1) \text{ mm/day} \qquad (5.12)$$

where Δ is the slope of the saturation vapour pressure curve for water at the mean air temperature in mm Hg/°F; and γ is the constant of the wet and dry bulb psychrometer equation (0.27 mm Hg/°F). This relationship assumes equality of the coefficients of water vapour and convective heat transfer, a requirement which is well met in windy and unstable conditions [25], but which becomes less certain in conditions of strong insolation and light winds [106]. The ratio Δ/γ is dimensionless, and is, in effect, a factor which makes allowance for the relative significance of net radiation and evaporativity in total evaporation. For example, at 10°, 20°, and 30°C, the values of Δ/γ are 1.3, 2.3, and 3.9 respectively [126]. Thus, during those periods, e.g., summer months, when totals of evaporation are significantly high, the net radiation term is given more weight than the evaporativity term. Furthermore, in humid areas, H is in any case usually greater than Ea, so, with its greater size and its greater weighting, H tends to be the dominant term in the equation. The quantity E, as estimated in eq. (5.12) will vary according to the value assigned to the reflection coefficient r in eq. (5.10). Thus, if r is given the value 0.25, E will represent the PE from an extended short green crop, whereas, if r is given the value 0.05, E will represent E_o, the evaporation from an extended sheet of open water.

This formula is probably the most comprehensive approach to the estimation of PE which has so far been devised, and takes into account almost all of the factors which are known to influence it. Modifications have, however, been suggested from time to time in order to overcome some of the comparatively minor inadequacies of the Penman method. Many of these suggested modifications have concerned the radiation terms in eqs. (5.10) and (5.11). Glover and McCulloch [47], for example, proposed the alteration of (5.10) thus:

$$A = (1 - r)Ra(0.29\cos(\text{lat}) + 0.52n/N) \text{ mm/day} \qquad (5.13)$$

This proposal was based upon regression analysis of actual values of radiation and hours of sunshine and gives a better relationship over the range of latitude from 0^c to 60°. Later modifications, based on work by L. P. Smith at the Meteorological Office, were described in MAFF Technical Bulletin No. 16 [97].

It will be evident that if all or part of the available energy represented in eq. (5.9) is measured directly, then even short-period estimates of PE may be tolerably accurate. In this respect, one of three alternative techniques may be employed. First, a value of measured incoming radiation may be substituted for eq. (5.10) [99]. A second alternative is to replace the whole of the expression for H, i.e., eq. (5.9), by a value of measured net radiation [25]. Thirdly, in addition to net radiation, measurements may also be made of the heat flow through the soil by means of flux plates embedded beneath the surface, and used in conjunction with a radiometer. In this way, the complete energy balance is measured, and errors caused in the estimation of PE by the neglect of heat storage may be obviated. This, in turn, might correct a large part of the seasonal discrepancy which has been noted between measured and estimated results using the Penman formula [195], and which has been attributed to the fact that a larger proportion of the incoming radiation will be required to heat the soil in the spring than in the autumn, so that the proportion available for evaporation will be correspondingly lower.

It should be emphasized, however, that the insertion of measured radiation values in place of the Penman radiation terms will result in an improved estimate of PE only if the network of radiation instruments is sufficiently dense to enable a representative sampling of vegetation type having different albedo and surface temperatures. Otherwise, the gain in accuracy of the point measurement tends to be lost in the extrapolation to an areal estimate.

Like the Thornthwaite method, Penman's has achieved wide acclaim, partly in spite of, and partly because of, its inherent defects. The formula is quite complex, although this is of small significance now that the use of high-speed computers is so common c.f. [203] and generalized programs, having a wide application with regard to location and variety of input data, are available c.f. [23]. In addition, a number of graphical solutions have been proposed c.f. [135]. More important is the number of basic variables which need to be known. If net radiation is measured, air temperature, vapour pressure, and wind speed are the only other variables needed to estimate PE. If, however, net radiation is calculated from eq. (5.9), it would be necessary to know latitude, time of year, duration of bright sunshine, mean air temperature, mean vapour pressure, and mean daily wind speed. This extended data requirement obviously means that the formula is less widely applicable than Thornthwaite's, particularly in respect of historic data.

Nevertheless, the Penman method has been extensively used and 'has worked satisfactorily in a wide range of climates' [126]. For example, the formula has been found to give satisfactory results in the British Isles by Green [49], [50], [51], [52], Guerrini [54], Morgan [113], Pearl and others [118], Ward [195], and Winter and others [200]; in the Netherlands by Makkink [98], [99] and Ritjema [143]; in North America, by Gilbert and van Bavel [46], King [79], and Kohler [82]; and in Africa, by Damagnez and others [31].

5.17 The van Bavel method

Van Bavel [3] took the combination equation as used by Penman but improved it to the point where it contains no empirical constants or functions and is not restricted to grass or any other specified set of surface conditions other than that water supply is unrestricted. The van Bavel method gave excellent agreement for 24-hour totals and acceptable agreement for hourly values when measured and calculated evaporation from open water, wet soil, and well-watered alfalfa were compared.

The basic equation takes the following form:

$$E = [(\Delta/\gamma) \ (H/L) + Bv.da] / [(\Delta/\gamma) + 1] \ \text{cal cm}^{-2} \ \text{min}^{-1} \qquad (5.14)$$

where E is the potential evaporation or potential evapotranspiration, depending on the surface, Δ/γ is as defined in eq. (5.12), H is as defined in eq. (5.9) but in units of langleys, L is the latent heat of vaporization (583 ly/cm^{-3}), and da is the vapour pressure deficit in millibars at height z_a. Bv is a transfer coefficient for water vapour found from the equation

$$Bv = (0.01222u/(\ln z_a/z_0)^2)298/T_a \qquad (5.15)$$

where u is daily windspeed at 2 metres in km per day, z_a is the height above the surface where temperature, humidity and wind are measured, z_0 is a roughness parameter, and T_a is the temperature of the air.

A detailed description of the formula for computing purposes was provided by McGuinness and Bordne [96].

Actual evapotranspiration

5.18 Introduction

It will be clear, from earlier discussions in this chapter, that the measurement of actual evapotranspiration, whether directly by instrumental means or indirectly through formulae, will be extremely difficult. As soon as the constriction of adequate water supply, which is conveniently imposed by the concept of PE, is removed it becomes necessary to make allowance for all factors relating to the soil and plant cover. Thus all the factors which affect the capillary movement of moisture in the soil are now relevant, as also are the resistances imposed on the passage of water by various parts of the plant system in all climatological and hydrological situations.

Consideration of actual evapotranspiration must, therefore, recognize that every moment of the day brings a unique set of conditions to each square centimetre of the ground surface. It would thus seem clear that no formula which relies solely on meteorological or climatological data can even approximate the rate of evapotranspiration, except fortuitously. Instead, a more realistic approach is to calculate PE using one of the more reliable formulae and then reduce the values obtained according to the degree of soil moisture depletion. This was the approach proposed by both Thornthwaite [182] and Penman [122], which has already been discussed in some detail in section 5.7, and which was eventually adopted by the UK Meteorological Office in the mapping of soil moisture deficit and actual evapotranspiration [53].

Apart from this eminently sensible and widely used method there are three other main approaches to the measurement of actual evapotranspiration. These are first, the use of formulae specifically designed to allow for conditions when water availability is a limiting factor in the evapotranspiration process; second, the physical modelling of the evapotranspiration system, in the sense of water balance calculations for known areas or volumes of vegetation or vegetation-covered ground surface, in which evaporative loss is derived as the residual item in the water balance equation; and third, the measurement of the flux of moisture either within the plant system or above the evaporating vegetation surface. Each of these approaches will now be discussed.

5.19 Actual evapotranspiration formulae

The main formulae for actual evapotranspiration are either of a general climatological type, which can normally be successfully applied only to rather long periods of time, or are of a type which involves a theoretical modelling of the entire evapotranspiration process from soil to atmosphere, using combination energy-balance and vapour transport formulae and taking into account surface and other resistances of the evaporating vegetation surface.

Where the demand is simply for a value of accumulated actual water loss over long periods of time, e.g., one year, it has been found that formulae such as those of Turc [184] and Budyko [15] give comparatively satisfactory results. For shorter term estimates, however, a more detailed approach is necessary which takes into account the entire soil-plant-atmosphere process of evapotranspiration. In this connection it became clear, soon after its introduction and subsequent minor modifications, that the Penman formula was not only a successful means of estimating PE but also an excellent and readily adaptable model of the evapotranspiration process. Provided that the appropriate vegetation and soil factors could be incorporated, it held out considerable promise for the successful estimation of actual evapotranspiration. A number of workers in different countries have made important contributions in this field including, in Britain, H. L. Penman himself and J. L. Monteith and his collaborators, particularly G. Szeicz, whilst in the Netherlands the similar work of P. E. Rijtema has been equally rigorous and successful.

In each case, investigation proceeded from the premise that although in the Penman combination formula the simplifying assumption of a wet vegetation surface was used, when, as is normal, the vegetation surface is not wet the rate of evapotranspiration will be reduced because of the diffusion resistances imposed by the soil pores and the stomata. Penman and Schofield [131] presented the combination formula so as to include the stomatal resistance of the vegetation cover. Subsequently, a solution to the problem of estimating stomatal and other surface resistances was suggested by Monteith and Szeicz [112] who showed that the effective surface resistance of field crops could be estimated from their radiative temperature, and by Monteith [110] who showed that this resistance could also be calculated from temperature, vapour pressure, and wind profiles over the vegetation surface. Later, Monteith [111] presented a generalized version of the Penman formula incorporating the stomatal resistance factor, and Szeicz and others [172] and Szeicz and Long [171] showed that the mean aerodynamic resistance can be

119

calculated from windspeed and surface roughness by reference to vegetation height. Calculated and measured rates of actual evaporation for water, pine forest and for two agricultural crops were in good agreement in both southern England and California.

Rijtema [144] [147] [148] and Endrödi and Rijtema [40] presented a quantitative description and physical formulation of the entire evapotranspiration process from soil to atmosphere, including not only meteorological conditions but also such soil physical properties as soil moisture suction and hydraulic conductivity, and vegetational properties such as rooting characteristics, internal resistance to liquid flow from root surface to sub-stomatal cavities, suction in the leaf tissue in relation to stomatal reaction, ground coverage, and vegetation height. Rijtema showed that, by taking all these factors into account, actual evapotranspiration for weekly periods could be calculated using a combined aerodynamic and energy balance approach.

5.20 Water balance methods

A number of water balance methods, involving different scales of operation, have

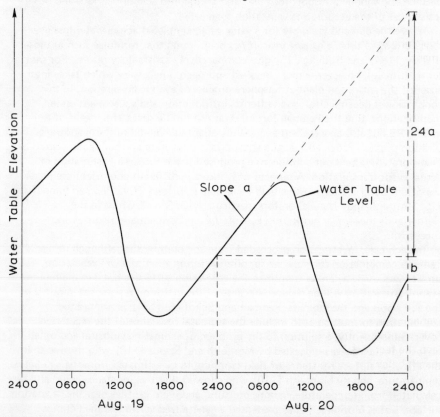

Fig. 5.5. *Fluctuations of a shallow water table reflect diurnal variations in evapotranspiration. (From an original diagram by H. C. Troxell,* Trans. Amer. Geophys. Union, **17**: 496-504, 1936.)

been applied to the measurement of actual evapotranspiration. These include lysimeters, water table fluctuations, catchment areas and small experimental watersheds, and finally, water balance calculations for the soil profile.

Lysimeters

Lysimeters, defined by Harrold [60] as small units of soil on which water balance values can be obtained, have been in use as hydrological instruments for a long period of time (c.f. section 4.12) during which there have been a number of important literature reviews by Kohnke and others [83], Pelton [120], Slatyer and McIlroy [159], and Harrold [59]. By definition, evapotranspirometers are a type of lysimeter but throughout this book a distinction is preserved between the evapotranspirometer used for measuring potential evapotranspiration and the lysimeter used for measuring actual evapotranspiration. In the latter case, the aim is normally to reproduce the natural soil profile within a watertight container, to induce a typical vegetation cover with a representative root system to develop, and to expose this sample surface to the same meteorological or climatological conditions as those experienced by the area being sampled. Then, evapotranspiration equals precipitation minus percolation through the lysimeter. The accuracy with which variations in soil moisture storage need to be assessed, usually by weighing the entire lysimeter tank, varies with the length of the period of measurement.

Water table fluctuations

The calculation of evapotranspiration from water table fluctuations was well exemplified in the classic paper by White [198], by the work of Troxell [183] on the floodplain of the Santa Ana river, and much later by Meyboom [104] in the Canadian prairies. In conditions where diurnal fluctuations of a shallow water table occur, White [198] suggested that evapotranspiration could be estimated from the following:

$$ET = Sy(24a + b) \qquad (5.16)$$

where ET is evapotranspiration in cm/day, Sy is the specific yield of the soil, a is the rate of rise of the water table (in cm/h) from midnight to 0400 hours, when evapotranspiration is assumed negligible, and b is the net change in water table elevation during the day (in cm), (see Fig. 5.5).

The main disadvantages of the method are the large number of influencing variables, apart from evapotranspiration, many of which, e.g., groundwater flow into and out of the area, are difficult to determine with a high degree of accuracy, and the fact that the method can be used only where the water table is at a shallow depth below the ground surface.

Catchment areas

An approximate measure of long-term evapotranspiration may be determined by considering the major water balance components of a river basin or catchment area, although more accurate results are possible using small experimental catchments in which detailed measurements are made of a wide range of hydrological variables. In such cases, the main thesis is that if the water balance equation $P - Q - E - \Delta S - \Delta G = 0$ (where P is precipitation, Q is streamflow, E is actual evapotranspiration, and ΔS and ΔG are changes in soil moisture and groundwater storage respectively) can be solved, then it is probable that the values of the individual components of that balance are accurate.

Clearly, this may not be so if discrepancies in the assessment of individual variables are fortuitously complementary, although this source of error can usually be guarded against by making specimen water balance calculations for different time periods within the same run of data. Since precipitation, streamflow, soil moisture, and groundwater measurements can almost certainly be made with a greater accuracy than the corresponding measurements or estimations of evaporative loss, it can be argued that the value of E which consistently gives the best result in the water balance equation is the most 'suitable'. Work of this type was presented by Dunin [33] and by Pegg [119] and a large number of such experiments were reviewed by Ward [197].

Soil moisture studies

An estimate of evapotranspiration can also be obtained by calculating the partial water balance for the soil profile, often within the context of an experimental plot study. The important work done by Thornthwaite and his colleagues on the climatic water balance [181] substantiated the close relationships between P, E, and ΔS. An approach through soil moisture measurement, and particularly the determination of successive soil moisture profiles, is attractive if only because, within the soil profile, one can obtain a direct measure of the amount of water withdrawn by the vegetation cover. Valuable work along these lines was reported by Stern [164] [165] [166] and by Burman and Loudon [16] and the method was discussed in general terms by Federer [41].

5.21 Moisture flux measurement

The third main approach to the measurement of actual evapotranspiration involves the calculation of the actual flux of moisture either immediately above the evaporating surface or within the plant system itself. In the former case the moisture flux has been investigated in a number of different situations.

Theoretical approach

A theoretical approach was developed in detail by Thornthwaite and Holzman [180] and later modified by Pasquill [116] [117] and Rider [140] [141]. This approach is based upon the fact that evapotranspiration into the lower layers of the atmosphere will tend to establish a moisture gradient from the evaporating surface into the overlying air, and that turbulent motion will tend to break that gradient down and so establish uniform moisture conditions above the evaporating surface. If, then, both the moisture gradient and the turbulent motion of the air can be accurately measured, the contribution of water vapour from the evapotranspiration process can be determined.

Leaf and plant chambers

In leaf chamber studies, such as those reported by Bierhuizen and Slatyer [7] and by Jarvis and Slatyer [75], independent, continuous and simultaneous measurements are possible of the water vapour and carbon dioxide exchanges between each leaf surface and the surrounding atmosphere under controlled conditions of visible and total radiation, air and leaf temperatures, and carbon dioxide and water vapour concentrations. Field measurements from an entire plant or collection of plants are possible in larger plant chambers, such as that developed by F. W. Went [1] [162] [163]. This is a clear plastic chamber, of 6 cubic decimetres volume, which may be placed over a small living plant in the field and within which changes in humidity are measured. Stark [162] [163] found the method accurate and was able to make more than 400 measurements daily.

Eddy correlation method

The accurate measurement of the flux of water vapour away from a vegetation surface, which is an essential feature of both leaf and plant chamber studies, is also the basic principle of the eddy correlation method, except that the latter is applied to natural, field conditions with no artificial constraints such as plastic chambers or controlled atmospheric conditions. The eddy correlation method was pioneered in Australia by Swinbank [170], who demonstrated that fluxes could be determined from the correlation of temperature and humidity fluctuations with the vertical component of wind velocity. In order to avoid laborious data computations it was necessary to develop an instrument in which the relevant calculations were carried out instantaneously. Once the feasibility and validity of this concept had been demonstrated [34] a small, portable field instrument was designed and subsequently called the *evapotron*. After preliminary testing, the instrument was used in extensive investigations [36]. By means of delicate sensing devices simultaneous measurements are made of the minute eddy fluctuations of humidity, wind, and temperature above the evaporating surface. This information may be fed directly into a computer and an output of net upward movement of water vapour from the evaporating surface obtained.

Dyer and Maher [35] discussed the limitations of the fine-wire sensors, particularly with regard to the rather slow response time of the wet bulb. Dyer and others [37], however, later described an improved instrument for measuring heat transfer, called the *fluxatron,* in which slow eddies not contributing to the eddy flux could be filtered out, which had a much lower power consumption and which could be used in the field by relatively unskilled personnel. It was intended to extend its use to evaporation measurement when an alternative to the fine-wire sensor was found for the humidity measurement. Such an alternative, a barium fluoride film humidity sensor, developed by Jones and Wexler [76], was described by Goltz and others [48] who concluded that the sensor had sufficiently rapid response to allow reliable eddy correlation measurements of vapour flux within a metre of the surface, in comparison with the four metres necessary for the evapotron. McGavin and others [95] reported the development of a microwave evapotron in which a modified microwave refractometer and a vertical sonic anemometer were combined to measure the vertical transport of water vapour in field conditions. In terms of evaporation, the accuracy was within 0.02 cm/day and the resolution within 0.002 cm/day.

Clearly, such instruments are still in the process of active development, although eventually the eddy correlation method will probably come into standard, widespread use. Ironically, this is likely to raise almost as many problems as it solves, particularly in connection with the representativeness of the measuring sites and the degree of replication required, since the more sophisticated the measured data are, the less likely are they to be broadly representative of surrounding conditions. Little work has been done on the application of point measurements of evapotranspiration to larger areas, e.g., entire river basins [114].

Moisture flux within the plant

Although, strictly speaking, not a method of measuring evapotranspiration, in the sense that no numerical value for evapotranspiration is derived, measurements of moisture flux within the plant have yielded much valuable information about variations of transpiration with time. The classical approach for woody vegetation is the heat-pulse technique reported by Huber and Schmidt [69] and subsequently developed and modified by a number of workers [169]. The basic principle is that heat as a pulse is injected into a tree stem and its rate of dissipation measured at closely spaced points both upstream and downstream. Subsequent developments, sometimes equally applicable to non-woody vegetation have involved the tracing of injected dyes or radioactive materials [78].

References

1. ASHBY, W. C., in F. W. Went, *Experimental Control of Plant Growth,* Chronica Botanica Co., Waltham, Mass., 23: 301-306, 1957.
2. BAUMGARTNER, A., Energetic bases for differential vaporization from forest and agricultural lands, in Sopper, W. E. and H. W. Lull (Eds.), *Forest Hydrology,* Pergamon, Oxford, pp. 381-389, 1967.
3. BAVEL, C. H. M. van, Potential evaporation: The combination concept and its experimental verification, *WRR,* 2: 455-467, 1966.
4. BAVEL, C. H. M. van, Further to the hydrologic importance of transpiration control by stomata, *WRR,* 4: 1387-1388, 1968.
5. BAVEL, C. H. M. van, L. J. FRITSCHEN, and W. E. REEVES, Transpiration of sudangrass as an externally controlled process, *Science,* 141: 269-270, 1963.
6. BERNSTEIN, L., The needs and uses of water by plants, in *Water,* USDA Yearbook, pp. 18-25, 1955.
7. BIERHUIZEN, J. F., and R. O. SLATYER, An apparatus for the continuous and simultaneous measurement of photosynthesis and transpiration under controlled environmental conditions, *CSIRO Div. Land Res. Reg. Sen. Tech. Pap.* 25: 1-16, 1964.
8. BIERHUIZEN, J. F., A. A. ABD EL RAHMAN, and P. J. C. KUIPER, The effect of nitrogen application and water supply on growth and water requirement of tomato under controlled conditions, *Inst. for Land and Water Management Res., Tech. Bull.,* 16, 1960.
9. BLANEY, H. F., Consumptive use requirements for water, *Agric. Eng.,* 35: 870-880, 1954.

10. BLANEY, H. F., and W. D. CRIDDLE, Determining water requirements in irrigated areas from climatological and irrigation data, *USDA Soil Cons. Serv., Tech Pap.* 96, 1950.
11. BLANEY, H. F., and K. V. MORIN, Evaporation and consumptive use of water empirical formulas, *Trans. AGU,* **23**: 76-83, 1942.
12. BOSIAN, G., Relationships between stomatal aperture, temperature, illumination, relative humidity and assimilation determined in the field by means of controlled-environment plant chambers, in Functioning of terrestrial ecosystems at the primary production level, *UNESCO Natural Resources Res. Series,* **5**: 321-328, 1968.
13. BRIGGS, L. J., and H. L. SHANTZ, Hourly transpiration rate on clear days as determined by cyclic environmental factors, *J. Agric. Res.,* **5**: 132-133, 1916.
14. BRUNIG, E. F., Stand structure, physiognomy and environmental factors in some low-land forests in Sarawak, *Bull. Internat. Soc. Trop. Ecol.,* **11**: 26-43, 1970.
15. BUDYKO, M. I., *The heat balance of the earth's surface,* Transl. by N. A. Stepanova, Office of Technical Services, U.S. Dept. of Commerce, Washington, D.C., 1958.
16. BURMAN, R. D., and T. L. LOUDON, Evapotranspiration and microclimate of irrigated pastures and alfalfa under high altitude conditions, *Trans. ASAE,* **11**: 123-125 and 128, 1968.
17. CARTER, D. B., Climates of Africa and India according to Thornthwaite's 1948 classification, *Publ. in Climat.,* **7**: 455-479, 1954.
18. CARTER, D. B., The water balance of the Lake Maracaibo basin, *Publ. in Climat.,* **8**: 209-227, 1955.
19. CARTER, D. B., The water balance of the Mediterranean and Black Seas, *Publ. in Climat.,* **9**: 123-174, 1956.
20. CARTER, D. B., The average water balance of the Delaware basin, *Publ. in Climat.,* **11**: 249-270, 1958.
21. CHANG, J. H., An evaluation of the Thornthwaite classification, *Ann. Assoc. Amer. Geog.,* **49**: 24-30, 1959.
22. CHANG, J. H., On the study of evapotranspiration and water balance, *Erdkunde,* **19**: 141-150, 1965.
23. CHIDLEY, T. R. E., and J. G. PIKE, A generalized computer program for the solution of the Penman equation for evapotranspiration, *J. Hydrol.,* **10**: 75-89, 1970.
24. COCHRANE, N. J., Observable evapotranspiration in the basin of the River Thames, *QJRMS,* **85**: 57-59, 1959.
25. COLE, J. A., and M. J. GREEN, Measurements of net radiation over vegetation and of other climatic factors affecting transpiration losses in water catchments, *IASH Cttee. for Evaporation, Publ.* 62: 190-202, 1963.
26. CRAFTS, A. S., H. B. CURRIER, and C. R. STOCKING, *Water in the physiology of plants,* Chronica Botanica, Waltham, Mass., 1949.
27. CRIDDLE, W. D., Consumptive use of water and irrigation requirements, *J. Soil and Water Cons.,* **8**: 207-212, 1953.
28. CSIRO, *Annual Report 1968-69,* Div. Irrig. Res., CSIRO, Griffith, N.S.W. 75 pp., 1969.
29. CSIRO, *Annual Report 1969-70,* Div. Irrig. Res., CSIRO, Griffith, N.S.W., 77 pp. 1970.
30. CSIRO, *Annual Report 1970-71,* Div. Irrig. Res., CSIRO, Griffith, N.S.W., 78 pp., 1971.
31. DAMAGNEZ, J., C. RIOU, O. de VILLELE, and S. EL AMMAMI, Estimation et mesure de l'evapotranspiration potentielle en Tunisie, *IASH Gen. Ass. Berkeley, Cttee for Evaporation Publ.* 62, 1963.
32. DOUGLASS, J. E., Effects of species and arrangement of forests on evapotranspiration, in Sopper, W. D. and H. W. Lull (Eds.), *Forest Hydrology,* Pergamon, Oxford, pp. 451-461, 1967.
33. DUNIN, F. X., The evapotranspiration component of a pastoral experimental catchment, *J. Hydrol.,* **7**: 147-157, 1969.
34. DYER, A. J., Measurements of evaporation and heat transfer in the lower atmosphere by an automatic eddy-convection technique, *QJRMS,* **87**: 401-412, 1961.
35. DYER, A. J., and F. J. MAHER, Automatic eddy-flux measurement with the evapotron, *J. Appl. Met.,* **4**: 622-625, 1965.
36. DYER, A. J., and W. D. PRUITT, Eddy-flux measurements over a small, irrigated area, *J. Appl. Met.,* **1**: 471-473, 1962.
37. DYER, A. J., B. B. HICKS, and K. M. KING, The fluxatron—a revised approach to the

measurement of eddy fluxes in the lower atmosphere, *J. Appl. Met.,* 6: 408-413, 1967.

38. EAGLESON, P. S., *Dynamic Hydrology,* McGraw-Hill, New York, 462 pp:, 1970.
39. EHRLER, W. L., and C. H. M. van BAVEL, Leaf diffusion resistance, illuminance and transpiration, *Plant Physiol.,* 43: 208-214, 1968.
40. ENDRODI, G., and P. E. RIJTEMA, Calculation of evapotranspiration from potatoes, *Neth. J. Agric. Sci.,* 17: 283-299, 1969.
41. FEDERER, C. A., Measuring forest evapotranspiration—theory and problems, *USDA, Forest Service Res. Paper* NE-165, 25 pp., 1970.
42. FRITSCHEN, L. J., and C. H. M. van BAVEL, Energy balance as affected by height and maturity of sudangrass, *Agron. J.,* 56: 201-204, 1964.
43. GARDNER, W. R., Internal water status and response in relation to the external water regime, in Slatyer, R. O. (Ed.), *Plant Response to Climatic Factors,* pp. 221-225, UNESCO, Paris, 1973.
44. GARDNER, W. R., and C. F. EHLIG, The influence of soil water on transpiration by plants, *JGR,* 68: 5719-5724, 1963.
45. GEE, G. W., and C. A. FEDERER, Stomatal resistance during senescence of hardwood leaves, *WRR,* 8: 1456-1460, 1972.
46. GILBERT, M. J., and C. H. M. van BAVEL, A simple field method for measuring evapotranspiration, *Trans. AGU,* 35: 937-942, 1954.
47. GLOVER, J., and J. S. G. McCULLOCH, The empirical relation between solar radiation and hours of sunshine, *QJRMS,* 84: 172-175, 1958.
48. GOLTZ, S. M., C. B. TANNER, G. W. THURTELL, and F. E. JONES, Evaporation measurement by an eddy correlation method, *WRR,* 6: 440-446, 1970.
49. GREEN, F. H. W., A year's observations of potential evapotranspiration in Rothiemurchus, *J. Inst. Wat. Eng.,* 10: 411-419, 1956.
50. GREEN, F. H. W., Problems raised by the operation of, and the results from, a small network of British evaporation measuring stations, *IASH Gen. Ass. Toronto,* 3: 444-451, 1957.
51. GREEN, F. H. W., Four years' experience in attempting to standardize measurements of potential evapotranspiration in the British Isles and the ecological significance of the results, *IASH Sympos. Hannoversch-Münden,* 1: 92-100, 1959.
52. GREEN, F. H. W., Some observations of potential evapotranspiration 1955-57, *QJRMS,* 85: 152-158, 1959.
53. GRINDLEY, J., Estimation and mapping of evaporation, in *Symposium on World Water Balance, IASH Publ.* 92: 200-213, 1970.
54. GUERRINI, V. H., Analysis of evapotranspiration observations at Valentia Observatory, August 1952—July 1956, *Dept. Transport & Power, Met. Serv., Tech. Note* 25, Dublin, 1957.
55. HAGAN, R. M., and M. L. PETERSON, Soil moisture extraction by irrigated pasture mixtures as influenced by clipping frequency, *Agron. J.,* 45, 1953.
56. HALKIAS, N. A., F. J. VEIHMEYER, and A. H. HENDRICKSON, Determining water needs for crops from climatic data, *Hilgardia,* 24: 207-233, 1955.
57. HANKS, R. J., L. H. ALLEN, and H. R. GARDNER, Advection and evapotranspiration of wide-row sorghum in the Central Great Plains, *Agron. J.,* 63: 520-527, 1971.
58. HARE, F. K., The evapotranspiration problem, Cyclostyled manuscript, 33 pp. Subsequently published in revised form as Thornthwaite and Hare (1965).
59. HARROLD, L. L., Measuring evapotranspiration by lysimetry, in Evapotranspiration and its role in water resources management, *Conf. Proc. Amer. Soc. Agric. Eng.,* pp. 28-33, 1966.
60. HARROLD, L. L., Evapotranspiration: A factor in the plant-soil-water economy, *The Progress of Hydrology,* University of Illinois, Urbana, Illinois, 2: 694-716, 1969.
61. HENDRICKS, D. W., and V. E. HANSEN, Mechanics of evapotranspiration, *Proc. ASCE, J. Irrig. Drainage Div.,* 88 (I.R.2): 67-82, 1962.
62. HESKETH, J. D., Photosynthesis: Leaf Chamber Studies with corn, Ph.D. Thesis, Cornell Univ., 1961.
63. HEWLETT, J. D., and W. L. NUTTER, *An Outline of Forest Hydrology,* University of Georgia Press, 137 pp., 1969.
64. HOPMANS. P. A. M., Types of stomatal cycling and their water relations in bean leaves,

Z. Pflanzenphys., 60: 242-254, 1969.

65. HOPMANS, P. A. M., Rhythms in stomatal opening of bean leaves, *Meded. Landb. Hogesch., Wageningen,* 71(3): 1-86, 1971.

66. HOWE, G. M., The moisture balance in England and Wales based on the concept of potential evapotranspiration, *Weather,* 11: 74-82, 1956.

67. HOWE, G. M., The water budget in Wales, *Nature in Wales,* 2: 297-306, 1956.

68. HOWE, G. M., Climate in relation to crop production, *Agric. Prog.,* 32: 1-15, 1957.

69. HUBER, B., and E. SCHMIDT, A compensation method for the thermoelectric measurement of slow sap movements (Translation TT 64-16015, Office of Technical Services, U.S. Dept. of Commerce, Washington, D.C.) *Ber. deutsch. Bot. Ges.,* 55: 514-529, 1937.

70. HYLCKAMA, T. E. A. van, The water balance of the earth, *Publ. in Climat.,* 9: 57-117, 1956.

71. HYLCKAMA, T. E. A. van, Modification of the water balance approach for basins within the Delaware Valley, *Publ. in Climat.,* 11: 271-291, 1958.

72. HYLCKAMA, T. E. A. van, Growth, development and water use by saltcedar (tamarix pentandra) under different conditions of weather and access to water, *IASH Gen. Ass. Berkeley, Cttee. for Evaporation, Publ. 62:* 75-86, 1963.

73. IDSO, S. B., Comments on paper by Richard Lee 'The hydrologic importance of transpiration control by stomata', *WRR,* 4: 665-666, 1968.

74. ITAI, C., A. RICHMOND, and Y. VAADIA, The role of root cytokinins during water and salinity stress, *Israel J. Bot.,* 17: 187-195, 1968.

75. JARVIS, P. G., and R. O. SLATYER, A Controlled-environment chamber for studies of gas exchange by each surface of a leaf, *CSIRO Div. Land Res. Tech. Pap.* 29, 1966.

76. JONES, F. E., and A. WEXLER, A barium fluoride film hygrometer element, *JGR,* 65: 2087-2095, 1960.

77. KANEMASU, E. T., and C. B. TANNER, Stomatal diffusion resistance of snap beans, II. Influence of light, *Plant Physiol.,* 44: 1542-1546, 1969.

78. KENWORTHY, J. B., Water and nutrient cycling in a tropical rain forest, in J. R. Flenley (Ed.), *Trans 1st Aberdeen-Hull Sympos. on Malasian Ecology,* Univ. of Hull. Dept. of Geog. Misc. Series. No. 11: 49-65, 1971.

79. KING, K. M., Pasture irrigation control according to soil and meteorological measurements, Ph.D. thesis Univ. Wisconsin, Madison, 70 pp., 1956. Reported by Chang (21).

80. KING, K. M., Evaporation from land surfaces, *Proc. Hydrol. Sympos. No. 2, Evaporation,* pp. 55-80, Queen's Printer, Ottawa, 1961.

81. KNOERR, K. R., Contrasts in energy balances between individual leaves and vegetated surfaces, in Sopper, W. E. and H. W. Lull, (Eds.), *Forest Hydrology,* Pergamon, Oxford, pp. 391-401, 1967.

82. KOHLER, M.A., Meteorological aspects of evaporation phenomena, *IASH Gen. Ass. Toronto,* 3: 421-436, 1957.

83. KOHNKE, H., F. R. DREIBELBIS, and J. M. DAVIDSON, A survey and discussion of lysimeters and a bibliography on their construction and performance, *USDA Misc. Publ.* 372, 68 pp., 1940.

84. KRAMER, P. J., *Plant and Soil Water Relationships,* McGraw-Hill, New York, 1949.

85. KRAMER, P. J., Plant and soil water relations on the watershed, *J. Forestry,* 50 1952.

86. KUIPER, P. C., The effects of environmental factors on the transpiration of leaves with special reference to stomatal light response, *Meded. Landb. Hogesch., Wageningen,* 61(7): 1-49, 1961.

87. LASSEN, L., H. W. LULL, and B. FRANK, Some plant-soil-water relations in watershed management, *USDA Circ.,* 910, 1952.

88. LEE, C. H. Transpiration and total evaporation, Chapter 8 in Meinzer, O. E. (Ed)., *Hydrology,* McGraw-Hill, New York, 1942.

89. LEE, R., The hydrologic importance of transpiration control by stomata, *WRR,* 3: 737-752, 1967.

90. LEE, R., Reply (to Idso, S. B. 1968), *WRR,* 4: 667-669, 1968.

91. LEE, R., Reply (to Bavel, C. H. M. van, 1968), *WRR,* 4: 1389-1390, 1968.

92. LEMON, E. R., Energy and water balance of plant communities, Chap. 5 in Evans, L. T. (Ed.), *Environmental Control of Plant Growth,* Academic Press, New York, 1963.

127

93. LINACRE, E. T., A note on a feature of leaf and air temperatures, *Agric. Met.,* **1**: 66-72, 1964.

94. LINSLEY, R. K., M. A. KOHLER, and J. L. H. PAULHUS, *Hydrology for Engineers,* McGraw-Hill, New York, 1958.

95. McGAVIN, R. E., P. B. UHLENHOPP, and B. R. BEAN, Microwave evapotron, *WRR,* **7**: 424-431, 1971.

96. McGUINESS, J. L., and E. F. BORDNE, A comparison of lysimeter-derived potential evapotranspiration with computed values, *USDA Agric. Res. Service, Tech. Bull,* 1452, 71 pp., 1972.

97. MAFF, Potential Transpiration, *Ministry of Agriculture, Fisheries and Food, Tech. Bull,* 16, 77 pp., HMSO, 1967.

98. MAKKINK, G. F., Testing the Penman formula by means of lysimeters, *J. Inst. Water Eng.,* **11**: 277–288, 1957.

99. MAKKINK, G. F., Ekzameno de la formulo de Penman, *Neth. J. Agric. Sci.,* **5**: 290-305, 1957.

100. MAKKINK, G. F., and H. D. J. van HEEMST, The actual evapotranspiration as a function of the potential evapotranspiration and the soil moisture tension, *Neth. J. Agric. Sci.,* **4**: 67-72, 1956.

101. MATHER, J. R. (Ed), The measurement of potential evapotranspiration, *Publ. in Climat.,* **7**, 225 pp., 1954.

102. MEIDNER, H., and T. A. MANSFIELD, *Physiology of Stomata,* McGraw-Hill, London, 179 pp., 1968.

103. METEOROLOGICAL OFFICE, *British Rainfall 1961,* HMSO, London, 1967.

104. MEYBOOM, P., Three observations on streamflow depletion by phreatophytes, *J. Hydrol.,* **2**: 248-261, 1965.

105. MILLER, D. H., The heat and water budget of the earth's surface, *Advances in Geophys.,* **11**: 175-302, Academic Press, New York, 1965.

106. MILTHORPE, F. L., The income and loss of water in arid and semi-arid zones, *Plant-Water Relationships in Arid and Semi-Arid Conditions,* 9-36, UNESCO, Paris, 1960.

107. MOLCANOV, A. A. [The water conserving and protective function of forests], *Arch. Forstw.,* **4**: 591-607, 1955. Reported by Penman [126].

108. MONTEITH, J. L., Evaporation at night, *Neth. J. Agric. Sci.,* **4**: 34-38, 1956.

109. MONTEITH, J. L., The reflection of short-wave radiation by vegetation, *QJRMS,* **85**: 386, 1959.

110. MONTEITH, J. L., Calculating evaporation from diffuse resistances, Chapter X, *Second Ann. Rept. USAEPG, Contract DA-36-039-SC-88334,* Univ. of California, Davis, 1963.

111. MONTEITH, J. L., Evaporation and environment, *Proc. Symp. Exptl. Biol.,* **19**: 205-234, 1965.

112. MONTEITH, J. L., and G. SZEICZ, Radiative temperature in the heat balance of natural surfaces, *QJRMS,* **88**: 378, 496-507, 1962.

113. MORGAN, W. A., Potential evapotranspiration as measured at Valentia Observatory over the period August 1952 to February 1962 and a comparison with values as computed by the Penman formula, *Dept. Transport and Power, Met. Serv., Tech. Note,* 29, Dublin, 1962.

114 MUSTONEN, S. E., and J. L. McGUINNESS, Lysimeter and watershed evapotranspiration, *WRR,* **3**: 989-996, 1967.

115. NEW SCIENTIST, Making irrigation in Britain a more exact science, *New Scientist,* **11**: 11, 1961.

116. PASQUILL, F., Eddy diffusion of water vapour and heat near the ground, *Proc. Roy. Soc. A,* **198**: 116-140, 1949.

117. PASQUILL, F., Some further considerations of the measurement and indirect evaluation of natural evaporation, *QJRMS,* **76**: 287-301, 1950.

118. PEARL, R. T., R. H. MATHEWS, L. P. SMITH, H. L. PENMAN, E. R. HOARE, and E. E. SKILLMAN, The calculation of irrigation need, *Min. Agric. and Fisheries Tech. Bull.,* **4**, 1954.

119. PEGG, R. K., Evapotranspiration and the water balance in a small clay catchment, in Taylor, J. A. (Ed), *The Role of Water in Agriculture,* Pergamon, Oxford, 1970.

120. PELTON, W. L., The use of lysimetric methods to measure evapotranspiration, *Proc.*

Hydrol. Symp. No. 2, Evaporation, pp. 106-127, Queen's Printer, Ottawa, 1961.

121. PENMAN, H. L., Natural evaporation from open water, bare soil and grass, *Proc. Roy. Soc. A,* **193**: 120-145, 1948.

122. PENMAN, H. L., The dependence of transpiration on weather and soil conditions, *J. Soil Sci.,* **1**: 74-89, 1949.

123. PENMAN, H. L., Experiments on the irrigation of sugar beet, *J. Agric. Sci.,* **42**: 286-292, 1952.

124. PENMAN, H. L., Evaporation over parts of Europe, *IASH Gen. Ass. Rome,* **3**: 168-176, 1954.

125. PENMAN, H. L., Evaporation: An introductory survey, *Neth. J. Agric. Sci.,* **4**: 9-29, 1956.

126. PENMAN, H. L., *Vegetation and Hydrology,* Commonwealth Agric. Bur., Farnham Royal, 1963.

127. PENMAN, H. L., Evaporation from forests: a comparison of theory and observation, in Sopper, W. E. and H. W. Lull (Eds), *Forest Hydrology,* Pergamon, Oxford, pp. 373-380, 1967.

128. PENMAN, H. L., in discussion of Penman, H. L., Evaporation from forests: A comparison of theory and observation, in Sopper, W. E. and H. W. Lull, (Eds), *Forest Hydrology,* Pergamon, Oxford, p. 380, 1967.

129. PENMAN, H. L., The role of vegetation in soil water problems, in P. E. Rijtema and H. Wassink (Eds), *Water in the Unsaturated Zone,* pp. 49-61, IASH-UNESCO, 1969.

130. PENMAN, H. L., and J. F. LONG, Weather in wheat: an essay in micrometeorology, *QJRMS,* **86**: 16-50, 1960.

131. PENMAN, H. L., and R. K. SCHOFIELD, Some physical aspects of assimilation and transpiration, *Sympos. Soc. Exp. Biol.,* **5**, 1951.

132. PENMAN, H. L., and others, Discussions of evaporation etc., *Neth. J. Agric. Sci.,* **4**: 87-97, 1956.

133. PHILIP, J. R., Plant water relations: Some physical aspects, *Ann. Rev. Plant Physiol.,* **17**: 245-268, 1966.

134. PRUITT, W. O., *Correlation of climatological data with water requirement of crops,* Dept. of Irrigation, Univ. of California, 1960.

135. PURVIS, J. C., Graphical solution of the Penman equation for potential evapotranspiration, *Mon. Wea. Rev.* **89**: 192-196, 1961.

136. RAO, B., The water balance of the Ohio River basin, *Bull. Amer. Met. Soc.,* **39**: 153-154, 1958.

137. RASCHKE, K., Die Stomata als Glieder eines schwingungsfähigen CO_2-Regelsystems, Experimenteller Nachweis an Zea Mays L., *Zeitschr. Naturforsch.,* **206**: 1261-1270, 1965.

138. RASCHKE, K., Der Einfluss von Rot- und Blaulicht auf die Offnungs- und Schliessgeschwindigkeit der Stomata von Zea Mays, *Naturwiss,* **54**: 72-73, 1967

139. RASCHKE, K., and U. KUHL, Stomatal response to changes in atmospheric humidity and water supply: experiments with leaf sections of Zea Mays in CO_2-free air, *Planta,* **87**: 36-48, 1969.

140. RIDER, N. E., Evaporation from an oatfield, *QJRMS,* **80**: 198-211, 1954.

141. RIDER, N. E., Water losses from various land surfaces, *QJRMS,* **83**: 181-193, 1957.

142. RIDER, N. E., and J. R. PHILIP, Advection and evaporation *IASH Publ.* **53**: 421-427, 1960.

143. RIJTEMA, P. E., Calculation methods of potential evapotranspiration, *Inst. Land Water Management Res. Tech. Bull.,* **7**, 1959.

144. RIJTEMA, P. E., *An analysis of actual evapotranspiration,* Agr. Res. Dept. 659, Pudoc, Wageningen, 107 pp., 1965.

145. RIJTEMA, P. E., Evapotranspiration, *Inst. for Land and Water Management Res. Tech. Bull.* 47, 1966.

146. RIJTEMA, P. E., Transpiration and production of crops in relation to climate and irrigation, *Inst. for Land and Water Management Res. Tech. Bull.* 44, 1966.

147. RIJTEMA, P. E., Derived meteorological data: Transpiration *Proc. UNESCO Symp. Agroclimatol. Methods, Reading, 1966,* pp. 55-72, 1968.

148. RIJTEMA, P. E., On the relation between transpiration, soil physical properties and crop

production as a basis for water supply plans, *Inst. Land and Water Mangt. Res., Tech. Bull.* 58, 1968.

149. RITCHIE, J. T., Model for predicting evaporation from a row crop with incomplete cover, *WRR,* **8**: 1204-1213, 1972.

150. ROSENBERG, N. J., H. E. HART, and K. W. BROWN, Evapotranspiration: Review of research, *Univ. of Nebraska, Agr. Expt. Sta., M.P.* 20, 78 pp. 1968.

151. RUSSELL, E. W., Water utilization by crops and the ability of the soil to supply water to them, *World Crops,* 2, 1950.

152. RUTTER, A. J., Forest hydrology, Informal discussion of the Hydrological Group, *Proc. ICE,* **44**: 293-295, 1969.

153. SANDERSON, M., Some Canadian developments in agricultural climatology, *Weather,* **5**: 381-388, 1950.

154. SANDERSON, M., Measuring potential evapotranspiration, Toronto, 1947, 1948, 1949, *Publ. in Climat.,* **7**, 1954.

155. SERRA, L., The hydrological check on a catchment area, *Soc. Hydrotechnique de France, Comte rendu des 3eme journees de l'hydraulique, Algers, 12-14 Apr. 1954,* pp. 29-35, 1954.

156. SHEPHERD, W., Some evidence of stomatal restriction of evaporation from well-watered plant canopies, *WRR,* **8**: 1092-1095, 1972.

157. SIBBONS, J. L. H., The climatic approach to potential evapotranspiration, *Adv. of Sci.,* **13**: 354-356, 1956.

158. SLATYER, R. O., *Plant-Water Relationships,* Academic Press, London, 366 pp., 1967.

159. SLATYER, R. O., and I. C. McILROY, *Practical Microclimatology* CSIRO, Australia and UNESCO, 1961.

160. SMIRNOV, V. V., and N. S. ODINOVKA, [The hydrological role of aspen woods], *Dokl. Acad. Nauk. SSSR.* **99**: 849-852, 1954.

161. SMITH, K., A long-period assessment of the Penman and Thornthwaite potential evapotranspiration formulae, *J. Hydrol.,* 2: 277-290, 1964.

162. STARK, N., The transpirometer for measuring the transpiration of desert plants, *J. Hydrol.,* **5**: 143-157, 1967.

163. STARK, N., Spring transpiration of three desert species, *J. Hydrol.,* **6**: 297-305, 1968.

164. STERN, W. R., Evapotranspiration of Safflower at three densities of sowing, *Aust. J. Agric. Res.,* 16: 961-971, 1965.

165. STERN, W. R., Seasonal evapotranspiration of irrigated cotton in a low-latitude environment, *Austr. J. Agric. Res.,* **18**: 259-269, 1967.

166. STERN, W. R., Soil water balance and evapotranspiration of irrigated cotton, *J. Agric. Sci.,* **69**: 95-101, 1967.

167. STEWART, D. W., and E. R. LEMON, The energy budget at the earth's surface: a simulation of net photosynthesis of field corn, *Res. and Develt. Tech. Rept., ECOM.* 2-68, I-6, 132 pp., 1969.

168. STIGTER, C. J., Leaf diffusion resistance to water vapour and its direct measurement, 1. Introduction and review concerning relevant factors and methods, *Mededlingen Landbouwhogeschool Wageningen,* 72-3, 47 pp., 1972.

169. SWANSON, R. H., Seasonal course of transpiration of lodgepole pine and Engelmann spruce, in Sopper, W. E. and H. W. Lull (Eds), *Forest Hydrology,* Pergamon, Oxford, pp. 419-434, 1967.

170. SWINBANK, W. C., The measurement of vertical transfer of heat and water vapour by eddies in the lower atmosphere, *J. Met.,* **8**: 135-145, 1951.

171. SZEICZ, G., and I. F. LONG, Surface resistance of crop canopies, *WRR,* **5**: 622-633, 1969.

172. SZEICZ, G., G. ENDRODI, and S. TAJCHMAN, Aerodynamic and surface factors in evaporation, *WRR,* **5**: 380-394, 1969.

173. TANNER, C. B., Factors affecting evaporation from plants and soils, *J. Soil and Water Cons.,* 12: 221-227, 1957.

174. TANNER, C. B., A. E. PETERSON, and J. R. LOVE, Radiant energy exchange in a corn field, *Agron. J.,* **52**: 373-379, 1960.

175. THORNTHWAITE, C. W., A contribution to the Report of the Cttee. on Transpiration and Evaporation, 1943-44, *Trans. AGU,* **25**: 686-693, 1944.

130

176. THORNTHWAITE, C. W., An approach toward a rational classification of climate, *Geog. Rev.,* **38**: 55-94, 1948.
177. THORNTHWAITE, C. W., Climate and scientific irrigation in New Jersey, *Publ. in Climat.,* **6**: 3-8, 1953.
178. THORNTHWAITE, C. W., A re-examination of the concept and measurement of potential evapotranspiration, *Publ. in Climat.,* **7** 1954.
179. THORNTHWAITE, C. W., and F. K. HARE, The loss of water to the air, *Meteorological Monographs,* **6**: 163-180, 1965.
180. THORNTHWAITE, C. W., and B. HOLZMAN, The determination of evaporation from land and water surfaces, *Mon. Wea. Rev.,* **67**: 4-11, 1939.
181. THORNTHWAITE, C. W., and J. R. MATHER, The Water Balance, *Publ. in Climat.,* **8**: 1-86, 1955.
182. THORNTHWAITE, C. W., and J. R. MATHER, Instructions and tables for computing potential evapotranspiration and the water balance, *Publ. in Climat.,* **10**: 185-311, 1957.
183. TROXELL, H. C., The diurnal fluctuation in the groundwater and flow of the Santa Ana River and its meaning, *Trans. AGU.,* **17**: 496-504, 1936.
184. TURC, L., Le bilan d'eau des sols: relation entre les précipitations, l'évaporation et l'écoulement, *Troisièmes journées de l'Hydraulique, Alger, 12-14 Avril,* 1954.
185. TURNER, N. C., Stomatal resistance to transpiration in three contrasting canopies, *Crop Sci.,* **9**: 303-307, 1969.
186. U.S. DEPARTMENT OF AGRICULTURE, Irrigation water requirements, *U.S.D.A., S.C.S., Engin. Div. Tech. Release* 21, 83 pp., 1967.
187. VEIHMEYER, F. J., Some factors affecting the irrigation requirements of deciduous orchards, *Hilgardia,* **2**: 125-284, 1927.
188. VEIHMEYER, F. J., Evaporation from soils and transpiration, *Trans. AGU,* **19**: 612-615, 1938.
189. VEIHMEYER, F. J., and A. H. HENDRICKSON, Soil moisture conditions in relation to plant growth, *Plant Physiol.,* **2**: 71-82, 1927.
190. VEIHMEYER, F. J., and A. H. HENDRICKSON, Does transpiration decrease as the soil moisture decreases?, *Trans. AGU,* **36**: 425-428, 1955.
191. VISSER, W. C., Soil moisture content and evapotranspiration, *Inst. for Land and Water Management, Tech. Bull.,* **31**, 1963.
192. VISSER, W. C., Moisture requirements of crops and rate of moisture depletion of the soil, *Inst. for Land and Water Management Res., Tech. Bull.,* 32, 1964.
193. VISSER, W. C., Progress in the knowledge about the effect of soil moisture content on plant production, *Inst. for Land and Water Management Res., Tech. Bull.,* **45**, 44 pp., 1966.
194. VRIES, D. A. de, and R. H. A. van DUIN, Some considerations on the diurnal variation of transpiration, *Neth. J. Agric. Sci.,* **1**: 27-39, 1953.
195. WARD, R. C., Observations of potential evapotranspiration (PE) on the Thames floodplain, 1959-60, *J. Hydrol.,* **1**: 183-194, 1963.
196. WARD, R. C., Measuring evapotranspiration; a review, *J. Hydrol,* **13**: 1-21, 1971.
197. WARD, R. C., Small Watersheds Experiments: An appraisal of concepts and research developments, *Univ. of Hull, Occ. Papers in Geography,* **18**, 254 pp. 1971.
198. WHITE, W. N., A method of estimating groundwater supplies based on discharge by plants and evaporation from soil, Results of investigation in Escalante Valley, Utah, *USGS Wat. Sup. Pap.* 659-A, 1932.
199. WIJK, W. R. van, and D. A. de VRIES, Evapotranspiration, *Neth. J. Agric. Sci.,* **2**: 105-118, 1954.
200. WINTER, E. J., P. J. SALTER, and G. STANHILL, Lysimetry at the National Vegetable Research Station, Wellesbourne, Warwicks, England, *IASH Publ.,* **49**: 44-53, 1959.
201. WRIGHT, S. T. C., and R. W. P. HIRON, (+) — Abscisic acid, the growth inhibitor induced in detached leaves by a period of wilting. *Nature,* **224**: 719-720, 1969.
202. YOUNG, A. A., and H. F. BLANEY, Use of water by native vegetation, *California Dept. of Public Works, Div. Water Resources Bull.,* 50, 1942.
203. YOUNG, C. P., A computer programme for the calculation of mean rates of evaporation using Penman's formula, *Met. Mag.,* **92**: 84-89, 1963.
204. ZELITCH, I., Stomatal control, *Ann. Rev. Pl. Phys.,* **20**: 329-350, 1969.

6. Subsurface Water - Soil Moisture

6.1 Introduction

Some part of the precipitation which reaches the ground surface may flow over the surface as overland flow, reaching drainage channels quite rapidly, and so making a direct addition to surface runoff, while the larger part will normally be absorbed by the surface layers. This latter portion may flow laterally, close to the surface, as interflow, or it may move downwards through the soil, subsoil, and rock layers, eventually becoming part of one or more of the moisture zones shown in Fig. 6.1. This simple, diagrammatic, cross-section of a river valley shows the four main zones in which water is normally found below the ground surface. Referring to the section A-B, it will be seen that precipitation first enters the *soil zone* which, as its name implies, comprises the water held in the comparatively thin layer of soil or weathered mantle at the ground surface. Water moves out of the soil zone, either by means of evapotranspiration to the atmosphere, or by means of downward percolation towards the water table. Immediately above the water table, is the *capillary fringe,* where a substantial number of pores are full of water as a result of suction forces operating against the force of gravity, and between this and the soil zone is the *intermediate zone,* where the movement of water is generally downwards.

Figure 6.1 clearly illustrates that the term *soil moisture* may refer to rather different conditions in different parts of a catchment area. Thus, on the valley flanks, the situation is that which has been described above, in which water drainage from the soil zone proper passes into the intermediate zone, and may or may not eventually reach the zone of saturation some tens, or perhaps hundreds, of feet below. A different situation may, however, exist in the floodplain areas of the catchment, where the capillary fringe often extends either into the soil zone or even to the ground surface itself, depending on the distance between the ground surface and the water table, and on the height of the capillary fringe. In these areas the upward flow of moisture from the underlying groundwater is likely to maintain a higher soil moisture content than on the valley flanks, where replenishment can come only from precipitation. A situation intermediate between these two extremes may be envisaged in the lower terraces or similar lowlying areas in the catchment, where the distance between the capillary fringe and the soil zone is such that seasonal fluctuations of the water table may, from time to time, bring the two into either contact or overlap.

132

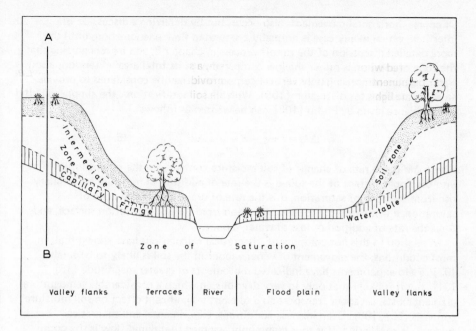

Fig. 6.1. *The zones of subsurface water.*

In the light of such varying circumstances throughout a catchment area, soil moisture is commonly regarded as comprising all the subsurface water in the *zone of aeration,* i.e., the unsaturated soil and subsoil layers above the water table which marks the upper surface of the *zone of saturation,* and it is in this sense that the term soil moisture is used in this chapter.

This distinction between the saturated and the unsaturated zones of subsurface water is an important one which emphasizes the basic differences between the zone of saturation, in which the pore spaces are almost completely filled with water and in which the pressure of water is equal to or greater than atmospheric pressure, and the zone of aeration in which the pore spaces are filled with both water and air, in which capillary suction forces predominate, and in which the pressure of water is less than atmospheric. Van Bavel [5] distinguished the domain of unsaturated water movement as a three-phase system, in order to stress the point that its essential characteristics derive from the variation with time of the proportion of gas and liquid in a solid matrix. However, in emphasizing the uniqueness of the three-phase domain, one must not overlook the essential continuity between surface water and groundwater, nor that this continuity is effected by the movement of water in the zone of aeration [37] [68], even though some workers have found it convenient to postulate a discontinuity between saturated and unsaturated flow conditions [38] [67].

From what has already been said, it will be apparent that soil moisture conditions are continually in a state of flux. The zone of aeration is, in fact, a zone of transition in which water is absorbed, held, or transmitted either downwards towards the water table, or upwards towards the soil surface and the atmosphere.

Of course, horizontal movements also occur, but by deferring a discussion of interflow, which in any case is normally a saturated flow phenomenon, until the more detailed discussion of the runoff process in chapter 8, and by recognizing that the zone of aeration is rather shallow compared with its total areal extension, a soil water balance comprising only vertical components may be considered to provide an adequate basis for discussion [100]. With slight modifications, the simple soil water balance given by Philip [100] can be written as follows:

$$\Delta M = f + c - d - e \pm \Delta v \qquad (6.1)$$

Where ΔM is the rate of change of soil moisture content, f is the rate of infiltration into the upper surface of the soil, c is the rate of addition of moisture by capillary rise from the zone of saturation, d is the rate of drainage into the zone of saturation, e is the rate of evapotranspiration from the soil-vegetation surface, and Δv is the rate of addition or loss of water vapour.

In relation to this last component, theoretical calculations have shown that, in most conditions, the movement of water vapour in the soil is likely to be small [92]. Field experiments have indicated movements of greater magnitude [15] [31], but it is likely that only in very dry soils, in which a considerable temperature gradient exists, is vapour transport able to exert a significant effect on soil moisture contents [55], [83]. Even in these conditions, experimental and theoretical evidence is conflicting. It is thus commonly assumed that liquid flow is the dominant process in moist, nearly isothermal soils [85].

In general terms, therefore, the aspects of soil moisture which are of most importance to the hydrologist concern the moisture content or wetness of the soil, and the ability of the soil to serve as a temporary storage reservoir for infiltrating precipitation, the vertical downward or upward movement of water through the unsaturated zone between the ground surface and the water table, and the entry of water through the surface layers of the soil. These aspects will now be discussed under the headings of *storage of soil moisture, movement of soil moisture,* and *infiltration.*

Storage of soil moisture

It has been remarked that if gravity were the only force to which water in the soil were subject, the soil would drain completely dry after rainfall—the only water to be found being that below the water table [122]. In such a situation, plant growth would be virtually impossible except in areas where rainfall occurred frequently. In other areas, devoid of vegetation, there would be little cohesion in the surface layers of the soil, so that these would tend to be eroded in large quantities during high winds. The fact that, in natural conditions, the soil always contains *some* moisture, even at the end of long dry periods lasting many months or even years, indicates that the forces which hold moisture in the soil must be very powerful.

6.2 Moisture retention forces

The nature of these forces has prompted numerous investigations during the past century, although it is only comparatively recently that their real nature and influence on soil moisture retention has become apparent. The main forces responsible for holding water in the soil are those of capillarity, adsorption, and osmosis.

Most of the water retained in comparatively moist, coarse-grained soil is held by *surface tension,* or *capillary forces* which develop at the interface between the soil air and the soil water (or more correctly, soil solution). Surface tension results from the fact that the molecules comprising a water surface are attracted mainly by the molecules within the water body and only infinitesimally by the water vapour molecules in the air. Films of water within the soil mass are thus held *in situ* by the surface tension force, which must be overcome if the air-water interface is to move. Since each soil particle is in contact with several others, it is evident that water will

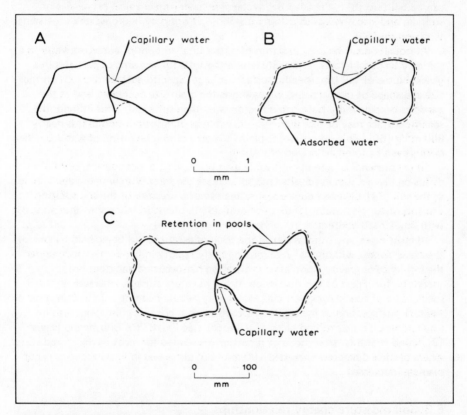

Fig. 6.2. *Modes of water retention: (A) Capillary wedges in pores; (B) capillary wedges in equilibrium with adsorbed water layer; and (C) capillary wedges, adsorbed water layer, and the retention of free-standing water on concave surfaces of coarse material.*

be held in the small pore spaces between them as in Fig. 6.2A. The force with which these small wedge-shaped films of water are retained will vary, just as the capillary force in a glass capillary tube varies, with the radius of the tube, the curvature of the surface meniscus, and with the viscosity and surface tension of the water. With a given viscosity and surface tension, therefore, water will be held more strongly in small pore spaces than in large ones, and more strongly, also, as the curvature of the water surface increases.

Various types of *adsorption* can occur, depending on the different phases which may come into contact, e.g., the adsorption of gases upon solids, of gases upon liquid surfaces, or of liquids upon solids, and the forces of attraction may differ, e.g., electrostatic or molecular. However, the adsorption of liquid soil water or soil water vapour upon the solid surfaces of soil particles is generally of an electrostatic nature, in which polar water molecules attach to the charged faces of the solids [60]. Since the attractive forces involved are only effective very close to the solid-liquid interface, only very thin films of water can be held in this way (see Fig. 6.2B), although some investigators believe that a thickness of 3-7 molecules or more can be built up [81], the adsorbed water film having mechanical properties of strength and viscosity which differ from those of ordinary liquid water at the same temperature [60].

Although occurring only as a thin film, the total amount of adsorbed water in a soil may be considerable, especially where the total surface area of particles in a given volume of soil (the specific surface) is large. Specific surface depends on the size and shape of the particles, increasing as the grain size decreases, and as the particles become more flattened and elongated than spherical. Thus in sand, the specific surface may be less than 1 m^2/g, whereas in montmorillonite it is nearly 800 m^2/g [60]. This, of course, explains the very strong retention of water by clays during even a prolonged period of drying.

In the presence of a semi-permeable membrane, e.g., a root surface, or of a diffusion barrier, dissolved salts tend to increase the force with which water is held in the soil [74] by an amount equal to the *osmotic pressure* of the soil solution. For this reason, less water tends to be available to plants in saline soils than in soils with smaller salt contents.

In most cases, any other water in the zone of aeration, not being held by one of the above forces, will normally be moving fairly rapidly in large interstices, under the influence of gravity, soon after infiltration has occurred and does not, therefore, form part of this discussion of soil moisture storage. However, in the special case of coarse rock particles, water may be retained not only as thin films on the rock surfaces and in the capillary-sized spaces at contact points, but also in small pools on the upturned faces of the rocks (see Fig. 6.2C). Boushi and Davis [8] found that this latter mode of retention accounted for most of the stored water where particle diameters exceeded 100 mm, but decreased in importance as particle diameter decreased.

6.3 Soil moisture energy relationships

As Hillel [60] pointed out, although classical physics recognizes that soil water, like any other body, may contain both kinetic and potential energy, the slow movement of water in the soil means that its kinetic energy, which is proportional to the square of the velocity, can generally be considered negligible, whereas its potential energy, resulting from position or internal condition, is of major importance in determining the state and movement of soil moisture.

The concept of *soil-moisture potential* relates the total potential energy of soil water to that of water in a standard reference state at atmospheric pressure. Thus, according to Aslying and others [3], the total potential of soil water may be defined as 'the amount of work that must be done per unit quantity of pure water in order to transport reversibly and isothermally an infinitesimal quantity of water from a pool of pure water at a specified elevation at atmospheric pressure to the soil water (at the point under consideration)'. Total soil-moisture potential has a number of components, including the gravitational potential and the osmotic potential, although by far the most important component is the *pressure potential.* Thus when soil is saturated, as below the water table, its hydrostatic pressure is greater than atmospheric pressure and the potential energy status of the soil water is, therefore, greater than that of the reference state, i.e., the pressure potential is positive and the water will tend to flow out of the soil into an adjacent free water body. In contrast when, as is normally the case in the zone of aeration, the soil is moist but unsaturated, the water is attracted to the soil particles by the capillary and adsorptive forces described in the preceding section. These forces thus act to lower the potential energy of the soil water below that of the reference state, thereby causing a *negative pressure potential* (or *suction*). In this situation the soil water will not be able to flow out of the soil into an adjacent free water body. On the contrary, the spontaneous tendency will be for the soil to take up water from such a free water body if placed in contact with it, much as blotting paper takes up ink [60].

Buckingham [12] first suggested the use of the term capillary potential to signify the negative pressure potential of soil water but this term is not satisfactory because, as has already been shown, adsorptive as well as capillary forces operate within the soil, and although the latter predominate in sandy soils the former predominate in clayey soils and at high suctions. Moreover, the separate effects of capillary and adsorptive forces on the negative pressure potential cannot easily be determined, since the capillary wedges are in a state of internal equilibrium with the adsorption films and neither can be changed without affecting the other [60]. In addition, the presence of solutes in soil water lowers its potential energy and must also be taken into account, particularly in considerations of water availability to plants. A more appropriate term for the total negative pressure potential of soil water is *matric potential,* which denotes the total effect resulting from the affinity of the soil solution to the whole matrix of the soil, including its pores and particle surfaces together [60], or simply soil-moisture potential or *soil-moisture suction.*

Application of the concept of soil-moisture potential has revolutionized the study of soil moisture and has made redundant some of the early arbitrary classifications of types of soil moisture, such as that of Briggs [11] who referred to hygroscopic, capillary, and gravitational water. Clearly, all soil moisture is influenced by gravity, and to most of it the laws of capillarity apply. Soil moisture cannot, then, be differentiated by type but only by its potential-energy status, which can be related to the amount of moisture stored in the soil and also to the rate and direction of soil moisture movement. With reference to moisture storage, as soil dries out, the moisture held least strongly by the attractive forces of capillarity and adsorption will be removed first, i.e., moisture in large pores and in thick films. As drying proceeds, the remaining moisture will be held with increasing attraction in successively smaller pores and thinner films. Because the attractive forces lower the potential energy of the soil water, i.e., increase its negative pressure potential or suction, there is a clear relationship between the amount of water retained in the soil and the soil moisture suction. In terms of its movement, soil moisture tends towards equilibrium energy conditions and therefore moves from a point of high potential energy to a point of lower potential energy, i.e., from low suction to high suction.

Soil moisture suction, or matric potential, may be expressed in a number of ways, including energy per unit mass, per unit volume or per unit weight, although the latter is the most convenient and the most frequently used. Energy per unit weight may be expressed in terms of units of hydrostatic pressure or, preferably, in terms of the equivalent hydraulic head, so that 1 atm is equivalent to a hydraulic head of 1033 cm. In expressing normal variations of soil moisture suction, one may be dealing with a negative head of as much as − 10 000 to − 100 000 cm of water, values which would clearly be unwieldy to write and difficult to plot on an arithmetic graph. It was partly for this reason that Schofield [113] suggested using the logarithm of the soil moisture suction, pF, which he defined as the logarithm of the negative head of water, in centimetres. By analogy with the pH acidity scale, the symbol p indicates the logarithm and the symbol F indicates the connection to the free energy of the soil water. A negative hydraulic head of 10 cm is therefore pF = 1 and a negative hydraulic head of 100 000 cm is pF = 5.

6.4 Soil moisture characteristics (Retention curves)

When a slowly increasing suction (or negative hydrostatic pressure) is applied to a completely saturated non-shrinking soil, such as a sand, the air-water interface, initially at the soil surface, begins to retreat below the soil surface and the moisture retentive influences of surface tension/capillarity associated with that air-water interface thereafter come increasingly into effect. As Childs [21] pointed out, large pore channels, in which only gentle interface curvatures (or menisci) can be maintained, will empty as low suctions while narrower pore channels, supporting interfaces of sharp curvature, will not empty until larger suctions are imposed. Hence, as the soil moisture suction is gradually increased, the soil moisture content is progressively reduced, the larger pores emptying at the lower suctions and the smaller pores at the higher suctions.

A similar relationship between soil moisture suction and soil moisture content is observed for shrinking, colloidal soils such as a clay. In this case, however, over much of the range of drying, interface effects cannot be held responsible since initially, as water is withdrawn from the interstices between the plate-like particles, the particles move closer together, thereby reducing the size of the interstice which remains full of water. Thus, as Childs [21] observed, withdrawal of water is accompanied by an equivalent reduction of overall volume, (i.e., by shrinkage) without the retreat of the air-water interface into the soil, and the suction is required to maintain this smaller volume by overcoming the mutual repulsion of the particles. Clearly, as more water is withdrawn and the particles move closer together, their mutual repulsion increases, and this accounts for the observed relationship between suction and moisture content.

At very high suctions the soil moisture content may be so reduced that, even in colloidal soils, the pore space is largely filled with air, and in this suction range the soil moisture content is increasingly due to adsorption, increasing soil-moisture suction being associated with a decreasing thickness of adsorbed-water film on the

soil-particle surfaces. In this situation, the size and shape of the pores is of little
importance compared with the specific surface of the soil material. Gardner [42]
suggested that at pF 4.18 the soil moisture content would represent about 10
molecular layers if distributed uniformly over the particle surfaces.

The curve obtained by plotting soil moisture content against suction is known as
a *soil-moisture characteristic,* using the terminology suggested by Childs [19], or
simply as a *soil-moisture retention curve.* Examples of moisture characteristics for
three types of soil are shown in Fig. 6.3. These curves exhibit to a greater or lesser

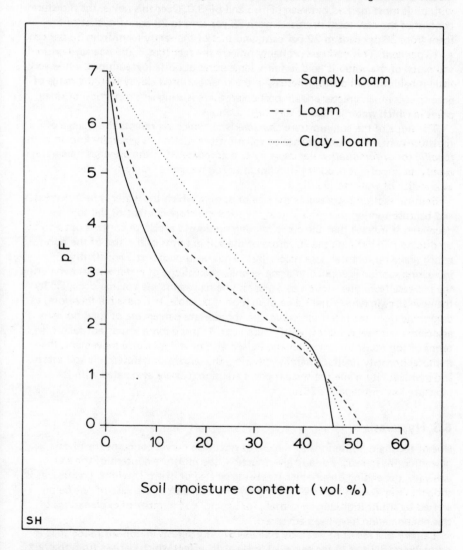

SH

Fig. 6.3. *Moisture characteristics for three soils. (From an original diagram by P. E. Rijtema, in*
Plant-Water Relationships, in Arid and Semi-Arid Conditions, *Proc. Madrid Symposium, 1959,*
UNESCO, Paris, 1960.)

degree a typical reclining S-shape which normally results from a comparatively flat central section. The clay-loam curve, however, has only a very shallow curvature in its upper section, reflecting the fact that, above pF 2.0, clay loses its moisture more gradually than sand, which in turn means that the clay-loam holds more water than the other two soils throughout most of the range of the graph. A further implication is that the final 20 per cent of moisture in the clay-loam is unavailable to the plant cover which is unable to remove moisture from the soil at suctions greater than about pF 4.2. The contrast in the rate of change of moisture content with pF is most marked between pF 1.8 and pF 3.0. Over this range, the moisture content of the clay-loam decreases from 38 per cent to 28 per cent, that of the loam from 36 per cent to 22 per cent, and that of the sandy loam from 36 per cent to 13 per cent. This contrast probably reflects the fact that in the coarser-textured soil much of the water is held in fairly large pores at quite low suctions, while very little is held in small pores; whereas, in the finer-textured clayey soil, the range of pore sizes is much greater and, in particular, there is a higher proportion of small pores in which water is held at very high suctions.

The slope of the soil-moisture characteristic, which represents the change of moisture content per unit change of soil-moisture suction, is generally known as the *specific* (or *differential*) *water capacity* and as indicated in the previous paragraph is clearly an important property in relation to soil-moisture storage and the availability of water to plants.

Because suction increases as the size of pore in which the water is held decreases, and because suction increases also as the total moisture content of the soil decreases, it follows that the curve relating suction to moisture content can also be used to help in defining the structure of the soil in terms of the size of the pores which make up its total pore space [82]. Thus, assuming that a constantly increasing suction is applied to a soil sample, the volume of water withdrawn during the increase from one suction to a greater one represents the volume occupied by the pore sizes in which that range of suction is possible. It is possible therefore, to determine from the retention curve, the appropriate percentage of total porosity associated with each successive pF increment. This is a more important factor in terms of soil moisture retention and indeed also of soil moisture movement, than the total porosity itself. Generally speaking, the larger pores determine soil aeration and permeability while the smaller pores are more closely associated with its moisture holding capacity [6].

6.5 Hysteresis

One of the main limitations to the use of retention curves concerns the phenomenon of hysteresis. For any given suction, the moisture content of the soil will vary, depending on whether it is being wetted or dried. It will be greatest, as is shown in Fig. 6.4, in the case of a drying soil, and least in the case of one being wetted by infiltration during rainfall, for example. A number of explanations of this phenomenon have been advanced.

Largely as a result of work by Haines [57], hysteresis in non-shrinking soils is normally attributed to the so-called 'ink-bottle' effect which derives from the fact that many pores have rather narrow necks or connections with neighbouring pores. It is argued that when the soil is drying, such pores may not empty until a high

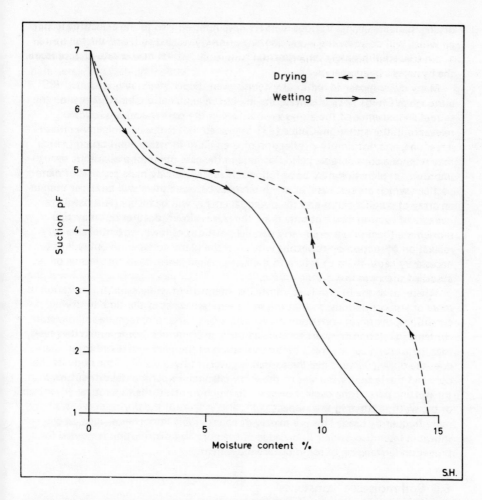

Fig. 6.4. *Moisture characteristics for a silty clay showing hysteresis. (From an original diagram by Woodman, J. H. and others [135].)*

suction, related to the attractive forces with which water is held in the narrow neck, has been reached; whereas, when the soil is being wetted, these pores will not fill until the suction has fallen to the much lower level associated with their widest diameter, at which point they will fill very rapidly. Later work [13], [120] indicated that, in fact, the delay in filling pores during wetting is more important than their retarded emptying during drying. The hysteresis effect is, in general, more pronounced in coarse-textured soils in the low-suction range, where pores may empty at a much larger suction than that at which they fill [60].

Other explanations of hysteresis have drawn attention to the fact that the contact angle of fluid interfaces on solids tends to be larger when the interface is advancing (wetting) and smaller when the interface is receding (drainage) [78]. A given water content will, therefore, tend to be associated with a greater suction in

141

drying than in wetting. Various workers have referred also to the effects of trapped air which will decrease the water content of newly wetted soil, and the failure to attain true equilibrium in experimental conditions, which may create or accentuate the hysteresis phenomenon.

Many clays appear to exhibit hysteresis even though remaining saturated [30], although as Childs [21] observed, the ink-bottle explanation, based as it is on the retreat and advance of the air-water interface in the pore space, can have no relevance in this situation. Childs [21] suggested that if the phenomenon in clays is a real one, and not simply a reflection of the failure to attain equilibrium (which may require a considerable period of time in the case of clays) a plausible, though unproven, explanation may be as follows. If initially a clay mass consists of plate particles which are not parallel, the interstices between them will be larger than in an array of parallel plates and the potential energy will be higher. A subsequent increase of suction may not only draw the plates closer together but may also reorientate them into more nearly parallel positions of lower potential energy. A relaxation of suction on rewetting will allow the plates to separate but will not necessarily cause them to return to their original alignments, so that less water is absorbed than was lost during drying.

Where, as in the British Isles, rainfall is intermittent while evapotranspiration is more or less continuous, so that the soil is exposed most of the time to drying conditions, the drying retention curve is normally used to determine soil moisture contents in relation to water balance and irrigation studies. Some authorities feel that the wetting curve gives a better indication of the pore size distribution than does the drying curve, since the former is governed by the size of the body of the pore and the latter by the size of the entry channel. Experimental difficulties with the wetting part of the cycle, however, do much to offset the theoretical advantage of the wetting curve in this respect [21], and as a result the drying curve is the more frequently used. This is a matter of considerable importance since, as will be shown in later discussions, knowledge of the pore size distribution is needed for a proper understanding of soil-moisture movement.

6.6 Soil moisture 'constants'

Despite the phenomenon of hysteresis, the form of the moisture retention curves which have been discussed above indicate that the energy status of soil-water is a continuous function of the soil moisture content [127] although, as was suggested earlier, different attractive forces may be dominant in, say, wet and dry conditions. Thus, it follows that in response to the changing balance between rainfall and evapotranspiration and the variable proximity of underlying groundwater, the energy status of soil-water and the moisture content of a soil are in a continual state of flux. But despite the fact that natural conditions seldom, if ever, permit soil moisture contents to reach a state of equilibrium, a number of arbitrary soil moisture 'constants' have, in the past, been described largely to facilitate comparisons between different soils. Even when these can be approximately located on a moisture characteristic, they have little intrinsic significance.

The constants most frequently referred to are the hygroscopic coefficient, the wilting point, the moisture equivalent, the field capacity, the maximum capillary capacity, and the saturation or maximum water capacity. Some of these must be determined for arbitrarily sized samples in specified laboratory conditions while others, more usefully, refer to natural field conditions. To the hydrologist, the most useful of these constants are, first, the *field capacity,* which has been defined as 'the amount of water held in soil after excess water has drained away and the rate of downward movement has materially decreased' [128] and which is normally taken to refer to the range of slowly changing moisture content once excess water has drained from a saturated or near-saturated soil and second, the *wilting point,* which may be defined as that range of moisture content in which permanent wilting of plants occurs. These two constants, therefore, define the approximate limits of the water in the soil which is available for plant use between about pF 3.0 and pF 4.2 and, therefore, also the potential depletion of each increment of rainfall to the soil.

Clearly, these values are not 'constants' but rather subjectively determined variables, whose magnitude depends upon a variety of changing climatic, vegetation, and soil-hydrological characteristics. Although untenable in theory, both have proved valuable in practice, and in relation to field capacity in particular, van Bavel [5] suggested that, in broad generalizations pertaining to drainage, irrigation, and watershed hydrology, the idea of a retentive capacity is essential and useful. This being so, it is rather surprising to find some investigators [48] [60] disproving their theoretical basis at such inordinate length. In any case these 'constants' will undoubtedly continue in general use.

6.7 Measurement of soil moisture content

From what has already been said about the relationships between soil moisture content and soil moisture suction, it will be apparent that the measurement of soil moisture may be approached in two main ways. Either a direct measurement of the actual water content of the soil may be made, or the water content may be derived indirectly through a measurement of the soil moisture suction.

Direct measurement of water content or wetness

One of the most accurate and most frequently used ways of measuring the water content of the soil is the *gravimetric method* which involves determining the weight loss from a number of oven-dried field samples obtained by coring or augering. A most useful review of the gravimetric method, in terms of equipment and methodological problems and of sampling requirements, was provided by Reynolds [105] [106].

The use of *lysimeters* represents an extension of the gravimetric method whereby changes in the moisture content of a block of soil are determined by recording changes in the weight of the soil block [53]. Aspects of this method have already been discussed in section 5.20.

Increasing use is now being made of the *neutron probe* in the direct measurement of soil moisture content. One of the earliest versions of this instrument was described by Gardner and Kirkham [46], although subsequent development and modifications greatly increased its suitability for field use [64] [65] [115] [126] and there is now an extensive and growing literature concerned with both theoretical and field measurement applications [25] [26] [111] [130].

In some cases a double tube system may be used, in which the source and counter are placed in separate access tubes a short distance apart. Gray and others [53] noted that one advantage of this system over the single tube system is that the source and counter tubes can be collimated, so that moisture contents over small depth intervals can be measured.

Measurement of soil moisture suction

Although only measuring its capillary component, *tensiometers* are normally used to indicate the suction force with which water is held in comparatively moist soil. In order to extend greatly the soil moisture suction range that can be measured, Peck and Rabbidge [91] proposed the use of an osmotic tensiometer or osmometer, i.e., an apparatus very similar to that described by Klute and Peters [79] but filled with an osmotic solution instead of water. Laboratory tests indicated the possibility of measuring soil moisture suctions as high as pF 4.3.

One of the great advantages of the tensiometer is that it may be used for field measurements *in situ*. For the direct measurement of soil moisture suctions higher than about pF 2.93, however, other than by using the type of experimental tensiometer which has just been described, it is necessary to make laboratory use of *porous plates* or *pressure membrane* apparatus, such as that described by Richards and Fireman [109] and Richards [108].

An alternative means of extending soil moisture suction measurements beyond the range of the normal tensiometer involves the use of buried *porous blocks*. The soil in which the block is buried has a suction appropriate to its moisture content and this suction will be transmitted to the buried block which then assumes a moisture content appropriate to the transmitted suction, in accordance with its own moisture characteristic [21].

Other methods

Other methods may be appropriate in particular circumstances, e.g., the water balance method developed principally by Thornthwaite and Mather [123] [124] [4] [138]. There are also possibilities of monitoring soil moisture over large areas using remote-sensing techniques [28] [34].

Movement of soil moisture

Soil moisture moves in response to a number of forces such as gravity, suction forces, and the vapour pressure and temperature gradients, and because of the fact that gravity is not necessarily the dominant force, unsaturated movement may be in any direction. There is, however, a tendency for the main controlling forces to operate either from the surface or from the bottom layers of the zone of aeration with the result that vertical movements predominate. The moisture involved in these movements may be in the form of either liquid water or water vapour, and it is frequently difficult to distinguish their relative importance. No clear distinction between these two phases will, therefore, be made in the following discussion, although in a later section (6.12) the role of water vapour transport will be discussed in relation to those conditions in which it is known to be important.

6.8 Principles of unsaturated flow

Klute [78] observed that soil moisture movement may be considered from three viewpoints—the molecular, the microscopic, and the macroscopic. At the molecular level, flow is considered in terms of the behaviour of the water molecules and explanations might be based upon statistical mechanical concepts whereas, at the microscopic level, principles of fluid mechanics might be used to calculate the behaviour of soil moisture within the pores. Almost inevitably, however, the development of theory must proceed at the macroscopic level, since we can observe neither the behaviour of the individual molecules nor the fluid velocities and pressures within individual pores. In the macroscopic approach the porous medium is treated as being uniform, all variables are assumed to be continuous functions of space and time [78], and the description of soil moisture movement assumes proportionality between the flux of water and the appropriate potential energy gradient [33]. Thus, Darcy's law gives the macroscopic velocity of flow (i.e., the average velocity) through a given cross-section of porous medium although, in fact, the flow occurs at a higher velocity through the more restricted cross-section of the actual interstices and pore spaces only.

According to Philip [102], the foundations of the theory of soil moisture movement were laid by Buckingham [12] and later workers such as Richards [107], Childs and Collis-George [23], Klute [77], and Philip [96] [97] [98] developed the approach sometimes known as *diffusion analysis,* which was reviewed in detail by Philip [101]. The three essential concepts which form the basis of the diffusion approach are (a) that soil moisture movement is a response to a total hydraulic head gradient which includes both suction and gravitational components,

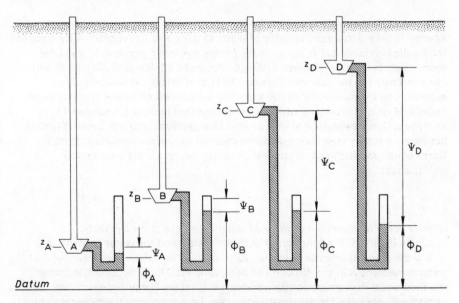

Fig. 6.5. *The concept of total potential. (From an original diagram by D. M. Gray, D. I. Norum, and J. M. Wigham, in Gray, D. M. (Ed.),* Handbook on the Principles of Hydrology, *Canadian National IHD Committee, Ottawa, 1970.)*

145

(b) that Darcy's law is valid for non-saturated flow, and (c) that soil moisture suction and conductivity are unique functions of the moisture content. Each of these basic concepts will now be briefly examined.

As indicated in section 6.3, water moves as a liquid in unsaturated soils, under the combined influence of gravity and suction, from a point of high total potential to a point of lower total potential. The total potential Φ, then, comprises the moisture potential (or suction) Ψ and the gravitational potential $-z$, thus:

$$\Phi = \Psi - z \qquad (6.2)$$

where the potentials are expressed in cm of water and z is positive downward.

The way in which the gravitational and moisture potentials combine to affect soil moisture movement is illustrated by the bank of tensiometers in Fig. 6.5. In this case, the soil moisture contents at A and B are equal (i.e., $\Psi_A = \Psi_B$) but moisture will move from B to A because the total potential at B is greater (i.e., $\Phi_B > \Phi_A$). Or, again, the soil is drier at C than at B (i.e., $\Psi_B > \Psi_C$ (N.B. values are negative)) but since the total potentials at B and C are equal, there will be no moisture movement between the two. Finally, provided that the soil at D is considerably drier than at C (the situation shown), moisture will move upward against the gradient of gravitational potential because $\Phi_C > \Phi_D$.

The second basic concept of the diffusion analysis approach to soil moisture movement is that Darcy's law is valid for non-saturated flow. Darcy [29] confirmed that, in saturated conditions the flow of water through a porous medium is proportional to the hydraulic gradient and the conductivity, so that

$$v = -k \nabla i \qquad (6.3)$$

where v is the macroscopic velocity of water in cm/s, k is a proportionality constant (hydraulic conductivity) in cm/s, and ∇i is the hydraulic gradient. It was later shown, particularly by Richards [107], Childs and Collis-George [22] [23] and more recently by, for example, Watson [131] that this law is basically also applicable to the movement of soil moisture in unsaturated conditions. Although occasional reports several of which were summarized by Swartzendruber [117], imply non-Darcy behaviour in unsaturated flow, according to van Bavel [5] there has been no totally convincing demonstration of significant deviations from the Darcy equation which can, therefore, be simply expressed for unsaturated conditions as

$$v = -k \nabla \Phi \qquad (6.4)$$

where v is the macroscopic velocity of water in cm/s, k is the hydraulic conductivity which is a function of both the matrix and the water content, and $\nabla \Phi$ is the gradient of total potential. As in eq. 6.3, the minus sign indicates that water movement is in the direction of decreasing total head or potential. In fact, the theoretical flow equations derived from this simple Darcy statement are exceedingly complex for the unsaturated case, largely because the movement of water at any point usually results in a change of water content at that point which, in turn causes a change in hydraulic conductivity and suction at that point [53].

146

Notwithstanding the general applicability of Darcy's law to unsaturated flow, certain basic conditions must first be satisfied. In the first place, the porous body must be large compared with its microstructure so that it can safely be regarded as uniform. Childs [21], for example, suggested that it would be most hazardous to apply the law to the flow of water to a plant root from a specific cylinder of surrounding soil occupying a volume of only a cubic centimetre or so. Secondly, the velocity of flow must be sufficiently small, and in this connection Childs [21] observed that the Darcy law may not be safely applied if the Reynolds number exceeds unity. It is most unlikely that such a value would even be approached in any case of natural soil moisture movement.

The third basic concept of diffusion analysis is that both the moisture potential (or suction) and the hydraulic conductivity are unique functions of the soil moisture content, although this can never be precisely true because of the effects of

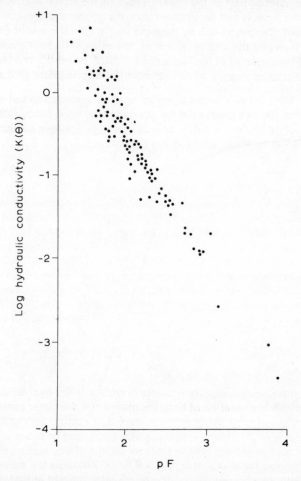

Fig. 6.6. *Relationships between hydraulic conductivity and pF for a heavy clay soil. (From an original diagram by Wind [133].)*

hysteresis. Relations between suction and moisture content were discussed in detail in sections 6.3 through 6.5. The relations between hydraulic conductivity and moisture content are such that whereas in saturated conditions k may be regarded as more or less constant for any given material, in unsaturated conditions k will vary considerably with soil moisture content (Θ), i.e., $K(\Theta)$, as is shown in Fig. 6.6, in which data are plotted for a heavy clay soil. It can be seen that at high moisture contents hydraulic conductivity is high, reaching a maximum at or near saturation, but that it decreases rapidly with moisture content. The main reasons for this marked decline in hydraulic conductivity are that, in a saturated soil all the pore spaces are filled with water, and therefore form an effective part of the moisture conducting system, since the movement of liquid moisture can take place only through existing films of water on and between the soil grains. By definition, however, in an unsaturated soil, some of the pores are filled with air and, therefore, act as a non-conducting part of the system, reducing the effective cross-sectional area available for flow. It will be apparent that, the greater the decrease in soil moisture content, the greater will be the reduction in effectiveness of the conducting system and the smaller, therefore, the value of hydraulic conductivity.

In terms of the variation of hydraulic conductivity, however, the effects of moisture content must be considered in association with the pore size, and pore size

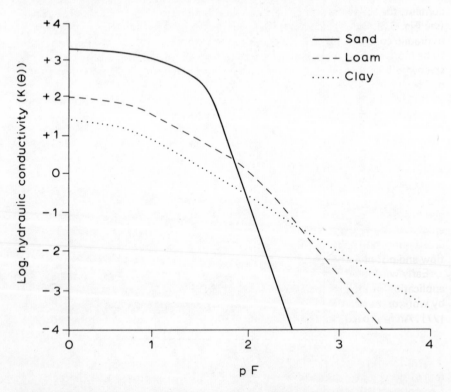

Fig. 6.7. *Relationship between hydraulic conductivity and pF for three soil types. (From an original diagram by Wind [133].)*

distribution of the soil. At high soil moisture contents, conductivity is more or less directly related to soil texture, and increases as the latter becomes coarser. This relationship is to be expected, because water will obviously be transmitted more easily through large water-filled pores than through small ones. As the moisture content decreases, and the suction increases, however, this relationship between conductivity and texture is reversed (see Fig. 6.7) so that in dry conditions, clay soils are likely to have higher conductivities than loamy or sandy soils. Again, this relationship is to be expected, since the finer soils will have more water-filled pores at high suctions and, therefore, a larger cross-sectional area through which flow can take place, than will coarser soils in which only a small proportion of the pores will contain water at high suctions. Finally, in shrinking soils, the increased suction which accompanies drying reduces the size of the pores that remain full of water (see also section 6.4) and this further helps to reduce the hydraulic conductivity.

Temporarily ignoring the effects of gravity, the main implications of the above discussion are that, in normal circumstances, there will be a movement of soil moisture from moist soil into dry soil or, in other words, from an area of low suction to an area of high suction. Where the nature of the soil profile results in the juxtaposition of soils of markedly different texture having the same initial moisture content, there will usually be a movement of moisture from the coarse soil—sand, for example—to the fine soil—clay, for example—because, for a given moisture content, the soil water will be held at lower suctions in the sand than in the clay (see Fig. 6.3). Finally, in view of the relationship between soil moisture content and hydraulic conductivity, it is apparent that the movement of soil moisture will tend to be more rapid in moist soils than in dry, and that the drier the soil becomes, the steeper will be the suction gradient necessary to induce any appreciable moisture movement.

It has already been stated that the theoretical unsaturated flow equations derived from the simple Darcy statement are exceedingly complex. In addition, it is necessary to measure the change of suction and moisture content with depth and time in order to apply the equations. In some cases it has proved possible to simplify the solution of the unsaturated flow process by resorting to diffusion analysis and by transforming the flow equation so that the movement (flux) of soil moisture is related to the gradient of water content (Θ), or wetness, rather than to the gradient of suction (Ψ). Analogous solutions in the fields of chemical diffusion and heat flow have provided the basic models [14] [27], although it must be emphasized that the use of diffusion terminology in soil moisture problems can be grossly misleading. Clearly, the process of soil moisture movement is one of mass flow and not one of molecular diffusion.

Early work by Childs [17] [18] and Nicholson and Childs [88] on the application of the diffusion equation to water movement in clay soils was followed by work on horizontal moisture movement by Kirkham and Feng [75] and Klute [77]. An improved method derived by Philip [98] made solution of the diffusion equation much easier and was closely followed [99] by an extension of the procedure to include the effects of gravity.

The analogy with heat flow helps to illustrate the basic principles. Thus the rate of flow of heat is proportional to the temperature gradient, which may be transformed into a gradient of heat content by reference to the specific heat (i.e., the ratio between the increase of heat per unit mass and the increase of temperature). In the case of soil moisture, the rate of flow depends in part on the gradient of suction (Ψ), which may be transformed into a gradient of water content (Θ) by reference to the specific water capacity (c) of the soil where

$$c(\Theta) = \frac{d\,\Theta}{d\,\Psi} \qquad (6.5)$$

The specific water capacity is thus the rate of change of water content, per unit volume, with change of suction and is, therefore, simply the slope of the moisture characteristic which changes more or less continuously with changing moisture content. The specific water capacity is thus a function of moisture content, i.e., $c(\Theta)$.

Childs and Collis-George [23] introduced a function called the diffusivity, D, which is defined as the ratio of the hydraulic conductivity to the specific water capacity. Since we have already seen that both hydraulic conductivity and specific water capacity are functions of soil moisture content, it is clear that the diffusivity is also such a function so that

$$D\,(\Theta) = \frac{k(\Theta)}{c\,(\Theta)} \qquad (6.6)$$

Darcy's law can thus be rewritten for the one-dimensional case as

$$v = -D(\Theta)\frac{\delta\Theta}{\delta\,x} \qquad (6.7)$$

where $\frac{\delta\Theta}{\delta\,x}$ is the gradient of moisture content in direction x and other terms are as previously defined. In this form, the equation is mathematically identical with the first equation of diffusion [60]. For flows in two or three dimensions, the equation is most conveniently written in vector notation.

This is clearly where the third basic concept of the diffusion approach to soil moisture movement, namely that both suction and hydraulic conductivity are unique functions of soil moisture content, has formerly imposed considerable limitations, and has largely restricted attention to flows in which all parts of the system were either always wetting or always drying [78]. Comparatively recent work, however, has been quite successful in incorporating the phenomenon of hysteresis into the diffusion equation for soil moisture movement [21] [37] [112] although, according to Hillel [60], the diffusion approach still fails if the hysteresis effect is appreciable, or in the case of a layered soil profile, or in the presence of thermal gradients. because in these conditions, soil moisture movement bears no simple or even consistent relation to the decreasing gradient of moisture content, and may even occur in the opposite direction. In addition, Philip [102] discussed some of the problems involved in extending the classic theories of water movement to swelling soils.

The main advantages in using the diffusion approach are that the range of variation of diffusivity is smaller than that of conductivity and that soil moisture content and its gradient are often easier to measure, either in the laboratory or the field, and to relate to volume fluxes, than are soil moisture suction and its gradient [60].

Finally, a brief mention should be made of the fact that, in addition to the mass flow of the soil solution, soil moisture movement in the liquid phase may also result from thermal and osmotic gradients [15] [50] [78] [103].

6.9 Soil moisture movement during infiltration

While infiltration into a dry soil is taking place, as , for example, during heavy rainfall or overhead irrigation, the immediate surface layers will be saturated, and there will be a decrease in moisture content with depth into the soil. In these conditions, therefore, both the suction gradient and the gravitational gradient, as well as adsorptive forces, encourage the downward movement of moisture into the soil profile.

Some of the classic experimental work on the entry of water into the soil in this situation was performed by Bodman and Colman [7], [24], who found that the wetted portion of a column of soil into which infiltration is taking place is comprised of a number of zones, which are illustrated in Fig. 6.8. The saturated zone, as its name implies, is just a shallow saturated layer, a centimetre or so in thickness, at the ground surface. Immediately below this, is another shallow zone, only a few centimetres in thickness, in which the moisture content decreases very rapidly from top to bottom and which is, therefore, known as the transition zone. Below this, again, is the transmission zone, through which water from the upper two zones is transmitted, with little or no change in moisture content, to the underlying wetting zone which, like the transition zone, has fairly steep moisture gradients, and where the moisture content also changes appreciably with time. Finally, at the base of the wetting zone is the sharply defined wetting front, which is characterized by very steep moisture gradients, and which marks the limit between the wetted soil above and the dry soil below. Provided that the supply of water to the soil surface from rainfall or irrigation continues, the wetting front advances steadily downwards into the unwetted soil as a result of the passage of water through the transmission zone.

It follows, therefore, that variations in the quantity of rainfall and infiltration result in variations in the depth of wetted soil rather than in continued increases in moisture content in the surface layers. Indeed, during infiltration and percolation, the only part of the soil profile in which the moisture content changes significantly with time is the wetting zone and the wetting front. The saturated zone, of course, remains saturated, the moisture gradient in the transition zone remains fairly constant, and so, too does the moisture content and suction in the transmission zone. Thus, with continuing infiltration, the transmission zone becomes longer and the wetting zone and the wetting front move farther downward into the soil. This is illustrated more clearly in Fig. 6.9, where the shaded areas show the increase in soil moisture content over the initial soil moisture values at 0 hours in the upper 150 cm of a soil profile, during continuous infiltration. The main features of these graphs are the rapid establishment of a soil moisture gradient in the upper 60 cm of the profile which changes only slightly after a period of five hours has elapsed, and the fact that, in the lower part of the profile, the rapid downward extension of the wetted zone is a considerably more marked feature than its slowly increasing moisture content.

A development of the soil profile during infiltration similar to that described above was subsequently derived from theoretical considerations by Philip [100],

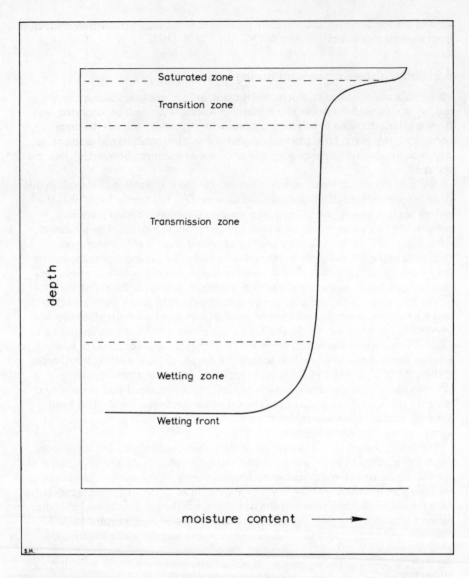

Fig. 6.8. *Moisture zones during infiltration. (From an original diagram by G. B. Bodman and E. A. Colman,* Soil Sci. Soc. Amer. Proc., **8:** *116-122, 1943 and adapted by permission of the publisher.)*

although, in fact, his analysis had to be modified slightly in order to accommodate the transition zone. Figure 6.10 shows typical moisture profiles during infiltration into a light clay, as computed by Philip [100], with the approximate time in hours shown on each profile; the similarities with the moisture profile shown in Fig. 6.9 are immediately apparent.

From the above discussion, it will be seen that, in the early stages of infiltration

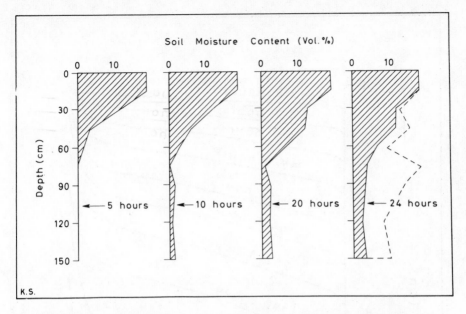

Fig. 6.9. *Variations of soil moisture content with depth during infiltration. The broken line represents complete saturation. (From an original diagram by D. R. Nielsen,* Proc. Seepage Symposium, *Phoenix, Arizona, 1963, pp. 47-51, 1965).*

into a dry soil the suction gradient is likely to be the most important factor determining the amount of infiltration and downward movement of moisture, and since, at first, the suction gradient is large, initial rates of movement are likely to be rather high. This initial rapid penetration of the moisture profile is clearly illustrated in Fig. 6.10. However, as the lengthening, wetted part of the soil profile attains approximately the same moisture content, that of the transmission zone, the suction gradient decreases, the gravitational gradient becomes relatively, more important, and in these conditions the rate of infiltration and downward movement will be much lower. This is again clearly illustrated by the time-located profiles in Fig. 6.10. As this trend continues, the suction gradient in the upper part of the profile eventually becomes negligibly small so that the constant gravitational gradient remains as the only force moving water downward in the transmission zone. Since the gravitational gradient has a value of unity (the gravitational head decreasing by 1 cm for each centimetre depth below the surface) it follows that the downward moisture flux tends to approach the hydraulic conductivity as a limiting value [60].

Infiltration under field conditions will naturally be a more complex process than that represented by the simple conceptual and mathematical models which have been described above. Klute [78] referred to some of the reasons for deviations between theoretical and experimental values, including non-uniformity of initial water content and soil properties, hysteresis, changes of various soil and boundary conditions with time, and the existence of divergent rather than one-dimensional flow, and concluded that much still remains to be done in the field application of infiltration theory.

153

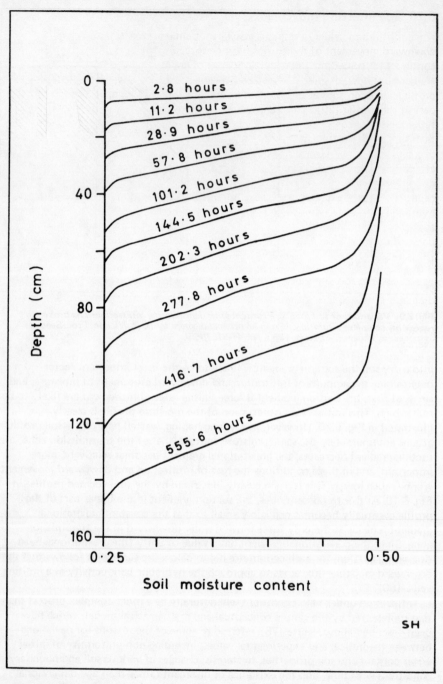

Fig. 6.10. *Computed moisture profiles during infiltration into a light clay. (From an original diagram by Philip [100].)*

6.10 Soil moisture movement after the cessation of infiltration

When infiltration ceases, a redistribution of the moisture in the soil begins. The downward movement of moisture, and therefore, also, of the wetting front, continues for some time under the influence of gravity and the existing suction gradient but now, of course, moisture movement into the lower part of the soil profile can occur only at the expense of the moisture in the upper, initially wetter, part of the profile [20]. The transmission zone which exists during infiltration subsequently becomes a draining zone [82] in which the water content and therefore, also, the non-saturated hydraulic conductivity, especially in sandy soils, decrease with time. Accordingly, moisture movement is gradually reduced until, after a few days it becomes very slow, at which stage the water content of the draining zone takes a long time to change and the soil is said to be at 'field capacity'. Thus, field capacity is closely related to the reduction in non-saturated hydraulic conductivity resulting from decreasing moisture content, and so is more applicable to conditions in sandy soils than to those in clays, because the large pores of the sand soon empty, and conductivity decreases rapidly as a consequence [82].

Figure 6.11 clearly illustrates the sequence of moisture redistribution, described above, in a soil from which no evaporation loss occurred during the period of drainage. The shaded areas represent that portion of the initial infiltration which remains at various levels in the soil profile at specific time intervals after the cessation of infiltration, and the figures at the bottom of the profiles indicate the average rate of water movement at the 150 cm level. The concept of field capacity is well illustrated by the comparatively uniform and slowly changing moisture contents of profiles 4 to 7.

A similar illustration is provided in Fig. 6.12 which shows successive moisture profiles in a fine sandy loam soil during redistribution, in the absence of evaporation, of two irrigations of 10 cm each. Field capacity is represented in the upper 60 cm of the soil profile by the slowly changing moisture content between days 2 and 33. Other moisture profiles in the same paper by Gardner and others [48], showing the redistribution of smaller amounts of irrigation, provide more convincing evidence of the authors' contention that this particular soil '. . . exhibits nothing resembling a distinct field capacity value'.

In natural conditions, however, evaporation will normally take place from the soil after the cessation of rainfall or irrigation, and will result in a drying-out of the surface layers. In this way, a suction gradient is created which encourages the upward movement of moisture from the draining zone and which, therefore, acts as an additional factor in reducing the downward movement of water. Normally the reduction in conductivity is the dominant factor, but the imposition by evaporation of a reverse suction gradient in the upper layers of the soil often encourages the early development of the quasi-equilibrium conditions represented by the term field capacity. This is illustrated in Fig. 6.13 in which are plotted moisture profiles developed during the redistribution of moisture in slate dust (comparable to a clay soil), after the cessation of infiltration. The duration and total amount of infiltration was smaller in the case of profile 1 than for profile 2, and it is interesting to see the effect of this contrast on the pattern of moisture movement. Both the rate of downward movement and the depth of penetration are smaller in profile 1, but the time taken to attain field capacity is considerably greater than in

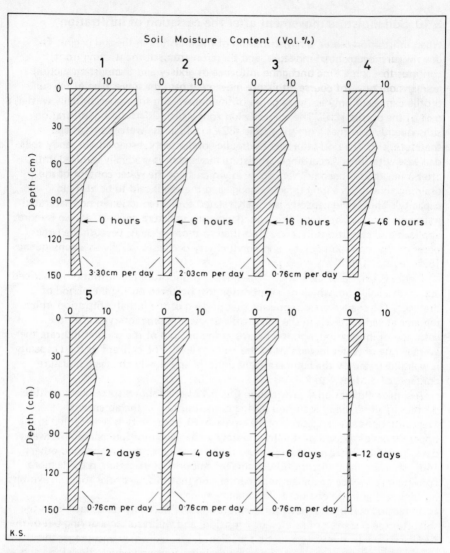

Fig. 6.11. *Variation of soil moisture content with depth during drainage. The values between pairs of profiles indicate average rates of water movement at the 150-cm depth. (From an original diagram by D. R. Nielsen,* Proc. Seepage Symposium, *Phoenix, Arizona, 1963, pp. 47-51, 1965.)*

the case of profile 2, where an almost constant moisture content has been established after only 24 hours.

An interesting comparison of the progress of moisture distribution in three situations, redistribution without evaporation, simultaneous evaporation and redistribution, and evaporation only, is provided in Fig. 6.14. The curves of the simultaneous processes resemble those for distribution alone, except for the obvious evaporation effect in the surface layers. In particular, the lower portions of the curves indicate

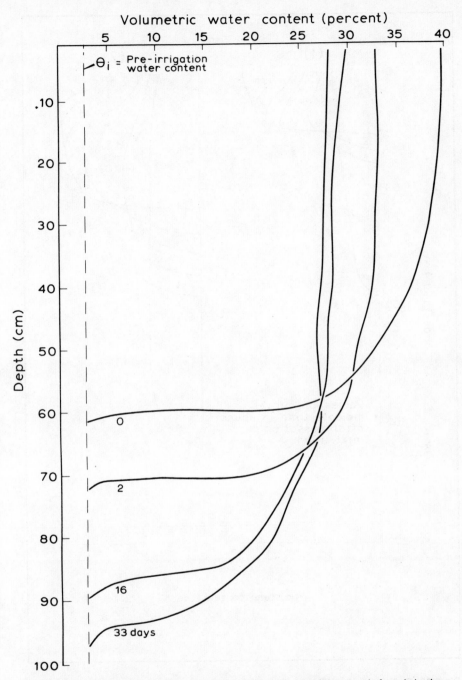

Fig. 6.12. *Successive moisture profiles during the redistribution of the second of two irrigations of 10 cm each in a fine sandy loam soil. (From an original diagram by W. R. Gardner, D. Hillel, and Y. Benyamini,* Water Resources Res., *6, p. 857, Fig. 3, 1970. © American Geophysical Union.)*

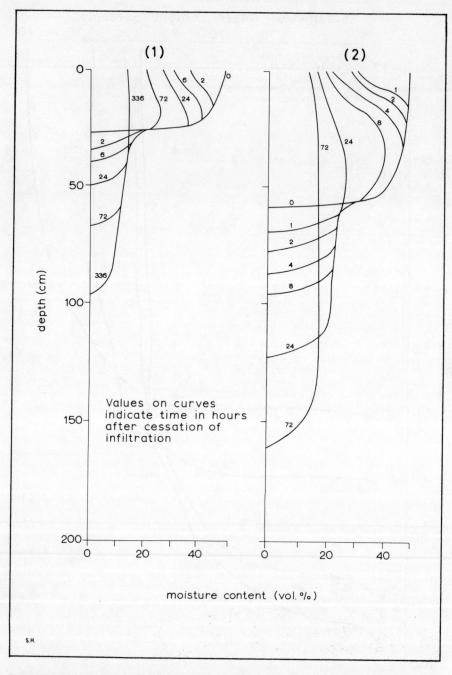

Fig. 6.13. *Profiles showing the redistribution of moisture after the cessation of infiltration into slate dust (similar to a clay soil). (From an original diagram by E. G. Youngs,* Soil Sci., **86**: 202-207, 1958. © *1958, The Williams and Wilkins Co., Baltimore, Md. 21202, USA.)*

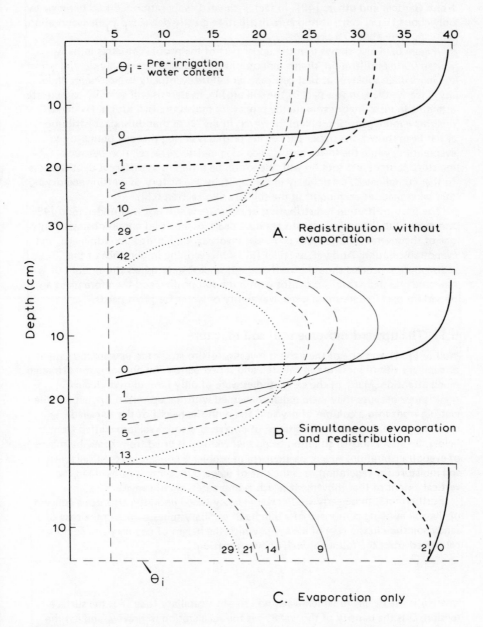

Fig. 6.14. *Successive moisture profiles in soil columns in three situations following an irrigation of 50 mm. Values indicate number of days after irrigation. (From an original diagram by W. R. Gardner, D. Hillel, and Y. Benyamini,* Water Resources Res., **6,** *p. 1149, Fig. 1, 1970. © American Geophysical Union.)*

that evaporation had little effect on the shape and rate of advance of the wetting front. Gardner and others [49], in fact, estimated evaporation reduced drainage by only about 10 per cent, although redistribution greatly detracted from evaporation, reducing it by about 75 per cent.

Figure 6.14 also shows that the upper part of the profile, which is initially wetted during infiltration, drains monotonically, though increasingly slowly, whilst the immediate sublayer at first gets moister before eventually beginning to drain dry. Clearly, then, in this part of the soil profile, hysteresis will severely complicate attempts to measure or estimate the process of moisture distribution. Hysteresis presents a more general problem, however, in the sense that during redistribution, as has been shown, the upper part of the profile is drying through drainage and evaporation, while the lower part is becoming moister. The relation between moisture content and suction will, therefore, be different at different depths but is further complicated, particularly in clay soils, by the history of wetting and drying that takes place at each point in the soil (see also section 6.5).

The post-infiltration redistribution of moisture is still not well understood [48] and most of the experimental studies have dealt, in idealized conditions, with only one of the several simultaneous processes involved, e.g., infiltration, drainage, and evapotranspiration. And yet, as Hillel [60] observed, the importance of the redistribution process should be self-evident, since it determines the amount of moisture retained at various depths within the soil profile, and therefore plays an important part in determining the availability of water for plant needs.

6.11 The upward movement of soil moisture

Wollny [134] was one of the earliest investigators to study the upward movement or capillary rise of moisture in the soil, and subsequently the phenomenon attracted much attention. Many of the early studies were of only limited hydrological significance because they were much concerned with the upward movement of the wetting front into a column of dry soil, and with the height of the so-called capillary fringe. In natural conditions, of course, dry soil is seldom wetted from below; the more normal situation being that wet soil is dried from above by means of evapotranspiration. Again, particularly in problems concerning the availability of soil moisture for vegetation, it is the rate of upward movement rather than the vertical extent of that movement which is of greatest significance.

Furthermore, these early studies were largely based upon the erroneous concept of the soil as being comprised of a bundle of capillary tubes (wider in the case of a sandy soil than in the case of a clay) in which the height of capillary rise could be calculated from the following well-known equation:

$$h = \frac{2T}{gDr} \qquad (6.8)$$

where h is the height to which water will rise in a capillary tube, T is the surface tension, D is the density of the water, g is the acceleration of gravity, and r is the radius of the tube. Earlier discussion in this chapter has shown this concept of the soil to be totally inadequate, and has emphasized that, in fact, water movement takes place through films of water in the irregularly shaped and variously sized inter-particle spaces, and that the speed and direction of movement are largely determined by the hydraulic conductivity and by the combined gradients of suction and gravity.

The effect of evaporation and transpiration (referred to briefly in the preceding section) is to create a suction gradient which rapidly becomes greater than the opposing gravitational gradient and which, therefore, encourages the movement of water towards the soil surface or the root zone.

Hillel [60] referred to the recognition, under constant external conditions, of two fairly distinct stages in the drying process, the first of which is characterized by a fairly constant rate of evaporation governed largely by external and soil surface conditions, and the second of which is characterized by a declining rate of evaporation governed largely by the ability of the soil profile to deliver moisture to the evaporating surface (see Fig. 6.15). In natural conditions, which are rarely constant, this two-fold distinction can only be approximate. Thus, at first (Fig. 6.15A) as the soil becomes drier, so the suction gradient increases but, at the same time, the films through which water movement takes place become thinner and fewer in number. The effect of the latter in reducing the non-saturated hydraulic conductivity tends to have a greater influence on water movement than the increased suction gradient. The rate of moisture movement, therefore, decreases, usually only slowly when the soil is wet, but very rapidly once drying has progressed some way [95]. In the second stage (Fig. 6.15B) this reduced moisture movement, and therefore reduced evaporation, is further encouraged by the decreasing moisture gradients which develop as the deeper soil layers lose moisture by continued upward movement.

It is at this stage that water vapour movement (see next section) in the dry upper layers of the soil will begin to occur at a significant rate and, in fact, the true evaporating surface will move down into the soil. Gardner [40] suggested, however,

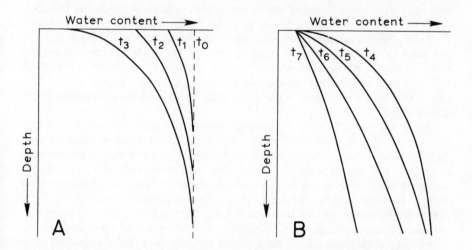

Fig. 6.15. *The changing moisture profile in a drying soil. (A) In the first stage of drying the gradients toward the surface become steeper; (B) in the second stage the moisture gradients decrease as the deeper layers lose moisture by continued upward movement. (From an original diagram by Hillel [60].)*

161

that existing data showed that the depth of evaporation will not generally move down more than a few centimetres below the soil surface, but that the depth will be greater for coarse-textured soil than for fine-textured soils.

The amount of moisture which will move upwards through the soil under the influence of evapotranspiration will be considerably affected by a number of 'external' factors, including the rate of drying of the soil surface, the density and depth of the root system, and the depth of the water table.

Penman [93], [94], [95] showed that the amount of moisture lost from a soil surface by evaporation will be affected by the *rate of drying* of the surface layers. If the drying occurs comparatively slowly, as during the spring in the British Isles, for example, evaporation losses will be similar to those from an open water surface exposed to the same meteorological conditions, indicating that the movement of moisture through the soil profile is able to keep pace with the removal of moisture from the surface, i.e., stage A in Fig. 6.15 has been maintained for a comparatively long time. If, on the other hand, drying occurs rapidly, as it does during warm summer periods, upward movement from below is unable to keep pace with surface losses and, in these conditions, the hydraulic conductivity of the surface layers may be so markedly reduced that, despite the fact that soil moisture contents a few centimetres below the surface are still high, further evaporation losses are negligible. In this situation the initial drying stage A is of very short duration. In hot, sunny conditions, therefore, bare soil may tend to be self-mulching and this, in turn, reduces the total amount of upward movement of soil moisture which takes place, although doubts about the effectiveness of a high initial rate of evaporation in reducing total cumulative water loss were raised by Gardner and Hillel [45] and Hillel [60].

Similar principles are applicable to the drying imposed by *plant roots* within the soil profile. Penman [95] suggested that it was more satisfactory to consider root behaviour in terms of 'foraging' ability rather than in terms of the ability to 'suck' water from any great distance. Similarly Visser [129] considered that around each root and rootlet a cylinder of soil may be distinguished, from which water is extracted by the root, while the moisture outside this cylinder is extracted by other roots. Evidently, the denser and deeper the root system is, the smaller will be the radius of each contributing cylinder and the higher, therefore, the rate of abstraction by the root which is likely to be satisfied by moisture movement within the cylinder.

One of the most important influences, however, on the amount and rate of moisture movement is the *depth to the water table.* Where the water table is close to the ground surface, drainage and redistribution of moisture after irrigation has ceased will continue until an equilibrium is reached, in which the soil moisture suction in centimetres at any point corresponds approximately to the height of that point in centimetres above the water table. In natural conditions, and with a plant-covered ground surface, however, this equilibrium stage is never attained, because suctions are increased in the surface layers as a result of the removal of moisture by evapotranspiration. Now, in this situation the upward movement of soil moisture may be considerable, because as long as the suction in the surface layers exceeds the depth to the water table, water will move upward through the soil [40].

In most cases, hydrologists are more concerned with the *rate* of upward

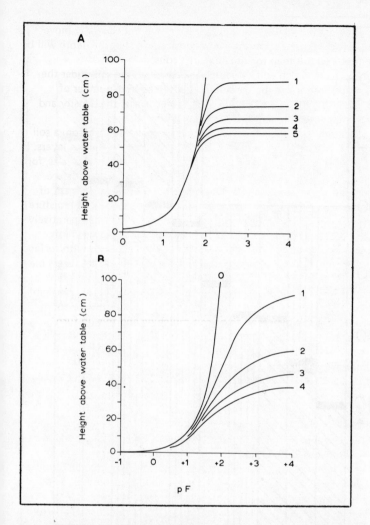

Fig. 6.16. *Maximum capillary movement in millimetres per day related to height above a water table and soil moisture suction for (A) a coarse-textured soil and (B) a fine-textured soil. (From original diagrams by Wind [133].)*

movement than with the fact that some movement takes place, although very slowly, and, in this respect, water table depth has been shown to be a vital factor. The results of experiments by Wind and his associates in the Netherlands [133], which confirm earlier theoretical work by Gardner [39], are summarized in Fig. 6.16 in which maximum rates of capillary movement in a coarse-textured (A) and a fine-textured soil (B) are related to water table and soil moisture conditions. It can be seen that, particularly in the case of the coarse-textured soil, the curves become almost horizontal at the higher suctions, indicating that the maximum value of capillary flow is dependent on height above the water table rather than on the suction imposed at the soil surface. The fact that the soil profile can so limit the

163

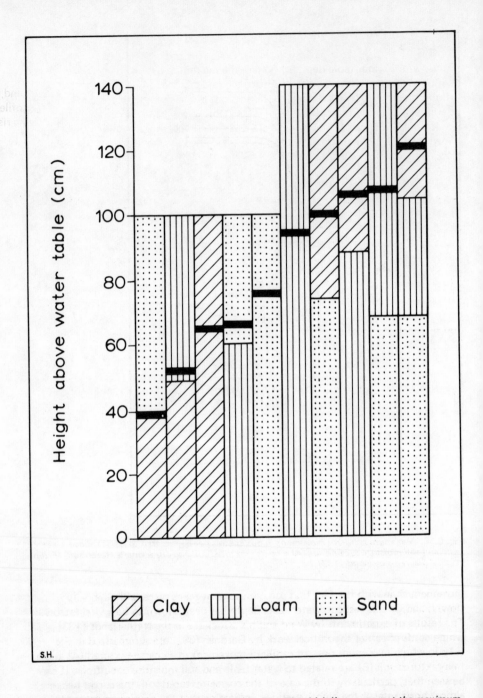

Fig. 6.17. *Capillary rise through different soil profiles. The thick lines represent the maximum height above the water table at which a capillary flow of 2 mm per day can be attained. (From an original diagram by Wind [133].)*

rate of evaporation, through the medium of hydraulic conductivity, was described by Hillel [60] as a remarkable and useful feature of the unsaturated flow system. Thus, with a water table depth of 60 cm, the maximum capillary rise is about 5 mm per day whereas, with a depth of 90 cm, the rate declines to about 1 mm per day. In the fine-textured soil the horizontal sections of the curves are not well developed, although the influence of water table depth can still be clearly seen.

It can be deduced from Fig. 6.16 that, with an appropriate suction gradient, very small rates of capillary flow may occur at a considerable height above the water table and, in fact, laboratory experiments have shown that measurable movement may occur over vertical distances of at least 7 metres [44]. Especially in arid and semi-arid areas, therefore, where rates of evaporation are high, such movements may lead to damaging accumulations of salt at the ground surface, even where the water table occurs at a considerable depth [133]. It nevertheless remains true that, as far as plant activity is concerned, the total amount of moisture that can move through the soil at a useful rate is effectively limited, at one extreme, by the conditions obtaining when the soil is at field capacity, and, at the other, by decreasing non-saturated hydraulic conductivity as the soil dries out.

It was earlier shown (see page 148 and Fig. 6.7), however, that hydraulic conductivity is determined by soil texture as well as by moisture content, and, accordingly, the rate and effectiveness of capillary movement will be much affected by this factor, and particularly by the juxtaposition of different textural horizons in the soil profile. In a previous reference to Fig. 6.7, it was noted that, in wet conditions, the conductivity of clay soils is rather low, but that, in dry conditions, clays have higher conductivities than sandy soils. Thus, as Wind [133] pointed out, where clay overlies sand in a soil profile having a high water table, the sand will tend to be moist and the clay dry, and so conditions favour a high rate of capillary movement throughout the profile. If sand overlies clay, with similar water table conditions, capillary movement will be less effective. This relationship is illustrated in more detail in Fig. 6.17, from which it will be seen that the highest rates of capillary movement occur when the soil texture becomes gradually coarser with depth below the surface.

6.12 The movement of water vapour

The movement of water vapour in the soil may in certain circumstances be quite important, particularly, for example, where a shallow surface layer has dried out while the underlying soil mass remains comparatively moist. Vapour movement through the soil profile results from three main factors: a pressure gradient, a vapour pressure gradient, and a temperature gradient, the latter being the most generally important.

Under the influence of a temperature gradient, there tends to be a movement of water vapour from warm soil to cold soil, an effect which has been frequently demonstrated under laboratory conditions [56] [72] [73] [76] [80] [121], although only a few successful field measurements have been reported [31]. With a given temperature gradient, water vapour movement through the soil profile will tend to occur more rapidly in open-textured soils than in those with a fine texture and low porosity and will tend to be considerably more important in dry soils, having a large number of air-filled pores, than in moist soils.

6.13 Moisture movement and the soil profile

In humid areas, where rainfall exceeds evaporation, the downward movement of moisture is dominant and normally results in the leaching out of soluble minerals and nutrients and, eventually, in the development of a typical iron- and aluminium-rich podsol profile. In severe cases, a hardpan or iron pan may develop in the B horizons, giving rise to impeded drainage, and perhaps having far-reaching hydrological consequences in terms of drainage, infiltration, and runoff.

In dry areas, where evaporation exceeds rainfall, the downwashing of soluble minerals is not likely to be important below the normal depth of penetration of rainfall in the surface layers. Indeed, in such areas there may be a predominant upward movement of moisture so that soluble materials are concentrated in the surface layers of the soil, giving rise to a typically lime-rich pedocal profile. Various investigators have given attention to the problems involved in the movement of salts in the profile during and after irrigation [9] [10] [43] [90].

Infiltration

The term 'infiltration' was used by Horton [69] to describe the process whereby water soaks into, or is absorbed by, the soil. Horton laid some stress upon the limitations imposed on infiltration by surface and surface cover conditions and thus implicitly distinguished between infiltration and percolation, the latter term being used to describe the downward flow of water through the zone of aeration towards the water table, the former being restricted to the entry of water through the surface layers of the soil. Clearly, however, these two phenomena are closely related, as was emphasized by Childs [21], who suggested that the rate of infiltration may be considered either as a consequence of the hydraulic conductivity and suction gradient at the soil surface, following Darcy's law, or as the rate of increase of total moisture storage within the soil profile.

Although in nature, therefore, the precipitation-infiltration process is the ultimate source of moisture in the zone of aeration and in this sense precedes both its storage and movement, in this discussion its consideration has been deferred to the end of the present chapter on the grounds that it can only be properly understood as a *function* of moisture storage and movement.

6.14 Infiltration capacity

One aspect of infiltration which has long been considered important in hydrology is the *infiltration capacity* of the soil surface. This was defined by Horton [69] as the maximum rate at which rain can be absorbed by a soil in a given condition. The usefulness of this concept has often been questioned on the grounds that since the actual infiltration rate will equal the infiltration capacity when the latter is exceeded or equalled by the rainfall intensity and, in all other cases, will equal the rainfall intensity, when allowance is made for interception and surface storage, the term infiltration capacity is redundant [53] and could be replaced by the term infiltration rate. In the present context, however, the two terms will be distinguished partly because *infiltration rate* is often used to imply that infiltration is proceeding at a rate lower than the infiltration capacity [125], and partly because the relationship between rainfall intensity and the rate of infiltration varies depending on whether rainfall intensity exceeds the infiltration capacity. Thus, when the rainfall intensity is lower than the infiltration capacity of a soil, all the falling rain not held as surface storage will infiltrate into the soil so that there will be a *direct* relationship between the rate of infiltration and the intensity of rainfall. When, however, rainfall intensity exceeds the infiltration capacity, the foregoing relationship breaks down and may, indeed, be replaced by an *inverse* relationship between infiltration and rainfall intensity. This is normally the case when an increase in rainfall intensity is reflected in an increase in raindrop size and consequently in an increase in their compacting force as the drops strike the ground surface.

In many circumstances, infiltration capacity is important in determining the disposition of precipitation falling upon a catchment area. Thus the relationship between rainfall intensity and infiltration capacity determines how much of the falling rain will flow over the ground surface, possibly directly into streams and rivers, and how much will enter the soil where it may move laterally as interflow or be retained for some period of time before being either passed downwards as percolation or returned to the atmosphere by the processes of evaporation and transpiration. The reasons for variations of infiltration capacity in both time and space, and the methods by which it may be determined, therefore merit careful consideration.

6.15 Factors affecting infiltration

Repeated field observations have confirmed that the infiltration capacity decreases with time through a period of rainfall until a more or less constant value is reached. Horton [69], [70], [71], considered that this reduction in infiltration was more probably controlled by factors operating at the soil surface than by the flow process within the soil profile [41] although subsequent work, much of it at a comparatively early stage of development, has shown that the reverse is more likely to be true. The main factors affecting infiltration will now be discussed, recognizing that in some circumstances Horton's view will be correct and that infiltration is limited by soil surface and surface cover conditions, that in other circumstances infiltration will be dependent upon the rate of downward movement of water through the soil profile, and that in all circumstances infiltration will be affected by the quality of the infiltrating water.

6.16 Soil surface conditions

In many cases soil surface conditions impose an upper limit to the rate at which water can be absorbed, despite the fact that the capacity of the lower soil layers to receive and to store additional infiltrating water remains unfilled. In general infiltration will tend to be reduced by surface compaction, the washing of fine particles into surface pores and by frost and to increase as the depth of standing water on the surface, the number of suncracks, and the surface slope increase. Cultivation may either increase or decrease infiltration capacity.

6.17 Surface cover conditions

The nature of the surface cover is also an important influence on the infiltration

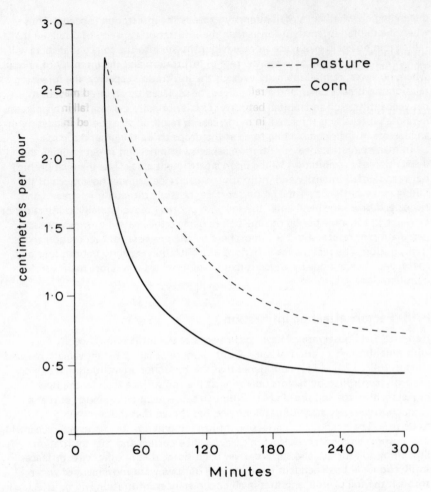

Fig. 6.18. *Infiltration rates under pasture and corn. (From an original diagram by G. W. Musgrave,* Water, *USDA Yearbook for 1955, pp. 151-159, 1955.)*

process. Thus a vegetation cover tends to increase infiltration in comparison with areas of bare soil not only by retarding surface water movement but also by reducing raindrop compaction. Most experimental evidence indicates that infiltration is higher beneath forest than beneath grass although the presence of ground litter has a more pronounced effect on the infiltration rate than does the main vegetation cover itself. For arable crops grains tend to be intermediate between row crops and grass (see Fig. 6.18) in their effects on infiltration [86]. Snow may have a similar effect to ground litter and urban areas effectively reduce the infiltration capacity to zero over considerable areas.

6.18 Flow conditions in a homogeneous soil mass

At this point it must be emphasized that water cannot continue to be absorbed by the soil surface faster than it is being transmitted through the underlying soil profile. The capacity of the soil surface to absorb water may suddenly be doubled, but no increase in infiltration rates will occur unless transmission rates through the soil profile are correspondingly large. Clearly, the distinction between conditions in the surface and subsurface layers is very arbitrary so that, in fact, while the rate of water supply to the surface is less than the infiltration capacity, water infiltrates as fast as it is supplied, i.e., the supply rate determines the infiltration rate. Once the supply rate exceeds the infiltration capacity, however, the surface becomes saturated, the limiting value of surface conductivity is attained, and the infiltration rate is thereafter determined by the surface moisture gradient [21], i.e., the infiltration process becomes profile-controlled [60].

Vertical infiltration will normally occur under the combined influence of the gradients of suction and gravity. In the case of a homogeneous uniformly dry profile which is suddenly saturated at the surface, the suction gradient acting in the surface layer will initially be very steep, partly because of purely capillary forces and partly because, at this stage in particular, the adsorptive forces, responsible for attracting and holding a thin film of water on the surfaces of the soil particles, play an important part, especially in fine grained soils having a large specific surface (see section 6.2). As wetting proceeds downward into the profile, however, the suction gradient is reduced as the adsorbed films approach their maximum thickness, the capillary pore spaces become filled with water, and the percolating water encounters increased resistance to flow due to reduced extent or dimension of flow channels and the increased length of the supply channels from the surface. The infiltration rate thus tends to settle down to a steady, gravity-induced rate which approximates the saturated hydraulic conductivity [60].

Infiltration equations

Recognition that infiltration involves both storage and transmission of water has formed the basis of a number of attempts to calculate mass infiltration such as that by Holtan [66] in which the infiltration rate was defined as a function of the exhaustion of soil moisture storage:

$$f = a(S - F)^n + f_c \qquad (6.9)$$

where f is the rate of infiltration, S is the storage potential of the soil expressed as the volumetric difference between pore saturation and the wilting point, F is the accumulated infiltration, f_c the constant rate of infiltration after prolonged wetting, and a and n are constants for a particular soil in a given condition.

The course of infiltration from a saturated surface as a function of time, as it has been described above, is usually termed the law of infiltration [21]. A number of formulae expressing this law have been proposed, all of which aim to express the rate of infiltration as a function of time from initial surface saturation, and to account for the rapid initial decrease of infiltration with time and, in the case of uniform soils, its asymptotic approach to an ultimate constant value [21].

One of the earliest attempts was that by Green and Ampt [54] in a classic paper on the flow of air and water through soils. Starting with Poiseuille's law for flow through capillaries, they derived an equation relating the depth of wetting to time:

$$(P/S)t = I - (a + \Psi) \cdot \ln \left[1 + I/(a + \Psi)\right] \qquad (6.10)$$

where P is the permeability of the soil to water, S is the porosity, t is the time, I is the distance to which the water has penetrated, a is the height of water on the soil surface, and Ψ is the soil-moisture suction at the wetting front.

This approach assumes that the soil-moisture suction at the wetting front remains constant, and that the front itself is a precisely defined surface which separates uniformly saturated soil behind it, of uniform hydraulic conductivity, from uniformly unsaturated, and as yet uninfluenced, soil beyond it [21]. Despite the fact that such a moisture distribution is virtually impossible in a natural soil, the equation was satisfactorily applied by the authors and by a number of subsequent workers [63] [118].

Horton [71] presented an equation which was essentially similar to one proposed earlier by Gardner and Widtsoe [47]. This approach, which Childs [21] described as 'intuitive', recognized that in many natural decay processes the rate at which a variable approaches an ultimate value is proportional to the amount by which it differs from that ultimate value. The resulting empirical equation takes the form:

$$F = f_c t + de^{-at} \qquad (6.11)$$

in which F is the accumulated infiltration, f_c is the fairly constant rate of infiltration after prolonged wetting, e is the base of natural logarithms, and a and d are constants. As with most empirical formulae, opinion about its applicability and usefulness is inevitably varied [21].

Undoubtedly, one of the most important contributors in this field has been J. R. Philip who in 1957 and 1958 published a series of papers on the theory of infiltration which developed some of his earlier work and which laid the foundations for a veritable barrage of subsequent publications. Philip showed that, when water is applied to a soil with a uniform initial moisture content, the solution of the Richards soil moisture diffusion equation can be expressed as an infinite power series thus:

$$F = St^{1/2} + At + Bt^{3/2} + \ldots \qquad (6.12)$$

where F is the accumulated infiltration, t is the time, and $S, A, B,$ and the other coefficients of the individual terms in the series are functions of both the soil-moisture diffusivity and the initial and surface moisture contents of the soil. For vertical infiltration the equation converges rapidly and in many cases only the first two terms are necessary.

As Gardner [41] pointed out, Philip explained the effect of various factors on the infiltration rate in terms of S (the sorptivity), which is proportional to the difference between the initial and the saturated moisture content of the soil, and to the square root of the soil-moisture diffusivity. S reduces approximately linearly as the initial moisture content is increased, and increases as the depth of surface-ponded water is increased, although both effects tend to disappear with the passage of time and, under all conditions of infiltration into a uniform soil, the same final value of infiltration is approached asymptotically (see Fig. 6.19).

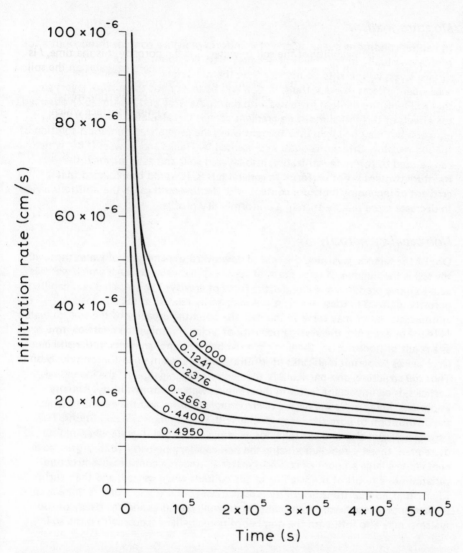

Fig. 6.19. *Computed relationship between infiltration and soil moisture content. The numbers on the curves denote the volumetric moisture content. (From an original diagram by Philip [100].)*

6.19 Flow conditions in a non-homogeneous soil mass

The foregoing discussions have considered infiltration only into a homogeneous soil of uniform moisture content. Clearly, natural conditions are considerably more complex in terms of the variability of both the moisture profile and the soil profile itself. Although it is not always possible to take account of this complexity in theoretical calculations of infiltration, some of the main factors will now be qualitatively discussed.

171

Moisture gradient

In natural conditions the existence of a uniform moisture content throughout a soil profile is virtually impossible. Immediately after precipitation, for example, the surface layers will tend to be moister than the subsurface layers, while the subsequent effects of evapotranspiration will be to dry out the surface layers so that soil moisture content increases with depth. As Gray and Norum [52] observed, the effects of this initial moisture gradient cannot be calculated using the Philip equation but may be taken into account using the generalized numerical solution of the soil moisture diffusion equation presented by Hanks and Bowers [58], which can be used to compute infiltration into layered soils and soils in which the moisture content is not uniform. In general terms, it would be expected that a gradient of increasing moisture content with depth would cause the infiltration rate to decrease more rapidly than in a uniformly dry profile.

Non-capillary porosity

One of the factors governing the rate of downward transmission of water through the soil is the number of large pores of super-capillary size through which water is able to move predominantly under the force of gravity. The so-called non-capillary porosity, defined in these terms, is not necessarily related to total porosity, although the latter may serve to indicate the potential capacity of the soil for water [114]. For example, the average porosity of a clay soil might be relatively low but, as a result of flocculation, there might exist between the larger granules sufficient large spaces to permit high rates of infiltration to be maintained for long periods. Thus *soil structure,* and particularly the degree of aggregation of the individual particles, is an important factor in determining water transmissibility. Equally important, however, is the degree of *structural stability,* and the extent to which the distribution of pore sizes may change with varying moisture conditions. The large granules in clay and silt soils, for example, often disintegrate when moistened by rainfall, thereby rapidly reducing the non-capillary porosity [87]. Again, soils containing a large amount of colloidal material, undergo considerable structural variations as a result of the swelling of the colloids when wetted, and their shrinking when dried, so that transmissibility is least in wet conditions. *Biotic factors,* such as the presence of earthworms and other burrowing creatures and the decay of root systems, may also influence the number of non-capillary spaces within the soil.

The nature of the soil profile

The stratification of a soil profile often results in a considerable variation of hydraulic conductivity with depth. In many humid areas, for example, leaching of minerals and fine particles from the surface layers and their accumulation at greater or lesser depths below the surface often results in a marked decrease in the number of large pores in the zone of accumulation. In extreme cases the formation of an iron pan or hard pan may result in zero permeability and consequently in the waterlogging of the surface layers. Although the surface soil tends generally to have a higher saturated hydraulic conductivity than the subsoil, this is not always the case and in fact any one soil horizon may limit the overall transmissibility of the complete profile.

Where in fact, a coarse layer of higher saturated hydraulic conductivity overlies a finer-textured layer the infiltration rate is initially controlled by the coarse layer but once the wetting front extends into the finer layer it is the latter which controls the rate of water movement and if infiltration is prolonged a perched water table may develop in the coarse soil, just above its boundary with the impeding finer layer [60]. In the opposite case, where fine material overlies coarse, the infiltration rate is again controlled initially by the upper layer and may, in fact, slow down when the wetting front reaches the underlying coarser material. This effect was observed by Miller and Gardner [84] and results from the fact that soil moisture suction at the wetting front may be too high to allow water to enter the larger pores of the coarse material until continued infiltration has reduced the suction to the appropriate value. Clearly then, a layer of sand in fine-textured soil may impede water movement through the profile rather than increase it [60].

In climatic conditions where evapotranspiration exceeds precipitation the net upward movement of water and solutes may result in surface deposition and the formation of a crusted or indurated surface layer. Such a layer may also result from compaction by raindrops [119], or from the breakdown of soil aggregates during wetting [59]. Even a thin surface crust can considerably impede infiltration and must, therefore, be taken into account in infiltration studies. Analyses of crust effects on infiltration were presented by Edwards and Larson [32] and by Hillel and Gardner [61] [62].

In high latitudes, and in some high altitude areas, frozen soils introduce a further complication to the entry of water into the soil profile. Some of the work in these conditions was reviewed by Gray and Norum [52] and by Gray and others [53] who emphasized that one of the important factors governing the rate of infiltration into frozen soils is the soil moisture content at the time of freezing, which determines the number and size of the ice-free pores. Thus, the wetter the soil is when freezing takes place the greater the number of ice-blocked pores and the lower the rate of infiltration, so that, when a saturated soil freezes, its intake rate will be virtually zero.

Other factors

Other factors may result in aspects of non-homogeneity in the soil profile. One which commonly affects the rate of infiltration to a greater or lesser degree is the entrapment of air bubbles in individual pores or groups of pores or the bulk compression of soil air if it is prevented from escaping from the profile when displaced by infiltrating water [1] [132] [137].

In lowlying, flat areas, for example, where the water table is close to the ground surface, infiltration may occur simultaneously over a wide area and considerable volumes of air may be temporarily trapped and compressed within the soil profile, and thus effectively limit its transmissibility. Although this effect is normally short-lived, and of comparatively little importance, Garner and others [51] noted its persistence for prolonged periods of several days or even weeks.

6.20 Characteristics of the infiltrating water

Since the viscosity of water and therefore the ease with which it may move through soil pore spaces varies with *water temperature,* the latter will tend to exert some influence upon infiltration rates. In terms of the effects of *water quality,* most water passing through the soil surface collects fine clay and silt particles and carries these in suspension into the soil profile, where blocking of smaller pore spaces may occur. Infiltration rates have also been found to vary when infiltrating water is contaminated by salts, particularly in very alkaline soils, because the salts affect not only the viscosity of the water but also the rate of swelling of colloids [87].

6.21 Time variations of infiltration capacity

It will be evident from the foregoing discussion that infiltration capacity varies from time to time according to the complex interactions of a large number of variables, the individual effects of which are difficult to assess.

Short-term variations

Figure 6.20A shows a typical curve of infiltration capacity plotted against time for a deep soil of reasonably uniform texture. The infiltration capacity is initially high, but falls very rapidly during the early stages of rainfall as a result of the combined effects of a number of factors such as surface compaction, inwashing of fine particles, colloidal swelling, the closing up of suncracks, increasing soil moisture content, and the marked reduction in the suction forces tending to pull water into the soil.

This rapid variation in infiltration capacity is frequently stressed in the literature but, in fact, the most important feature of Fig. 6.20A is the marked flattening out of the curve and the comparative uniformity of infiltration capacity during all but the early stages of rainfall. This state of semi-equilibrium, which closely approximates the saturated hydraulic conductivity, is sometimes reached within ten minutes, and almost always within an hour of the beginning of rainfall [36]. Where, however, the soil profile is complicated by the existence of a relatively impermeable layer at some distance below the surface, the curve of infiltration capacity versus time may resemble that shown in Fig. 6.20B. In this case, soon after the curve has flattened out there is a further sudden reduction of infiltration capacity, reflecting the fact that, when the available storage in the surface soil horizons has been filled, infiltration is governed by the rate at which water can pass through the layer of lower saturated hydraulic conductivity.

The infiltration curves of a frozen soil may, according to Gray and others [53], take three distinct shapes depending on conditions prevailing at the time of freezing or thawing. First, the intake rate may be reasonably constant with time at a low value (see Fig. 6.20C) where the soil is frozen when saturated or where an impervious ice layer develops on the surface during the melting period. Second, there may be an increase in the rate of infiltration with time (Fig. 6.20D), especially when a soil is frozen at a moisture content of about 70-80 per cent of field capacity. In this case, some of the meltwater is able to penetrate the soil and melt the ice-filled pores and as the ice progressively melts the infiltration rate increases, although eventually it will decrease again because of increasing soil moisture content. Third, there may be a decrease in the rate of infiltration with time, as shown in Fig. 6.20A, especially where the soil is frozen at low moisture contents.

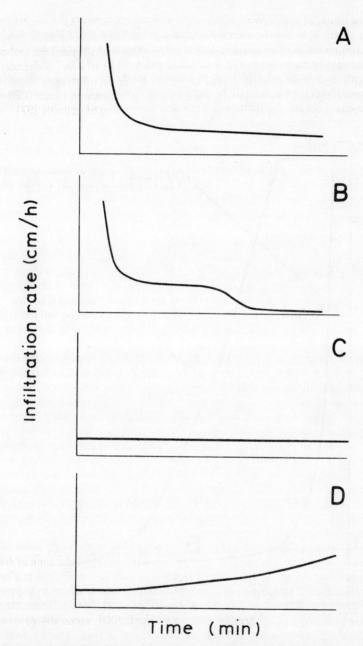

Fig. 6.20. *Variations of infiltration capacity with time in different situations: (A) uniform soil profile; (B) composite soil profile; (C) a frozen saturated soil or surface ice layer; (D) a frozen moist soil. ((A) and (B) from an original diagram by A. K. Turner, J. Hydrol., 1: 129-143, 1963 and (C) and (D) from an original diagram by D. M. Gray, D. I. Norum, and J. M. Wigham, in Gray, D. M. (Ed.),* Handbook on the Principles of Hydrology, *Canadian National IHD Committee, Ottawa, 1970.)*

The magnitude of the normal reduction in infiltration capacity during the early stages of a storm is, of course, directly reflected in the difference between the initial maximum infiltration capacity and the final minimum rate represented by the flat portions of the curve in Fig. 6.20A and will tend to vary with the type of soil and with its condition [69]. Thus, in a coarse, sandy soil, having a relatively high saturated hydraulic conductivity and where little compaction, inwashing, or internal swelling occurs, the difference between maximum and minimum infiltra-

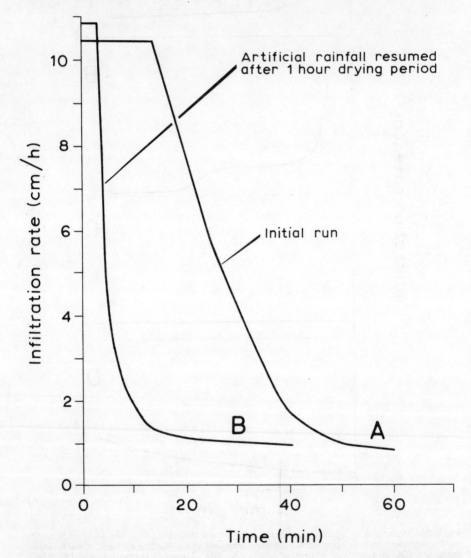

Fig. 6.21. *Average curves showing recovery of infiltration capacity after one hour's drying of a silt-loam soil. (From an original diagram by K. G. Reinhart and R. E. Taylor,* Trans. Amer. Geophys. Union, **35***: 791-795, 1954.)*

tion capacity may be negligibly small but will increase as the soil texture becomes finer. In the case of loamy soils, Horton [69] suggested that the difference may be between 2 and 3 cm. This has important hydrological implications, particularly for the runoff process in conditions where interflow is not a significant factor. If, in such cases, a constant rate of rainfall higher than the final infiltration capacity is assumed throughout a storm, it will be evident from curves A and B in Fig. 6.20 that shortly after rainfall begins, the amount of water moving over the ground surface towards the streams may increase considerably, causing stream levels and discharges to rise rapidly. Early investigators frequently attributed this effect to the saturation of the soil, a state which is rarely totally achieved except in locally restricted (although highly significant) areas of a catchment.

Where rainfall occurs as a sequence of showers, the effects of evapotranspiration in drying out the soil may help to increase infiltration capacity between showers. This is indicated in Fig. 6.21, in which curve A shows the infiltration rate during the initial application of artificial rainfall to a silt loam soil, while curve B shows the infiltration rate after the soil had been allowed to dry for one hour after the end of the first rainfall period. It can be seen that the infiltration capacity recovered considerably during the drying period and, in fact, exceeded the final rate of the initial curve for about ten minutes [104]. In showery conditions, therefore, total infiltration is likely to be somewhat higher for a given amount of rainfall than when the rain falls as one continuous storm. If, however, no drying of the soil occurs between showers, so that the soil moisture content at the beginning of each shower is high, total infiltration is not likely to vary much from that which would take place during continuous rainfall.

Seasonal variations

Various factors combine to encourage seasonal differences in infiltration capacities over a given area, and it is found that, generally speaking, these are higher during the summer months than during the winter months. Again, however, it is difficult to isolate the effects of individual factors, although certain trends are readily apparent. Thus, for example, soil moisture content is generally higher in winter and spring than in summer and autumn; earthworms are more active and vegetation roots more extensive during the summer; the vegetation cover is denser at this time of year and, of course, air and soil temperatures are higher. Apart from natural factors, however, account must also be taken of land use changes and agricultural patterns which in the British Isles, for example, result in a marked contrast between crop-covered land surfaces in the summer months, and bare plough-land during much of the winter and early spring. In high latitude and high altitude regions, these contrasts are further accentuated by widespread freezing during the cold season, which may cause the surface layers of the soil to become virtually impermeable.

References

1. ADRIAN, D. D. and J. B. FRANZINI, Impedance to infiltration by pressure build up ahead of the wetting front, *JGR*, **71**: 5857-5862, 1966.

2. AMERICAN SOCIETY OF CIVIL ENGINEERS, *Hydrology Handbook,* New York, 1949.
3. ASLYING, H. C., G. H. BOLT, W. R. GARDNER, J. W. HOLMES, R. D. MILLER, A. A. RODE, and E. G. YOUNGS, Soil physics terminology, *Inter. Soc. of Soil Sci. Bull.,* **23**: 7-10, 1963.
4. BAIER, W., and G. W. ROBINSON, A new versatile soil moisture budget, *Canad. J. Plast. Sci.,* **46**: 299-315, 1966.
5. BAVEL, C. H. M. van, The three-phase domain in hydrology, *Proc. Wageningen Symp. on Water in the Unsaturated Zone,* **1**: 23-32, UNESCO/IASH, 1969.
6. BAVER, L. D., Retention and movement of soil moisture, in Meinzer, O. E. (Ed.), *Hydrology,* McGraw-Hill, New York, 1942.
7. BODMAN, G. B., and E. A. COLMAN, Moisture and energy conditions during downward entry of water into soils, *Proc. SSSA,* **8**: 116-122, 1943.
8. BOUSHI, I. M. El, and S. N. DAVIS, Water-retention characteristics of coarse rock particles, *J. Hydrol.* **8**: 431-441, 1969.
9. BRESLER, E., A model for tracing salt distribution in the soil profile and estimating the efficient combinations of water quality and quantity, *Soil Sci.,* **104**: 227-233, 1967.
10. BRESLER, E., and R. J. HANKS, Numerical method of estimating simultaneous flow of water and salt in unsaturated soil, *Proc. SSSA,* **33**: 827-832, 1969.
11. BRIGGS, L. J., The mechanics of soil moisture, *U.S.D.A. Bur. Soils Bull,* No. 10, 1897.
12. BUCKINGHAM, E., Studies on the movement of soil moisture, *USDA Bur. Soils Bull.,* No. 38, 1907.
13. CARMAN, P. C., Properties of capillary-held liquids, *J. Phys. Chem.,* **57**: 56-64, 1953.
14. CARSLAW, H. S., and J. C. JAEGER, *Conduction of Heat in Solids,* OUP, London, 510 pp., 1959.
15. CARY, J. W., and S. A. TAYLOR, The interaction of the simultaneous diffusions of heat and water vapour, *Proc. SSSA,* **26**: 413-416, 1962.
16. CARY, J. W., and S. A. TAYLOR, The dynamics of soil water, II. Temperature and solute effects, in *Irrigation of Agricultural lands, Amer. Soc. Agron. Monograph* 11: Chap. 13, 1967.
17. CHILDS, E. C., The transport of water through heavy clay soils, I., *J. Agric. Sci.,* **26**: 114-127, 1936.
18. CHILDS, E. C., The transport of water through heavy clay soils, III, *J. Agric. Sci.,* **26**: 527-545, 1936.
19. CHILDS, E. C., The use of soil moisture characteristics in soil studies, *Soil Sci.,* **50**: 239-252, 1940.
20. CHILDS, E. C., Recent advances in the study of water movement in unsaturated soil, *Congress of Soil Science Proc.* 6: 265-274, Paris, 1956.
21. CHILDS, E. C., *An introduction to the physical basis of Soil Water Phenomena,* Wiley, London, 493 pp., 1969.
22. CHILDS, E. C., and N. COLLIS-GEORGE, Interaction of water and porous materials. *Soil geometry and soil water equilibria,* Proc. Faraday Soc., **3**: 78-85, 1948.
23. CHILDS, E. C., and N. COLLIS-GEORGE, The permeability of porous materials, *Proc. Roy. Soc.,* A 201: 392-405, 1950.
24. COLMAN, E. A., and G. B. BODMAN, Moisture and energy conditions during downward entry of water into moist and layered soils, *Proc. SSSA,* '9: 3-11, 1944.
25. COMMONWEALTH BUREAU OF SOILS, Bibliography of the determination of soil moisture using neutron probes (1963-51), Cyclostyled, The Bureau, Harpenden, Herts, 1963.
26. COMMONWEALTH BUREAU OF SOILS, Bibliography on The determination of soil moisture using neutron- and gamma ray probes (1967-63), Cyclostyled, The Bureau, Harpenden, Herts, 1967.
27. CRANK, J., *The Mathematics of Diffusion,* OUP., London, 347 pp., 1956.
28. CURTIS, L. F., Remote sensing for environmental planning surveys, in Barrett, E. C. and L. F. Curtis (Eds.), *Environmental Remote Sensing,* pp. 87-109, Arnold, London, 1974.
29. DARCY, H., *Les fontaines publiques de la ville de Dijon,* V. Dalmont, Paris, 1856.
30. ECKARDT, F. E., Eco-physiological measuring techniques applied to research on water relations of plants in arid and semi-arid regions, in *Plant-Water Relationships in Arid and Semi-Arid Conditions,* UNESCO, Paris, 1960.

178

31. EDLEFSEN, N. E., and G. B. BODMAN, Field measurements of water movement through a silt loam soil, *J. Amer. Soc. Agron.,* **33**: 713-731, 1941.
32. EDWARDS, W. M., and W. E. LARSON, Infiltration of water into soils as influenced by surface seal development, *Trans. ASAE,* **12**: 463, 465, 470, 1969.
33. ELRICK, D. E. and J. H. M. AALDERS, Mechanics of the movement of moisture and chemical substances in soils, in *Soil Moisture, Proc. Hydrology Sympos.* No. 6: 51-79, National Research Council of Canada, Ottawa, 1968.
34. EVANS, R., The time factor in aerial photography for soil surveys in lowland England, in Barrett, E. C. and L. F. Curtis (Eds.), *Environmental Remote Sensing,* pp. 67-86, Arnold, London, 1974.
35. FARRELL, D. A., and W. E. LARSON, Modelling the pore structure of porous media, *WRR,* **8**: 699-706, 1972.
36. FOSTER, E. E., *Rainfall and Runoff,* MacMillan, New York, 1949.
37. FREEZE, R. A., The mechanism of natural groundwater recharge and discharge, 1. One-dimensional vertical, unsteady, unsaturated flow above a recharging or discharging groundwater flow system, *WRR,* **5**: 153-171, 1969.
38. FUJIOKA, Y., and T. KITAMURA, Approximate solution to a vertical drainage problem, *JGR,* **69**: 5249-5255, 1964.
39. GARDNER, W. R., Some steady state solutions of the unsaturated moisture flow equation with application to evaporation from a water table, *Soil Sci.,* **85**: 228-232, 1958.
40. GARDNER, W. R., Soil water relations in arid and semi-arid conditions, in *Plant Water Relationships in Arid and Semi-Arid Conditions,* UNESCO, Paris, 1960.
41. GARDNER, W. R., Development of modern infiltration theory and application in hydrology, *Trans. ASAE,* **10**: 379-381, 390, 1967.
42. GARDNER, W. R., Availability and measurement of soil water, in *Water Deficits and Plant Growth,* **1**: 107-135, Academic Press, New York, 1968.
43. GARDNER, W. R., and R. H. BROOKS, A descriptive theory of leaching, *Soil Sci.,* **83**: 295-304, 1956.
44. GARDNER, W. R., and M. FIREMAN, Laboratory studies of evaporation from soil columns in the presence of a water table, *Soil Sci.,* **85**: 244-249, 1958.
45. GARDNER, W. R., and D. HILLEL, The relation of external evaporative conditions to the drying of soils, *JGR,* **67**: 4319-4325, 1962.
46. GARDNER, W. R., and D. KIRKHAM, Determination of soil moisture by neutron scattering, *Soil Sci.,* **73**: 391-401, 1952.
47. GARDNER, W. R., and J. A. WIDTSOE, The movement of soil moisture, *Soil Sci.,* **11**: 215-232, 1921.
48. GARDNER, W. R., D. HILLEL, and Y. BENYAMINI, Post-irrigation movement of soil water, I. Redistribution, *WRR,* **6**: 851-861, 1970.
49. GARDNER, W. R., D. HILLEL, and Y. BENYAMINI, Post-irrigation movement of soil water, 2. Simultaneous redistribution and evaporation, *WRR,* **6**: 1148-1153, 1970.
50. GARDNER, W. R., M. S. MAYHUGH, J. O. GOERTZEN, and C. A. BOWER, Effect of electrolyte concentration and exchangeable sodium percentage on diffusivity of water in soils, *Soil Sci.,* **88**: 270-274, 1959.
51. GARNER, D. E., J. K. DONALDSON, and G. S. TAYLOR, Entrapped air in a field site, *Proc. SSSA,* **33**: 634-635, 1969.
52. GRAY, D. M. and D. I. NORUM, The effect of soil moisture on infiltration as related to runoff and recharge, in *Soil Moisture, Proc. Hydrology Sympos.* No. 6: 133-153, National Research Council of Canada, Ottawa, 1968.
53. GRAY, D. M., D. I. NORUM, and J. M. WIGHAM, Infiltration and the physics of flow of water through porous media, Section V in Gray, D. M. (Ed.), *Handbook on the Principles of Hydrology,* National Research Council of Canada, Ottawa, 1970.
54. GREEN, W. H., and G. A. AMPT, Studies in soil physics. I. The flow of air and water through soils, *J. Agric. Sci.,* **4**: 1-24, 1911.
55. GURR, C. G., The movement of water vapour in soil, *Austr. Conf. in Soil Sci.,* **2**: 431-434, 1953.
56. HADLEY, W. A., and R. EISENSTADT, Thermally actuated moisture migration in granular media, *Trans. AGU,* **36**: 615-623, 1955.

57. HAINES, W. B., Studies in the Physical properties of soils. V. The hysteresis effect in capillary properties, and the modes of moisture distribution associated therewith, *J. Agric. Sci.,* **20**: 97-116, 1930.
58. HANKS, R. J., and S. A. BOWERS, Numerical solutions of the moisture flow equations into layered soils, *Proc. SSSA,* **26**: 530-534, 1962.
59. HILLEL, D., Crust formation in loessial soils, *Trans. Int. Soil Sci. Congr., Madison,* **7**: 330-339, 1960.
60. HILLEL, D., *Soil and Water,* Academic Press, New York, 288 pp., 1971.
61. HILLEL, D., and W. R. GARDNER, Steady infiltration into crust topped profiles, *Soil Sci.,* **108**: 137-142, 1969.
62. HILLEL, D., and W. R. GARDNER, Transient infiltration into crust-topped profiles, *Soil Sci.,* **109**: 69-76, 1970.
63. HILLEL, D., and W. R. GARDNER, Measurement of unsaturated conductivity and diffusivity by infiltration through an impeding layer, *Soil Sci.,* **109**: 149-153, 1970.
64. HOLMES, J. W., Calibration and field use of the neutron scattering method of measuring soil water content, *Aust. J. Appl. Sci.,* **7**: 45-58, 1956.
65. HOLMES, J. W., and K. G. TURNER, The measurement of water content of soils by neutron scattering: A portable apparatus for field use, *J. Agric. Eng. Res.,* **3**: 199-204,1958.
66. HOLTAN, H. N., A concept of infiltration estimates in watershed engineering, *USDA-ARS* 41-51, 25 pp., 1961.
67. HORNBERGER, G. M., and I. REMSON, A moving boundary model of a one-dimensional saturated-unsaturated, transient porous flow system, *WRR,* **6**: 898-905, 1970.
68. HORNBERGER, G. M., I. REMSON, and A. A. FUNGAROLI, Numeric studies of a composite soil moisture groundwater system, *WRR,* **5**: 797-802, 1969.
69. HORTON, R. E., The role of infiltration in the hydrologic cycle, *Trans. AGU,* **14**: 446-460, 1933.
70. HORTON, R. E., Analysis of runoff-plot experiments with varying infiltration capacity, *Trans. AGU,* **20**: 693-694, 1939.
71. HORTON, R. E., An approach toward a physical interpretation of infiltration-capacity, *Proc. SSSA,* **5**: 399-417, 1940.
72. HUTCHEON, N. B., and J. A. PAXTON, Moisture migration in a closed guarded hot plate, *Heating and Piping and Air Conditioning,* **24**: 113-122, 1952. (Quoted by Marshall) [82].
73. JENNINGS, J. E., P. R. B. HEYMANN, and L. WOLPERT, Some laboratory studies of the migration of moisture in soils under temperature gradients, *National Building Res. Inst., South Africa, Bull.,* **9**: 1952. (Quoted by Marshall) [82].
74. KIRKHAM, D., Soil physics, in Chow, V. T. (Ed.), *Handbook of Applied Hydrology,* McGraw-Hill, New York, 1964.
75. KIRKHAM, D., and C. L. FENG, Some tests of the diffusion theory, and laws of capillary flow, in soils, *Soil Sci.,* **67**: 29-40, 1949.
76. KIRSCHER, O., and H. ROHNALTER, Wärmeleitung und Dampfdiffusion in feuchten Guten, *Forsch. Verein Deut. Ing.,* **402**: 1-18, 1940. (Quoted by Taylor and Cavazza) [121].
77. KLUTE, A., A numerical method for solving the flow equation for water in unsaturated materials, *Soil Sci.,* **73**: 105-116, 1952.
78. KLUTE, A., The movement of water in unsaturated soils, *The Progress of Hydrology,* Univ. of Illinois, Urbana, Illinois, **2**: 821-888, 1969.
79. KLUTE, A., and D. B. PETERS, Hydraulic and pressure head measurement with strain gauge pressure transducers, *Proc. Wageningen Sympos. on Water in the Unsaturated Zone,* **1**: 156-165, 1969.
80. KUZMAK, J. M., and SEREDA, P. J., The mechanism by which water moves through a porous material subjected to a temperature gradient: 2. Salt tracer and streaming potential to detect flow in the liquid phase, *Soil Sci.,* **84**: 419-422, 1957.
81. LOW, P. F., Physical chemistry of clay-water interactions, *Advan. Agron.,* **13**: 269-327, 1961.
82. MARSHALL, T. J., *Relations between Water and Soil,* Commonwealth Agric. Bureaux, Farnham Royal, 1959.

83. MATHER, J. R., and J. K. NAKAMURA, The climatic and hydrologic factors affecting the redistribution of strontium in soils, *Publ. in Climat.*, **15**: 5-99, 1962.
84. MILLER, E. E., and W. H. GARDNER, Water infiltration into stratified soil, *Proc. SSSA*, **26**: 115-118, 1962.
85. MILLER, E. E., and A. KLUTE, Dynamics of soil water. Part 1 Mechanical forces, in *Irrigation of Agricultural Lands, Amer. Soc. Agron. Monograph* 11: 209-244, 1967.
86. MUSGRAVE, G. W., How much of the rain enters the soil? In *Water,* USDA, Yearbook of Agriculture, 1955. pp. 151-159, 1955.
87. MUSGRAVE, G. W., and H. N. HOLTAN, Infiltration, In Chow, V. T. (Ed.) *Handbook of Applied Hydrology,* McGraw-Hill, New York, 1964.
88. NICHOLSON, H. H., and E. C. CHILDS, The transport of water through heavy clay soils. II, *J. Agric. Sci.,* **26**: 128-141, 1936.
89. NIELSEN, D. R., Percolation of water in soil, Proc. Seepage Sympos., Phoenix, Arizona, 1963, *USDA Agric. Res. Service Publ.* 41-90: 47-51, 1965.
90. NIELSEN, D. R., and J. W. BIGGAR, Miscible displacement in soils. I. Experimental information, *Proc. SSSA,* **25**: 1-5, 1961.
91. PECK, A. J., and R. M. RABBIDGE, Direct measurement of moisture potential: A new technique, *Proc. Wageningen Sympos. on Water in the Unsaturated Zone,* 1: 165-170, IASH/UNESCO, 1969.
92. PENMAN, H. L., Gas and vapour movements in the soil. I. The diffusion of vapours through porous solids, *J. Agric. Sci.,* **30**: 437-462, 1940.
93. PENMAN, H. L., Laboratory experiments on evaporation from fallow soil, *J. Agric. Sci.,* **31**: 454-465, 1941.
94. PENMAN, H. L., Natural evaporation from open water, bare soil and grass, *Proc. Roy. Soc.* A **193**: 120-145, 1948.
95. PENMAN, H. L., *Vegetation and Hydrology,* Commonwealth Agric. Bureaux, Farnham Royal, 1963.
96. PHILIP, J. R., Some recent advances in hydrologic physics, *J. Inst. Eng. Aust.,* **26**: 255-259, 1954.
97. PHILIP, J. R., The concept of diffusion applied to soil water, *Proc. Nat. Acad. Sci. India (Allahabad),* A24: 93-104, 1955.
98. PHILIP, J. R., Numerical solution of equations of the diffusion type with diffusivity concentration dependent, *Trans. Faraday Soc.,* **51**: 885-892, 1955.
99. PHILIP, J. R., The theory of infiltration: 1. The infiltration equation and its solution, *Soil Sci.,* **83**: 345-357, 1957.
100. PHILIP, J. R., The gain, transfer, and loss of soil-water, in *Water Resources Use and Management,* Melbourne Univ. Press, Melbourne, 1964.
101. PHILIP, J. R., Theory of infiltration, *Adv. Hydrosci.,* **5**: 215-296, 1969.
102. PHILIP, J. R., Hydrostatics and hydrodynamics in swelling soils, *WRR,* **5**: 1070-1077, 1969.
103. PHILIP, J. R., and D. A. de VRIES, Moisture movement in porous materials under temperature gradients, *Trans. AGU,* **38**: 222-232, 1957.
104. REINHART, K. G., and R. E. TAYLOR, Infiltration and available water storage capacity in the soil, *Trans. AGU,* **35**: 791-795, 1954.
105. REYNOLDS, S. G., The gravimetric method of soil moisture determination, Part I. A study of equipment, and methodological problems, *J. Hydrol,* **11**: 258-273, 1970.
106. REYNOLDS, S. G., The gravimetric method of soil moisture determination, Part II. Typical required sample sizes and methods of reducing variability, *J. Hydrol.* **11**: 274-287, 1970.
107. RICHARDS, L. A., Capillary conduction of liquids through porous mediums, *Physics,* **1**: 318-333, 1931.
108. RICHARDS, L. A., Pressure-membrane apparatus, construction, and use, *Agric. Eng.,* **28**: 451-454, 1947.
109. RICHARDS, L. A., and M. FIREMAN, Pressure-plate apparatus for studying moisture sorption and transmission by soils, *Soil Sci.,* **56**: 395-404, 1943.
110. RIJTEMA, P. E., Evapotranspiration in relation to suction and capillary conductivity, in *Plant-Water Relationships in Arid and Semi-Arid Conditions,* Proc. of Madrid Symposium, 1959, UNESCO, Paris, 1960.

111. RIJTEMA, P. E., and H. WASSINK, (Eds.), *Water in the Unsaturated Zone, Proceedings of the Wageningen Symposium,* 2 vols., UNESCO/IASH, 1969.

112. RUBIN, J., Numerical method for analyzing hysteresis-affected post-infiltration redistribution of soil moisture, *Proc. SSSA,* **31**: 13-20, 1967.

113. SCHOFIELD, R. K., The pF of the water in soil, *Trans. 3rd Internat. Cong. Soil Sci.,* **2**: 37-48, 1935.

114. SHERMAN, L. K., and G. W. MUSGRAVE, Infiltration, In Meinzer, O. E. (Ed.), *Hydrology,* McGraw-Hill, New York, 1942.

115. STONE, J. F., D. KIRKHAM, and A. A. READ, Soil moisture determination by a portable neutron scattering moisture meter, *Proc. SSSA,* **19**: 419-423, 1955.

116. SWANSON, C. L. W., and J. B. PETERSON, The use of the micrometric and other methods for the evaluation of soil structure, *Soil Sci.,* **53**: 173-185, 1942.

117. SWARTZENDRUBER, D., The applicability of Darcy's law, *Proc. SSSA,* **32**: 11-18, 1968.

118. SWARTZENDRUBER, D., and M. R. HUBERTY, Use of infiltration equation parameters to evaluate infiltration differences in the field, *Trans. AGU,* **39**: 84-93, 1958.

119. TACKETT, J. L., and R. W. PEARSON, Some characteristics of soil crusts formed by simulated rainfall, *Soil Sci.,* **99**: 407-413, 1965.

120. TAYLOR, S. A., Principles of dry land crop management in arid and semi-arid zones, in *Plant-Water Relationships in Arid and Semi-Arid Conditions,* UNESCO, Paris, 1960.

121. TAYLOR, S. A., and L. CAVAZZA, The movement of soil moisture in response to temperature gradients, *Proc. SSSA,* **18**: 351-358, 1954.

122. TERZAGHI, K., Soil moisture and capillary phenomena in soils, in Meinzer, O. E. (Ed.), *Hydrology,* McGraw-Hill, New York, 1942.

123. THORNTHWAITE, C. W., and J. R. MATHER, The computation of soil moisture, in estimating soil tractionability from climatic data, *Publ. in Climat.,* **7**: 397-402, 1954.

124. THORNTHWAITE, C. W., and J. R. MATHER, The Water Balance, *Publ. in Climat.,* **8**: 1-86, 1955.

125. TURNER, A. K., Infiltration, runoff and soil classifications, *J. Hydrol,* **1**: 129-143, 1963.

126. UNDERWOOD, N., C. H. M. van BAVEL, and R. W. SWANSON, A portable slow neutron flux meter for measuring soil moisture, *Soil Sci.,* **77**: 339-340, 1954.

127. VEIHMEYER, F. J., and N. E. EDLEFSEN, Interpretation of soil-moisture problems by means of energy-changes, *Trans. AGU,* **18**: 302-318, 1937.

128. VEIHMEYER, F. J., and A. H. HENDRICKSON, Methods of measuring field capacity and wilting percentages of soils, *Soil Sci.,* **68**: 75-94, 1949.

129. VISSER, W. C., Moisture requirements of crops and rate of moisture depletion of the soil, *Inst. for Land and Water Management Res. Tech. Bull.* **32**, 1964.

130. VISVALINGAM, M., and J. D. TANDY, The neutron method for measuring soil moisture content—a review, *J. Soil Sci.,* **23**: 499-511, 1972.

131. WATSON, K. K., An instantaneous profile method for determining the hydraulic conductivity of unsaturated porous materials, *WRR,* **2**: 109-116, 1966.

132. WILSON, L. G., and J. N. LUTHIN, Effect of air flow ahead of the wetting front on infiltration, *Soil Sci.,* **96**: 136-143, 1963.

133. WIND, G. P., Capillary rise and some applications of the theory of moisture movement in unsaturated soils, *Inst. for Land and Water Management Res. Tech. Bull.,* **22**, 1961.

134. WOLLNY, E., Untersuchungen über die kapillare Leitung des Wassers in Boden, *Forsch. Geb. Agr. Phys.,* **8**: 206-220, 1885.

135. WOODMAN, J. H., I. M. FENWICK, and K. RUSSAM, An investigation of moisture conditions under a road in Southern Rhodesia, *Proc. 3rd Regional Conf. for Africa on Soil Mechanics and Foundation Eng.,* **1**: 57-60, 1963.

136. YOUNGS, E. G., Redistribution of moisture in porous materials after infiltration. II, *Soil Sci.,* **86**: 202-207, 1958.

137. YOUNGS, E. G., and A. J. PECK, Moisture profile development and air compression during water uptake by bounded porous bodies. I. Theoretical introduction, *Soil Sci.,* **98**: 290-294, 1964.

138. ZAHNER, R., Refinement in empirical functions for realistic soil-moisture regimes under forest cover, in Sopper, W. E. and H. W. Lull (Eds.), *Forest Hydrology,* pp. 261-274, Pergamon, Oxford, 1967.

182

7. Subsurface Water-Groundwater

Subsurface water other than groundwater itself formed the subject matter of chapter 6 which considered, under the general heading of soil moisture, water in the soil zone, in the intermediate zone, and in the capillary fringe. The discussion of groundwater in the present chapter, therefore, completes the survey of subsurface water by extending the study into the zone of saturation where nearly all interstices are filled with water and where, as a result, the retention and movement of water are governed by the laws of saturated flow. Again, however, one must qualify any apparently rigid differentiation between saturated and unsaturated conditions by stressing the essential continuity between all types of subsurface water and indeed also between surface and subsurface water.

Rocks which yield significant quantities of water [159] are known as *aquifers,* although nearly all rocks contain some water, however small in amount. The surface layers of the earth's crust thus comprise a great reservoir through which water moves very slowly, on what may be an extremely brief or a very long journey underground, and which acts as a vast natural regulator in the hydrological cycle, absorbing rainfall which would otherwise reach streams and rivers very rapidly as surface runoff, and maintaining stream flow during long dry periods during which no surface runoff occurs.

The occurrence of groundwater

7.1 Sources of groundwater

Groundwater is derived from a number of sources and, although the relative importance of these was much debated in the past, it is now clear that virtually all groundwater is composed of precipitated atmospheric moisture which has percolated down into the soil and subsoil layers, and that only a minute proportion can be attributed to other sources.

The water derived from the atmosphere is normally termed *meteoric* and may reach the zone of saturation either directly or indirectly. In the former case, rain infiltrates into the soil layers and moves downward through the zone of aeration to the underlying groundwater. The fact that this now so obvious and simple relationship between groundwater and the atmosphere was for a long time not generally accepted, probably arose partly from a general lack of understanding of the behaviour of subsurface water and partly from the fact that surface catchment areas frequently do not correspond with an underlying groundwater catchment, so that accretion to the groundwater underlying a particular surface catchment may occur from a much smaller or a much wider area than that of the surface catchment itself. A striking example concerns the River Itchen, in Hampshire, where the groundwater catchment area of 430 km^2 is approximately 20 per cent greater than that of the surface catchment [69]. In large artesian basins the dissociation between water on the surface and water in the zone of saturation would have been magnified by the large distances (frequently several hundreds of kilometres) which may separate recharge areas of infiltrating precipitation from the point at which groundwater emerges from a spring or well.

Another way in which groundwater may be derived directly from atmospheric moisture, is by means of the condensation of water vapour from air circulating through pores and interstices, although it has been shown that this can only be of importance in the immediate surface layers of the soil, and that the total amount of water involved is negligibly small [104].

Atmospheric moisture is also involved indirectly in that proportion of the groundwater which is derived from influent seepage from lakes, rivers, and man-made channels, especially in arid and semi-arid areas, and from the oceans themselves. Another minor source of groundwater comprises the connate water trapped in some sedimentary and volcanic rocks at the time of their origin. Finally, a relatively minute quantity of new juvenile groundwater is added periodically to the zone of saturation from deep down within the earth's crust.

7.2 Geological background

Only a small proportion of the total zone of saturation will be comprised of rocks which store and transmit significant quantities of water and which are, therefore, considered to be aquifers as previously defined. A rock which contains no interconnected openings or interstices and, therefore, neither absorbs nor transmits water may be referred to as an *impermeable bed* or *aquifuge*, [14], while a formation which, although porous and capable of absorbing water slowly, will not transmit it in appreciable quantities to a spring or well, but may be important in the regional movement of groundwater may be referred to as a *semi-permeable bed* or *aquiclude* [14]. Apart from the fact that there is a considerable diversity of, sometimes conflicting, terminology currently in use amongst groundwater hydrologists, it will be clear that some of the above terms lack precision. Thus a formation may be classified as an aquifer in one groundwater situation but as an aquiclude in another.

The expression 'groundwater is mostly rock' emphasizes that, even in high-yielding aquifers, water can occur only in the voids or interstices *between* solid rock particles and fragments, and thus underlines the fact that the fundamental geological factor affecting the occurrence of groundwater is the nature of these interstices, and particularly their size, shape, and distribution, both areally and in depth, through the zone of saturation.

7.3 Interstices

Walton [169] observed that although most interstices are small and interconnected, some are cavernous in size, and others are sufficiently isolated that there is little opportunity for water movement from one interstice to another. Since the nature of the interstices in a water-bearing rock is almost entirely determined by its geology and geological history, knowledge of the latter is clearly essential for a proper understanding of the storage and movement of groundwater.

Interstices may be classified according to one or more of their main characteristics of shape, size, and mode of formation. Thus, *intergranular* spaces vary widely in size from the minute voids between the component particles of, say, clays, shales, and slates on the one hand, to the large spaces between the pebbles of a well-sorted, unconsolidated valley gravel on the other. *Massive* spaces are those which occur between large blocks of rock such as fractures, joints, and bedding planes, which are sometimes enlarged by solution. The most frequently used classification, however, is based upon mode of origin, and considers original and secondary interstices [100]. *Original* interstices, as the term implies, were created at the time of origin of the rock in which they occur; in sedimentary rocks, therefore, they coincide with the intergranular spaces discussed above, while in igneous rocks, where they normally result from the cooling of molten lava, they may range in size from minute intercrystalline spaces to large caverns [1]. *Secondary* interstices result from the actions of subsequent geological, climatic, or biotic factors upon the original rock. Joints and faults, enlarged perhaps by weathering and solution, are the most common, so that this group closely resembles the massive spaces discussed above. Interstices of this type are most common in old, hard, crystalline rocks, and over large areas of Africa, for example, form the main channels for the storage and movement of groundwater [79]. Several types of interstice are shown in Fig. 7.1.

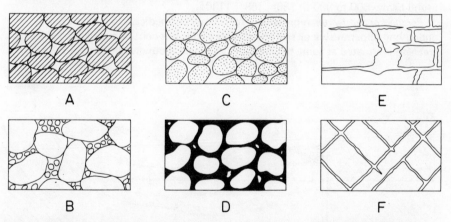

Fig. 7.1. *Types of interstice: (A) between relatively impermeable well-sorted particles, (B) between relatively impermeable unsorted particles, (C) between permeable particles, (D) between grains which have been partially cemented, (E) formed by solution along joints and bedding planes in carbonate rocks, and (F) formed by fractures in crystalline rocks. (From an original diagram by Meinzer [101]. With acknowledgements to the US Geological Survey.)*

7.4 The disposition of groundwater

The foregoing discussions have indicated the sources of groundwater and the potential of the subsurface strata to absorb and retain that water. By definition, however, only in the zone of saturation are the majority of the interstices full of water. It will be necessary, therefore, to discuss the boundaries of the zone of saturation, and thereby consider the form and manner in which bodies of groundwater occur. The normal simple distinction between *unconfined, confined,* and *perched groundwater,* tends to over-emphasize the differences between the three types of groundwater body, and is often difficult to apply even in simple geological conditions. In areas of complex geology the terms become almost meaningless. Despite these disadvantages, however, the terms are used in the present chapter as a convenient basis for discussion.

In all cases, the lower limit of groundwater represents a zone in which the interstices are so few and so small that further downward movement is virtually impossible. The boundary is thus frequently formed by a stratum of very dense rock, such as clay, or slate, or granite, or by the upper surface of the parent rock where the groundwater body occurs only in a surface deposit of weathered material. Alternatively, the gradual compression of strata with depth, as a result of the increasing weight of the overlying rocks, gives rise eventually to a zone in which the interstices have been so reduced in both size and number that downward movement is effectively limited. The depth at which this occurs will naturally depend on the nature of the water-bearing rock, so that the lower limit of the zone of saturation would be higher where a dense granite outcropped at the ground surface than where a porous sandstone extended to great depths. Nevertheless, the number of interstices tends generally to decrease with depth, and most authorities suggest that, although in theory the lower limit to which groundwater will penetrate is the zone of rock flowage [158], little groundwater has in fact been found below 600 to 900 m [36], [58], [180].

Perched groundwater represents a special unconfined case in which the underlying impermeable or semi-permeable bed is not continuous over a very large area and is situated at some height above the main groundwater body. In many

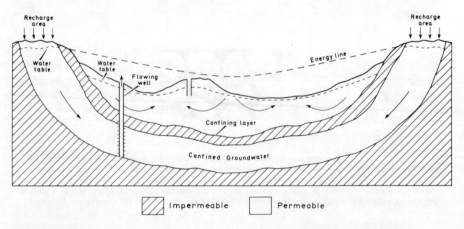

Fig. 7.2. *Simplified section of a groundwater basin showing confined and unconfined groundwater. (Based on an original diagram by P. A. Domenico,* Concepts and Models in Groundwater Hydrology, *McGraw-Hill, New York, 1972.)*

areas the first unconfined groundwater encountered is of such a type and thus constitutes a more or less isolated body of water whose position is controlled by structure or stratigraphy [34]. In addition, water percolating through the zone of aeration after heavy rainfall may be regarded as a temporary perched water body [36] [80], and will be discussed more fully in chapter 8, in relation to interflow.

The upper boundary of the zone of saturation varies according to whether the groundwater is confined or unconfined (see Fig. 7.2). In the case of *unconfined groundwater,* this boundary is normally known as the water table, which was defined by Tolman as the contact plane between unconfined groundwater and the capillary fringe [161], and which, therefore, approximates to the level where the porewater pressure is equal to atmospheric pressure [97] [129].

It is well known that the water table tends to follow the contours of the overlying ground surface, although in a more subdued form. Assuming a similar amount of infiltration from rainfall over both high and low ground, the amplitude of relief of the water table surface again depends largely upon the texture of the material comprising the zone of saturation. In the case of a very open-textured rock, such as well-jointed limestone, groundwater will tend to move through the interstices at such a rate that it is able rapidly to find its own level, and thus form a more or less horizontal surface. On the other hand, with a fine-textured rock, groundwater movement will be so slow that water will still be draining towards the valleys from beneath the higher ground when additional infiltration from subsequent precipitation occurs, so that its height is built up under the latter areas. This tendency is magnified by the fact that precipitation normally increases with relief.

In the case of *confined groundwater* (see Fig. 7.2), the upper boundary of the water body is formed by an overlying impermeable bed, and in the case of semi-confined groundwater by an overlying semi-permeable bed [34]. The distinction between unconfined and confined groundwater is often made because of hydraulic differences between the flow of water under pressure and the flow of free, unconfined groundwater. Hydrologically, however, the two form part of a single, unified system. Thus, all confined aquifers have an unconfined area through which accretions to the groundwater occur by means of infiltration and percolation, and in which a water table, as defined above, represents the upper surface of the zone of saturation. Furthermore, the confining impermeable bed rarely forms an absolute barrier to groundwater movement so that there is normally some interchange, and therefore a degree of hydraulic continuity, between the confined groundwater below the confining bed and the unconfined groundwater above it. Indeed, attention has already been drawn to the relative sense in which terms such as aquifer and aquiclude must be used and to the fact that a rock forming an aquiclude in one situation may form an aquifer in another. Thus relatively permeable sands overlying highly permeable gravels might constitute an aquiclude although the same sands overlain by a less permeable clay would constitute a confined aquifer.

Since the water table in the unconfined groundwater area, through which recharge takes place, is situated at a higher elevation than the confined area of the aquifer, it will be apparent that the groundwater in the latter area is under a pressure equivalent to the difference in hydrostatic level between the two. If, then, the pressure is locally released, as, for example, by sinking a well into the confined

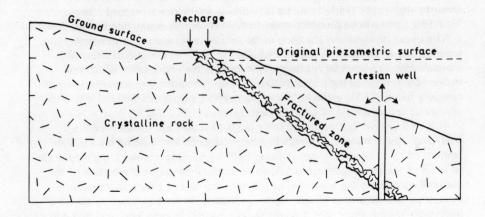

A

Ground surface

Recharge

Original piezometric surface

Artesian well

Crystalline rock

Fractured zone

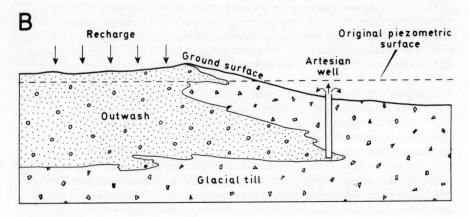

B

Recharge

Original piezometric surface

Ground surface

Artesian well

Outwash

Glacial till

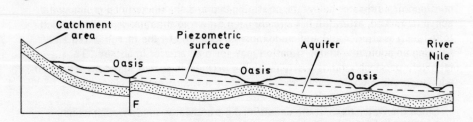

C

Catchment area

Piezometric surface

Aquifer

River Nile

Oasis

Oasis

Oasis

F

Fig. 7.3. *Examples of artesian conditions in (A) fractured crystalline rocks, (B) glacial deposits, and (C) in faulted and folded sedimentary rocks. (A) and (B) are based on original diagrams by Davis and De Wiest [34] and (C) on an original diagram by A. Holmes,* Principles of Physical Geology *(2nd Ed.), Nelson, London, 1965.)*

aquifer, the water level will theoretically rise in the well to the height of the hydrostatic head, i.e., the water table, in the recharge area minus the height equivalent of any energy losses resulting from friction between the moving groundwater and the solid matrix of the aquifer between the point of recharge and the point of withdrawal. The imaginary surface to which water rises in wells tapping confined aquifers is called the *piezometric surface* [169].

The normal textbook diagram of confined or artesian conditions (c.f. Fig. 7.2) illustrates the fairly simple synclinal basin situation which has been described above. This undoubtedly reflects the fact that some of the best known, and certainly some of the most spectacular, artesian conditions have been found in areas of gently folded sedimentary strata such as the type area in the province of Artois in northern France, the London Basin in England, or the great artesian basins of East Central Australia and the Great Plains of the United States. Early wells in these last two basins encountered water with sufficient pressure to flow upward in columns more than 45 m high although, of course, the head subsequently diminished rather rapidly [34]. Artesian conditions have also been found, however, in fissured and fractured crystalline rocks (Fig. 7.3A), particularly when these are overlain with relatively impervious superficial deposits, and also in alternating glacial and other superficial deposits of sand and clay, such as those in Wisconsin, USA [36] (Fig. 7.3B). Natural artesian springs may result from faulting in areas of folded sedimentary rocks. Thus, Fig. 7.3C represents a schematic section across part of the Sahara, showing that oases may occur where water under hydrostatic pressure escapes through fissures and other lines of weakness at faults and anticlines, or where the desert surface has been worn down almost to the level of the confined aquifer.

Groundwater Storage

Aquifers are commonly regarded as both reservoirs for groundwater storage and pipelines for groundwater movement, although as Nace [116] observed, the pipeline analogy is rather less apt. Except in massive interstices groundwater movement is very slow so that although groundwater is generally in continuous motion, its detention time in aquifers is long compared with the detention of water in river channels and commonly ranges from a few minutes to a few decades [116], although estimates of about 20 000 years have been made for artesian water in Australia [170] and several hundreds of years in extensive highly permeable aquifers such as the Columbia lava plateau [116]. Consideration will now be given to the main features of groundwater storage, and particularly to aquifer characteristics affecting it, to its role in the groundwater balance, and to aspects of storage changes in unconfined and confined aquifers.

7.5 Porosity

The amount of groundwater stored in saturated material depends upon its porosity. The porosity of a rock or soil is defined as its property of containing interstices [100] and is normally expressed in terms of the percentage of the total volume of' material which is represented by its interstices. Confusion occasionally arises in the case of, say, a well-jointed crystalline rock, between the porosity of the rock material itself (which may be very low) and the porosity of the stratum or formation which it comprises (which may be relatively high). According to the definition given above, however, all interstices, whether original or secondary, are involved in the concept of porosity, so that joints, bedding planes, and fractures, including those greatly enlarged by solution and weathering, must be included as part of the total interstitial volume.

From an earlier discussion on the distribution of interstices, it will be apparent that the main factors affecting porosity will be the shape, arrangement, and degree of sorting of the constituent particles, and the extent to which modifications arising from solution, cementation, compaction, and faulting have occurred. Most of these influences are self-explanatory and, accordingly, only a brief comment will be made upon the effect of particle arrangement and sorting.

Irregularity of particle shape normally has a more important effect on porosity than particle size; indeed, with perfectly spherical, even-sized particles, porosity is dependent only upon the degree of compaction, irrespective of the size of particle, as is shown by Fig. 7.4 where, with the loosest possible packing, the porosity is

 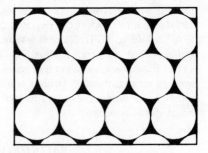

Fig. 7.4. *The effect of grain packing on porosity. See text for explanation.* (From an original diagram by A. Holmes, Principles of Physical Geology *(2nd Ed.), Nelson, London, 1965.)*

47.6 per cent, and with the closest possible packing, is 26 per cent [58]. With variously shaped particles, however, the porosity may be either increased or decreased. An even greater influence is exerted by the degree of sorting of the particles because, with a large range of particle sizes, the interstices between large fragments will be partially filled with smaller ones, and the porosity correspondingly reduced in comparison with material composed of uniform grains. The larger the size range, the greater, generally speaking, is the decrease in porosity likely to be.

The combined effect of the various factors noted above is summarized in Fig. 7.5, which shows the normal range of porosity values for a number of different types of material. Initially, it might seem strange that clay, which so often forms an aquiclude, has a very high porosity, while good aquifers, like sandstone and limestone, have only low to medium porosities. Further consideration, however, reveals that although porosity determines how much water a saturated rock can hold, by no means all of this water will be readily available for movement in the context of the hydrological cycle. The proportion of groundwater which is potentially 'mobile' will depend partly on how many of the interstices contributing to the overall porosity are interconnected, since water locked up in enclosed

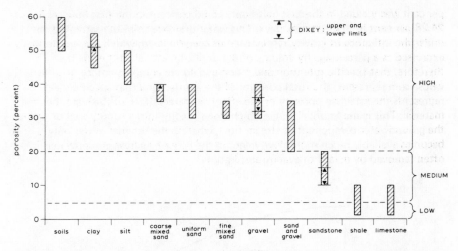

Fig. 7.5. *Porosity ranges for various natural deposits. (Based on data by Todd [159] and Dixey [36].)*

interstices however large, may be almost permanently out of circulation; and partly on the size of the interstices and, therefore, by implication, on the forces by which the water is retained in the rock.

7.6 Water retention forces in the zone of saturation

The forces retaining water in the zone of aeration have already been fully discussed (see pages 134-136), and the reader is referred to this discussion in order to avoid the necessity of restating the relevant parts of the argument here. It must suffice, at this stage, to note that in the zone of saturation the proportion of total ground-water which may be regarded as mobile will depend almost entirely on the size distribution of the interstices in the water-bearing material, being high in the case of, say, a clean gravel or a well-jointed limestone, and very low in the case of a clay, in which most of the water may be retained in very small interstices. These relationships are normally expressed in terms of specific retention and specific yield.

7.7 Specific retention and specific yield

Specific retention may be defined as the volume of water which a soil or rock will retain against the force of gravity if it is drained after having been saturated [148], and is normally expressed as a percentage of the total volume of the saturated material. The term is, therefore, analogous with field capacity, which is used when referring to soil moisture (see page 143), and is similarly imprecise in the sense that there is no fixed moisture content at which gravity drainage ceases. Rather, drainage is very rapid at first, it slows up after a short while, and then becomes extremely slow for a long period of time. Experiments by King [81], for example, showed that over a period of two-and-a-half years, the amount of water drainage from a coarse sand represented 26.88 per cent of its volume, and that of this, 10.68

191

per cent was yielded in the first half-hour, 15.56 per cent in the first hour, and 24.28 per cent in the first nine days. This water drainage from the saturated sand under the influence of gravity represented its *specific yield*, which is again normally expressed as a percentage by volume of the drained material. It is evident, therefore, that specific retention and specific yield are complementary qualities whose sum represents the total porosity of the aquifer, and that specific yield represents the effective porosity or the effective storage [87] of the water-bearing material. This is the 'mobile' water which flows readily into a supply well or forms the groundwater component of stream flow, whereas the retained water only becomes available much more slowly and, in the case of shallow groundwater, is often removed by means of evapotranspiration.

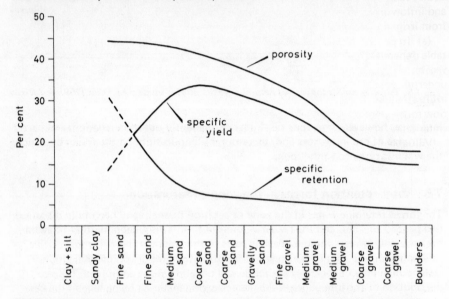

Fig. 7.6. *Diagram showing the relationship between specific retention, specific yield, and porosity for different types of material. (From an original diagram by Eckis [40].)*

The relationship between specific yield and retention, and porosity, for different types of material is shown in Fig. 7.6. It is clear that, as the texture of the material becomes coarser, and by implication the importance of larger interstices increases, the specific retention decreases, as does the total porosity. The specific yield curve is obtained by subtracting values of specific retention from values of total porosity.

7.8 The groundwater balance

It is evident that, if inflow into a groundwater system exactly equals outflow from that system, storage will remain constant; if recharge exceeds discharge, storage will increase while, if discharge is greater than recharge, storage will decrease. These relationships may be conveniently expressed in the form of the groundwater balance equation which, in simplified form, reads

$$\Delta S = Q_r - Q_d \qquad (7.1)$$

where ΔS is the change in groundwater storage, Q_r is recharge to groundwater, and Q_d is discharge from groundwater. The main components of this equation will now be discussed.

Recharge

In the final analysis, virtually all groundwater owes its existence, directly or indirectly, to precipitation. In detail, however, the main components of groundwater recharge are (a) infiltration of part of the total precipitation at the ground surface; (b) influent seepage through the banks and bed of surface water bodies such as ditches, rivers, lakes, and even the oceans; (c) groundwater leakage and inflow from adjacent aquicludes and aquifers; and finally, (d) artificial recharge from irrigation, reservoirs, spreading operations, and injection wells.

(a) In general terms, the proportion of precipitation infiltrating to the water table depends largely on characteristics of the precipitation itself, topography, vegetation characteristics, and on the type and structure of the soil and the underlying rocks. To the extent that recharge to groundwater resulting from infiltrating precipitation must traverse the unsaturated zone, and that groundwater flow must, therefore, be continuous with the unsaturated flow of water above the water table, earlier discussion of infiltration in chapter 6 is relevant here.

Valuable theoretical work by Freeze [43] assuming homogeneous, isotropic

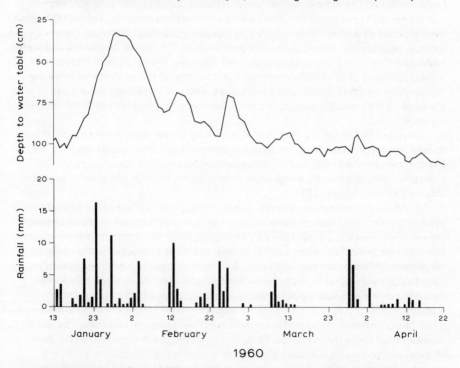

Fig. 7.7. *Graphs of rainfall and shallow water table data from the Thames floodplain, Berkshire, England.*

soils, indicated that recharge from infiltrating precipitation is likely to be greater for long-duration low-intensity rainfalls, shallow water table conditions, high antecedent soil moisture conditions, and soils with high hydraulic conductivity and/or low specific moisture capacity. Subsequent comparisons between laboratory data and field observations on the Canadian prairies [44] showed that, in fact, infiltration to the water table was isolated in both space and time, normally resulted only from very intense rainfalls and favourable conditions of water table depth, soil type, and antecedent soil moisture, and that spring snowmelt was the major source of groundwater recharge on the prairies, particularly in topographic depressions. The work also demonstrated that attempts to estimate recharge simply from a knowledge of soil texture and saturated hydraulic conductivity are likely to fail, and emphasized the need for detailed information on the functional relationships between suction, hydraulic conductivity, specific moisture capacity, and moisture content, since very small differences in these properties can account for large differences in the reaction of similar field soils to the same infiltration event [44].

Investigating the mechanism of recharge by infiltration, Horton and Hawkins [60] described a process of displacement whereby the water which is added to the water table during rainfall is not 'new' rainfall but previous stored rainfall which has been displaced downwards by successive bouts of infiltration. This process which was referred to as *translatory flow* by Hewlett and Hibbert [56], undoubtedly helps to explain the often rapid response of water tables to precipitation, even in low permeability material, especially when soil moisture content is within the range field capacity or wetter. Bonell [10], for example, discussed the rapid response of shallow groundwater in a boulder clay area and Fig. 7.7, based on rainfall and water table data from the Thames floodplain near Reading, shows that the increases in water table level after rainfall normally occur within 24 hours, although they are occasionally delayed by as much as two days. The role of soil moisture content is apparent in the reduced response of the water table to precipitation from winter to spring.

One result of translatory flow is that even deep water table levels may respond rapidly to precipitation even though rates of downward percolation are very low. Experiments by the Institute of Geological Sciences in the Lower Greensand of Hampshire indicated that average annual percolation through the zone of aeration may be less than 1m/yr. [71]

(b) Where groundwater occurs in direct contact with surface water bodies such as lakes, ponds, and streams, there will normally be a movement of water between the two water bodies. Either flow will take place from the stream to the groundwater body, in which case it is known as *influent seepage,* or the reverse movement, *effluent seepage,* will occur, in which case groundwater seeps into and adds to the volume and flow of the surface water body. The seepage relationship between surface and underground water is seldom static but changes with the changing levels of, say, a stream and the adjacent water table so that, in a matter of a few hours, influent seepage may supersede effluent seepage and then, in turn, be replaced once more by the latter. Thus, in Fig. 7.8A, the normal humid situation of effluent seepage is represented by water table profile (a). Occasionally, during periods of heavy rainfall or after snowmelt, the concentration of runoff in the stream channels raises stream levels above those of the adjacent water table (profile

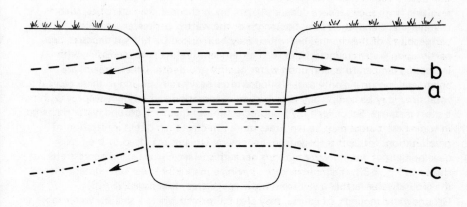

A

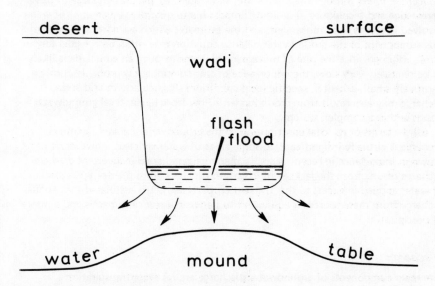

B

desert   surface

wadi

flash
flood

water   mound   table

Fig. 7.8. *(A) Relationships between groundwater and surface water in humid areas. See text for explanation. (B) Groundwater mound developed by influent seepage.*

(b)) resulting in groundwater recharge through influent seepage. Again, during the summer, withdrawals of groundwater by evapotranspiration may induce groundwater recharge, as shown in profile (c).

In arid and semi-arid areas, where evapotranspiration losses normally far exceed precipitation, the few perennial or semi-permanent streams which do exist are allogenic, and almost without exception lose water by influent seepage throughout the 'desert' sections of their courses. The rate and amount of groundwater recharge resulting from such seepage depends partly upon channel characteristics, such as channel shape and, therefore, the length of the wetted perimeter, and the permeability of the channel bed which may be reduced by fine silt deposits, and partly upon water characteristics such as temperature, seepage increasing with increasing temperature, and upon water quality and depth. Other streams are ephemeral, occurring only as flash floods after heavy rainfall, and in these cases the total flow may be completely absorbed by evaporation and influent seepage within a short distance. Schoeller [137] quoted estimates that such ground water recharge in some desert areas may be ten times more effective than direct infiltration of precipitation. Influent seepage from ephemeral streams may lead to the development of *groundwater mounds* beneath surface channels and depressions, as shown in Fig. 7.8B, the mounds often having a markedly lower salt content than the groundwater farther away from the seepage areas (see page 228) [8]. Groundwater mounds, of course, may also be formed where a shallow water table occurs in humid areas, as a result of the concentration of runoff in surface depressions.

(c) It has already been noted that aquicludes rarely form an absolute barrier to water movement, so that there is almost always some slow *drainage of water from adjacent aquifers* through the intervening aquicludes. At the present stage of our knowledge and techniques, it is almost impossible to quantify the amount of water involved in this type of movement, and the estimates which are normally made of this component of the groundwater balance equation may often be considerably in error, although since the rate of movement of water through an aquiclude is likely to be relatively very slow, the importance of this component is probably comparatively small, except in specific local conditions. Particularly in arid areas, recharge may also result from groundwater inflow from higher level groundwater basins within a complex system.

(d) In terms of its total contribution to the groundwater balance, artificial recharge is virtually insignificant; in some areas of water shortage, however, it has grown in importance in recent years. In many irrigated areas inadvertent artificial recharge results from the leakage of irrigation canals or from the deep percolation of water applied in excess to the irrigated crops. Nace [116] estimated that annual recharge from these sources is equivalent to between a few centimetres and a metre of precipitation.

Discharge

The main components of groundwater discharge are (a) evapotranspiration particularly in lowlying areas where the water table is close to the ground surface; (b) natural discharge by means of spring flow and effluent seepage into surface water bodies; (c) groundwater leakage and outflow through aquicludes and into adjacent aquifers; and (d) artificial abstraction.

(a) The role of evapotranspiration in groundwater discharge is relatively complicated and it is necessary to consider not only its indirect and direct effects upon variations of groundwater level, but also factors such as the effect of water table depth on the magnitude of evapotranspiration losses.

The main indirect effect of evapotranspiration concerns the reduction of soil moisture content, and the consequent effectiveness of rainfall, which have been discussed above. Short-term direct effects are evidenced in the diurnal fluctuation of shallow water tables which has been observed on numerous occasions during periods when evapotranspiration rates are high [51], [166], [177]. In valley bottom areas, particularly, losses of groundwater by evapotranspiration during the hottest part of the day may exceed the rate at which groundwater inflow from surrounding higher areas takes place and, as a result, the water table falls. During the evening and at night, when the evapotranspiration rate is much reduced, it will be exceeded by groundwater inflow, so that the water table level will recover. This regular diurnal rhythm will be maintained for much of the summer, although it will be interrupted by periods of rainfall and reduced evapotranspiration (see Fig. 7.9).

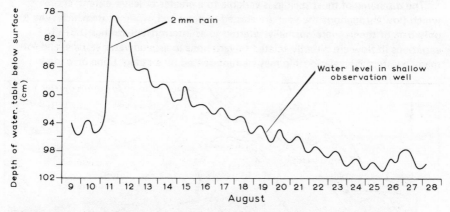

Fig. 7.9. *Diurnal variation of water table level resulting from evapotranspiration. (From an original diagram by P. Meyboom,* J. Hydrol., *2: 248-261, 1965.)*

Provided that the night-time recovery of water table level equals the daytime drawdown, the long-term effects of evapotranspiration may not be readily apparent. In the situation illustrated in Fig. 7.9, however, the daily drawdown exceeded the subsequent recovery for much of the time, with the result that, over a period of about a fortnight, the average water table level fell significantly. From earlier discussions of soil moisture movement in chapter 6, it will be apparent that this process cannot be continued indefinitely, because eventually the water table will drop to such a level that the capillary rise of moisture will be unable to satisfy the demands of the transpiring vegetation at the soil surface. The long-term effect of evapotranspiration in these circumstances would be an exponential reduction of groundwater levels similar to that shown by the composite curves in Fig. 4.8.

(b) Effluent seepage, either directly into surface water bodies, or at seepages and springs at the ground surface, is normally the major form of groundwater discharge and occurs where, for a variety of reasons, the upper surface of the zone of saturation intersects the ground surface. In humid areas where rainfall exceeds evapotranspiration losses, the normal relationship between surface water and groundwater is that depicted by profile (a) in Fig. 7.8A, where the water table slopes gently towards the stream, thereby facilitating the continuous discharge of groundwater into the surface channel. In this situation, the rate of seepage will obviously depend mainly on the available head, which will be represented by the difference in elevation between the stream surface and the adjacent water table. Seasonal modifications, represented by profiles (b) and (c), which result in influent seepage have already been discussed in relation to groundwater recharge.

Although major streams and other surface water bodies are normally also major sinks for groundwater discharge, there are many instances where broadly distributed *seepages* occur over the lower valley slopes as well [61] [162] [163]. *Springs* are often distinguished from seepage areas on the basis that a spring is a concentrated discharge of groundwater, which appears as a definite flow of water at the surface, whereas a seepage area indicates a slower movement of groundwater to the surface [159], although for the purposes of this discussion the difference is a matter of degree only.

The discharge of most springs is variable to a greater or lesser extent; those which flow throughout the year are classed as perennial, while those which flow for only part of the year are normally referred to as intermittent springs [88]. The variations in flow are directly related to variations in groundwater storage, and for many aquifers this relationship may be represented by a straight line on a log-log

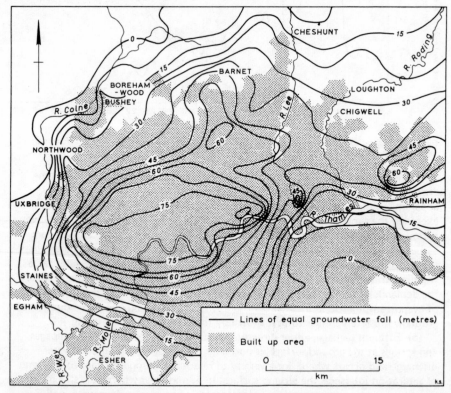

Fig. 7.10. *Decline of groundwater levels in the London area up to 1965. (From original maps by Water Resources Board [171].)*

plot. It follows, therefore, that spring flow from thick porous aquifers having a very large storage capacity will tend to be relatively constant throughout the year, since the volume of storage change will probably represent a comparatively small proportion of the total storage volume of the aquifer. Similarly, artesian spring flow from confined aquifers will not vary greatly from one season to another. In the case of thin aquifers, however, particularly veneers of superficial deposits such as scree or glacial gravels, changes of storage in relation to total storage capacity are likely to be large, so that spring flow tends to be very variable through the year, sometimes occurring only for short periods after rainfall in the case of, say, a scree foot spring.

(c) Groundwater leakage through aquicludes and between adjacent aquifers was briefly discussed in relation to groundwater recharge. Clearly a movement of groundwater which represents a source of recharge for one part of a groundwater system must necessarily represent a discharge from some other part of the system.

(d) Groundwater abstraction for supply purposes has had such a profound effect on storage and storage changes in some areas, that a brief comment on it seems unavoidable at this stage. If excessive abstraction takes place from a large number of wells over a long period of time, it will result in a gradual lowering of the water table, or piezometric surface, over a wide area, and this has, in fact, happened in many major groundwater supply areas. In the British Isles, the eastern part of the London Basin provides a classic example, and Fig. 7.10 illustrates the manner and extent in which water levels have fallen in the confined chalk aquifer. In other areas groundwater abstraction has resulted in the incursion of inferior quality, saline groundwater from adjacent aquifers or by direct seepage from the oceans along coastal areas.

7.9 Coefficient of storage

Now that the components of storage change have been described in largely qualitative terms, it remains to discuss the mechanism of storage change in both unconfined and confined conditions. The coefficient of storage, or storativity, of an aquifer is defined as the volume of water which an aquifer releases from, or takes into, storage per unit surface area of aquifer per unit decline or rise of head. It is, therefore, a ratio of a volume of water to a volume of aquifer [37]. This is illustrated in Fig. 7.11 in which the volume of water released from storage in the aquifer prism divided by the product of the prism's cross-sectional area and the decline in head results in a dimensionless number, i.e., the coefficient of storage. In unconfined conditions (Fig. 7.11B) the coefficient of storage approximates the specific yield (see page 191), provided that gravity drainage is complete, and normally ranges from about 0.02 to 0.30 [169]. In confined conditions (Fig. 7.11A), where no dewatering of the aquifer occurs, the volume of water released for a unit decline of the piezometric surface may be attributed to compression of the granular structure of the aquifer and to the expansion of the water itself. Although definite limits are difficult to establish, the storage coefficients of confined aquifers may range from about 0.000 01 to 0.001 [169].

Clearly, there are significant differences between the mechanism of storage changes in confined and unconfined conditions and some of the more important of these differences will now be discussed.

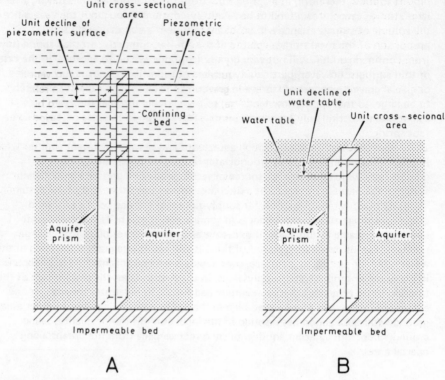

Fig. 7.11. *Diagram to illustrate the storage coefficient of (A) confined and (B) unconfined aquifers. (From an original diagram by Todd [159].)*

7.10 Storage changes in unconfined aquifers

It will be apparent from the foregoing discussions, that storage changes in unconfined conditions are relatively uncomplicated and may be directly reflected in variations of groundwater level. When recharge exceeds discharge, water table levels will rise, and when discharge exceeds recharge, they will fall. Since, in most natural circumstances, recharge to and discharge from the same groundwater system occur simultaneously, groundwater level fluctuations, in fact, reflect the net change of storage resulting from the interaction of these two components. The study and interpretation of water table fluctuations thus forms an integral part of the study of groundwater storage.

In many areas water table fluctuations tend to follow a fairly rhythmic seasonal pattern. In the British Isles, for example, high water levels occur during the winter months, and low levels during the summer months, and it is on the basis of this broad differentiation that the hydrological or water year has been delimited as beginning on the first day of October and ending on the last day of September. The water table levels illustrated in Fig. 7.12 refer to a chalk well in the Yorkshire Wolds, but may be considered as more generally typical. In some areas, such as Russia, where conditions are particularly suitable, considerable attention has been

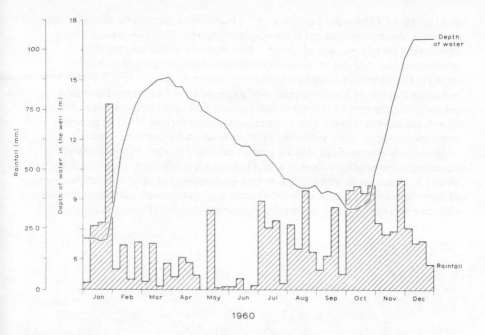

Fig. 7.12. *Rainfall and water table level in a chalk well in Humberside County, England. (Based on data supplied by the Yorkshire Regional Water Authority.)*

paid to the study of groundwater regimes, and to the extent to which the normal climatically determined regime may be modified by artificial factors such as irrigation and drainage [14] [143].

Seasonal fluctuations, reflecting as they do seasonal changes in storage, are normally of considerable interest to the hydrologist, in the sense that they represent large-scale contrasts in water availability for both natural and artificial uses. Short-term fluctuations, usually on a much smaller scale, may also be of relevance, however, in particular circumstances. Thus, in many coastal areas regular short-term fluctuations of water table level are associated with tidal movements since, if the sea level varies with a simple harmonic motion, a train of sinusoidal waves is propagated inland from the submarine outcrop of an aquifer [159]. Investigations of this phenomenon have shown that the groundwater fluctuations are of decreased amplitude, and lag behind the tidal fluctuations, the extent of the reduction and lag depending largely on distance from the sea and the ease with which groundwater can move through the aquifer.

Long-term irregular or secular fluctuations of water table level are also commonly experienced, and are often of considerable hydrological significance. These are mainly associated with secular variations of rainfall (see chapter 2, page 22). Numerous investigations of this phenomenon have been made during the past twenty-five years, especially in the United States, including classic analyses by Jacob of water levels and precipitation on Long Island [74].

However, the relationships between storage and water level are complicated by the fact that the latter responds to factors other than storage changes, for as

Meinzer noted, 'The water level in a well is sensitive to every force that acts upon the body of water with which the well communicates' [103]. Although this is particularly true in the case of confined groundwater it may also apply to unconfined groundwater in some circumstances. Thus, in the late nineteenth century, experiments by Baldwin Latham in the North Downs near Croydon, indicated that spring flow increased with a sudden fall of barometric pressure, presumably because of a slight increase in water table level as the pressure fell, and that spring flow decreased with a rise in barometric pressure [145]. The actual magnitude of water table variations must, however, have been very small.

Water-level change maps may be used to calculate changes in the saturated volume of an unconfined aquifer, and if the storage coefficient is known, the actual change in groundwater storage. Such maps are constructed by plotting the change of water levels in wells over a period of time. For short periods data from the same wells can normally be used but for longer periods initial and terminal water level

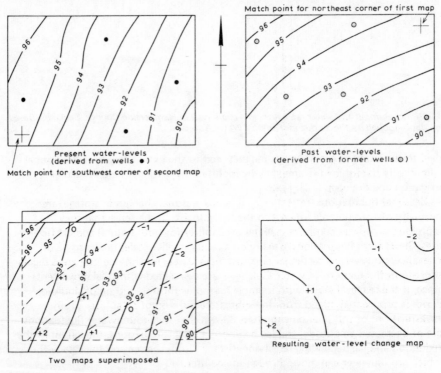

Fig. 7.13. *Diagram to illustrate the construction of a water-level change map by superimposing water-level contour maps. (From an original diagram by Davis and De Wiest [34].)*

data may be from different wells. In this case, the maps for present and past water levels may be superimposed and the water level changes at contour intersections transferred to a third map on which lines of equal water-level change can be drawn in [34]. This procedure is illustrated in Fig. 7.13 which also shows that water-level change maps may be useful in delimiting the local effects of recharge or discharge, which are often difficult to detect simply from a comparison of successive water table maps. Thus in Fig. 7.13 the effects of recharge, which show only as a deflection of the water table contours, are well-defined by the curved isopleths in the water level change map.

7.11 Storage changes in confined aquifers

In the case of confined aquifers the relationships between changes in piezometric level and changes of groundwater storage are greatly complicated by the compressibility and elasticity of the aquifers, whereas under water table conditions aquifer compressibility is virtually negligible with respect to the gravity drainage of the interstices. Before Meinzer's classic work on the Dakota sandstone of North Dakota, it was generally assumed that confined aquifers were rigid incompressible bodies and that changes in pore water pressure were not accompanied by changes in pore volume. Meinzer [102] demonstrated that the long-period excess of discharge by pumping from the sandstone over natural recharge could not be accounted for by the compressibility of the water alone and that, therefore, the pore space of the sandstone was presumably reduced to the extent of the unexplained excess volume of water. Meinzer's pioneer work was subsequently further developed by Theis [156] and by Jacob [73] who also drew attention to the water derived from the compression of adjacent and included clay beds in interbedded aquifers and more recently valuable contributions have been made, as Domenico [37] observed by Lohman [90] on elastic compression, Poland [127] on permanent compression, and Hantush [52].

Within a confined aquifer there are pressures in the solid phase at the points of contact between individual particles and there are pressures due to the contained water. The former are referred to as intergranular pressures or *effective stresses,* and the latter as pore-water pressures or *neutral stresses* and their sum represents the total vertical stress acting on a horizontal plane at any depth [37]. Since neutral stresses act equally on all sides of the solid particles of the matrix they do not cause the particles to press against each other, so that all measurable effects such as compression and expansion result exclusively from changes in effective stress, i.e., in order for an aquifer to undergo compression, there must be an increase in the grain-to-grain pressures within the matrix and, conversely, in order for it to expand, there must be a decrease in grain-to-grain pressure [37].

Thus when water is pumped from a confined aquifer, there is no change in the total stress so that the reduction in pore-water pressure must result in a proportional increase in that part of the total load carried by the grain structure. Owing to the reduced fluid pressure the water expands to the extent permitted by its elasticity, while compression of the aquifer results in reduced porosity and the consequent release of water from the saturated volume. If pumping ceases and the pressure is restored, the aquifer grains gradually return to their original position and the water is compressed [37]. If the aquifer is perfectly elastic and water levels in the recharge and discharge areas do not change, the original artesian head will ultimately be restored [169].

It will be clear from the foregoing and from earlier discussion of the coefficient of storage that comparatively small yields of water from a confined aquifer may be accompanied by large variations in piezometric level compared with the corresponding water-table fluctuation in an unconfined aquifer. It will also be evident that any variation of loading on a confined aquifer may result in fluctuations of the piezometric surface. Such variations of loading may result from barometric pressure changes [73] [120] [159] [167] [180], variable tidal and gravitational loading [49] [131] [132] [157], or, in certain circumstances, from the passage of a railway train [72], the occurrence of an earthquake [138], or a nuclear explosion [65], and in each case may provide valuable information about the elastic and storage properties of the aquifers concerned.

With a continuing excess of discharge over recharge to a confined aquifer the ever-increasing intergranular pressures and the resulting compression of the aquifer may ultimately result in the subsidence of the overlying ground surface. Subsidence may also result where agricultural drainage of peat lands, for example, leads to lower water tables and the associated oxidation and wind erosion of the exposed peat [160]. Substantial subsidence resulting from reductions of groundwater storage have been recorded in California [50] [127] [128], Texas [179], Georgia [33], and Nevada [38], while outside the United States important examples include Mexico City [30] [89] and London [178].

The amount of subsidence at any location depends upon the sub-surface lithology, the thickness of the compressible units and their storage characteristics, and upon the magnitude and duration of the decline in head. Subsidence is normally inelastic and permanent and in the worst-affected areas may amount to between one-tenth and one-twenty-fifth of the observed decline of piezometric level [169].

Just as water table maps may be used to throw light on storage changes in unconfined aquifers, so piezometric or potentiometric maps may be used in the case of confined aquifers, although since the piezometric surface is determined by both water pressure and the height of the aquifer, it is both misleading and erroneous to imply that there is *one* piezometric surface for a given aquifer. Normally, piezometric surfaces are much smoother than water tables since local changes in head will be more rapidly propagated, as a recharge or discharge wave, in confined conditions [34]. Used with caution, therefore, piezometric change maps may provide valuable information about storage changes.

Groundwater movement

Groundwater taking an active part in the hydrological cycle is in more or less continuous motion from areas of recharge to areas of discharge in response to the hydraulic conductivity of the water-bearing material and the prevailing hydraulic gradient (both of which are incorporated in the basic Darcy equation). Hydrological interest is largely concerned with the speed and direction of groundwater movement, and it is mainly upon these two factors that attention will be concentrated in the following pages. It is commonly remarked that not only is the rate of movement of groundwater very slow, compared with that of surface water, but also, that it is very variable. Buchan [17], for example, quoted rates of groundwater movement through permeable strata ranging from as low as a fraction of a millimetre per day, in some fine-grained pervious rocks, to as much as 5500 m per day, through fissured chalk in Hertfordshire, England.

The direction of groundwater movement is similarly variable since, like surface water, groundwater tends to follow the line of least resistance. Other things being equal, therefore, flow tends to be concentrated in areas where the interstices are larger and better connected, and the hydrologist's problem is to locate such areas, often from rather scanty geological information. Theoretical analyses commonly assume ideal, homogeneous and isotropic conditions so that results from them may be difficult to apply in field conditions. Most studies attempt to define more or less complete and independent flow systems, e.g., bounded by impermeable beds. In most real situations, however, flow systems are bounded by semi-permeable, rather than by completely impermeable beds, so that very complex and widespread flow systems develop. It is, therefore, encouraging to note that during the past decade in particular there has been a considerable growth of interest in the definition and analysis of regional groundwater flow systems and a corresponding decline in the trend towards increasingly sophisticated, idealized solutions to problems in the field of well hydraulics. Even so, simplifying assumptions are normally essential and while these may be reasonable for many real situations, it should be emphasized that important 'untypical' groundwater flow situations are found in, say, limestone [118] [125] [142] [149] [150] or volcanic rocks [34] or in areas of permafrost [20] [21] [34] [115].

7.12 Darcy's Law

Since the main features and limitations of the Darcy equation were discussed in section 6.8, in relation to unsaturated flow, the present discussion will be brief. Most groundwater movement takes place in small interstices so that the resistance to flow imposed by the material of the aquifer itself may be considerable with the consequence that the flow is laminar, i.e., with fluid particles moving along smooth paths and with successive particles following the same path or streamline. As the velocity of flow increases, especially in material having large pores, the occurrence of turbulent eddies dissipates kinetic energy and means that the hydraulic gradient becomes less effective in inducing flow. In very large interstices such as those found in many limestone and volcanic areas groundwater flow is almost identical to the turbulent flow of surface water.

The law which expresses the relationship between capillary or laminar flow and the hydraulic gradient was stated by Poiseuille [126] and is generally described in physics as *Poiseuille's Law*. Later, Darcy [31] confirmed the application of this law to the movement of groundwater through natural materials and, for hydrologists, the law has since become associated with his name.

Darcy's law for saturated flow may be written as in eq. (6.4)

$$v = -k \, \nabla \, \Phi \qquad (7.2)$$

where v is the macroscopic velocity of the groundwater, k is the saturated hydraulic conductivity and $\nabla \Phi$ is the gradient of total potential or head. Total potential is defined as

$$\Phi = \Psi + z \qquad (7.3)$$

where Ψ is the pressure head and z is the elevation head above a selected datum

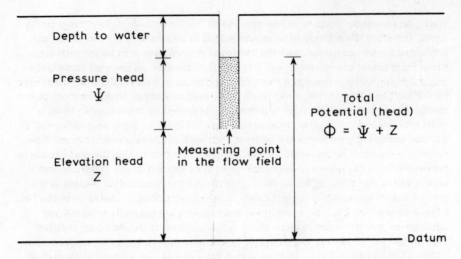

Fig. 7.14. *Diagram showing elevation head, pressure head, and total potential (head) for a point in the flow field. (From an original diagram by P. A. Domenico,* Concepts and Models in Groundwater Hydrology, *McGraw-Hill, New York, 1972.)*

level as shown in Fig. 7.14. Although Darcy described k as a coefficient dependent on the nature of the porous medium, it is accepted that the term also involves the physical properties of the water itself, in so far as these affect the rate of movement in saturated materials.

Over the years, the terminology relating to the flow of groundwater in general, and to Darcy's law in particular, became rather confused, and in 1952 it was recommended [129] that, since k in Darcy's equation refers to the characteristics of both the porous medium and the fluid, it should be termed hydraulic conductivity, while those characteristics of the porous medium itself which determine groundwater movement should be referred to as *intrinsic permeability.* The term hydraulic conductivity is thus virtually synonymous with the earlier term *coefficient of permeability* as defined by Meinzer [148].

As was emphasized in the discussion of unsaturated flow, the Darcy equation yields only a macroscopic velocity value through a cross-sectional area of solid matrix and interstices. Clearly, flow velocities through the interstices alone will be higher than the macroscopic value, and since the interstices themselves vary in shape, width and direction, the actual velocity in the soil is highly variable [57]. Furthermore, the actual flow path of a water particle through a given length (l) of aquifer exceeds the apparent length (l_a) because of the *tortuosity* of the flow-path, i.e., the ratio of the distance travelled by an average water particle around and between the grains of a porous medium to the measured length of the porous medium in the average direction of flow.

Another factor complicating the field application of the Darcy equation is that hydraulic conductivity is often markedly anisotropic, particularly in fractured and jointed rock. Anisotropic versions of the Darcy equation have been presented by a number of investigators [23] [91] [134] [146] [168].

Again, extreme flow velocities may result in deviations from the Darcy law. The effect of turbulence in modifying the relationship between the hydraulic gradient and high rates of flow has already been mentioned. At the other extreme, some investigators have claimed that in clay soils, with small pores and low hydraulic gradients, the very low flow rates are less than proportional to the hydraulic gradient [112] [117] [153]. A possible explanation is that much of the water in such material is strongly held by adsorptive forces and may be more rigid and less mobile than ordinary water [57] (see page 136).

Apart from minor deviations of this sort, however, the Darcy law can be successfully applied to virtually all normal cases of groundwater flow and is equally applicable to both confined and unconfined conditions. An understanding of most groundwater problems, however, demands information not only about the velocity of water movement but also about the *velocity of head transmission* which is usually many hundreds of times faster. The velocity of head transmission is proportional to the square root of the *hydraulic diffusivity (a)* which is the ratio of the hydraulic conductivity to the storage coefficient thus

$$a = \frac{km}{S} \qquad (7.4)$$

where k is the hydraulic conductivity, m is the saturated thickness of the aquifer, and S is the storage coefficient as defined in section 7.9. For confined conditions, therefore, this ratio depends not only on the conductivity of the aquifer material but also on its elastic properties [14]. The hydraulic diffusivity term is inherent in all groundwater flow equations and is frequently determined as a general hydrological parameter without separate determination of its constituent parts [14].

By itself, the Darcy law suffices to describe only steady flow conditions so that for most field applications the law must be combined with the mass-conservation law to obtain the general flow equation, or for saturated conditions, the Laplace equation. A direct solution of the latter equation for most groundwater flow conditions is generally not possible so that it is necessary to resort to various approximate or indirect methods of analysis, some of which will be referred to in a subsequent section (7.15).

7.13 Factors affecting hydraulic conductivity

Fundamental to the application of Darcy's law is a knowledge of the hydraulic conductivity of the saturated medium. The factors affecting hydraulic conductivity may be conveniently grouped into those pertaining to the water-bearing material itself, and those pertaining to the groundwater as a fluid.

One of the most important of the *aquifer characteristics* concerns the geometry of the pore spaces through which groundwater movement occurs. Although methods have been evolved to determine the pore size distribution of a porous medium directly [25], the difficulties in so doing have, for the most part, resulted in indirect approaches whereby the pore space geometry is related to factors such as grain size and shape, and the grain size distribution on the not always very sound assumption that there is a definable relationship between these properties and pore size distribution.

A second aquifer characteristic relates to the geometry of the rock particles themselves, particularly in respect to their surface roughness, which may have an important effect on the speed of groundwater flow.

Finally, an important influence is exerted by secondary geological processes such as faulting and folding, which may increase or decrease groundwater movement, depending on the lithology of the beds involved in these movements [63], secondary deposition which will tend to reduce the effective size of the interstices and, therefore, also the flow of water, and secondary solution in, for example, limestone, to which reference has already been made. Figure 7.15 illustrates the effect of factors such as these on groundwater movement in the chalk of East Anglia; the areas of high transmissibility (transmissibility is the product of hydraulic conductivity and the saturated thickness of the aquifer) tend to be

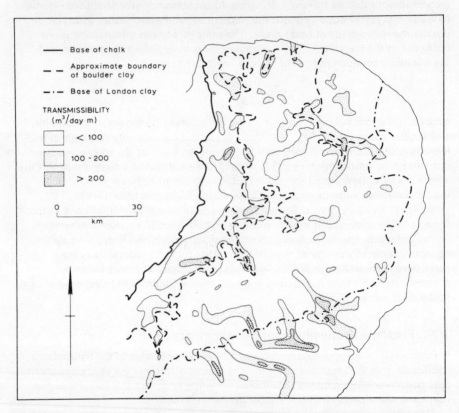

Fig. 7.15. *Regional variations in the transmissibility of the chalk in East Anglia. See text for explanation. (From an original diagram by Ineson [68].*

related to topographic valleys, which in turn are associated with fold or fault structures, or with increased fissuring, and the map suggests that these relationships may continue beneath relatively impermeable overlying deposits. In the London Basin, the effects of folding are even more pronounced, and here synclinal areas in the chalk are associated with low permeability as a result of compaction, while anticlinal areas frequently tend to encourage high rates of groundwater flow [68].

As Hillel [57] observed numerous attempts have been made to represent porous media by idealized theoretical models which are amenable to mathematical treatment. In a comprehensive review Scheidegger [135] noted that since most natural porous media are extremely disordered, they are unlikely to be adequately represented by such idealized models and suggested that a statistically based model might be preferable.

The effects of *fluid characteristics,* such as density and viscosity, on hydraulic conductivity, tend to be rather less important than the effects of the aquifer characteristics. Certainly, in relation to normal conditions of groundwater flow, the physical properties of the groundwater are likely to be influenced only by changes in temperature and salinity. Temperature, by inversely affecting the viscosity has a direct influence on the speed of groundwater flow which, for example, approximately doubles for an increase in temperature from $4.5°C$ to $32.5°C$. [1]. Since, however, most groundwater is characterized by relatively constant

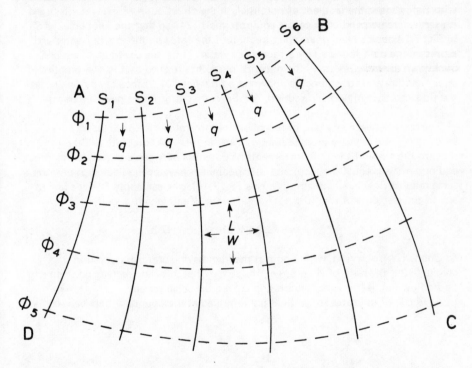

Fig. 7.16. *The flow net over a specified area ABCD. See text for explanation. (From an original diagram by Childs [24].)*

temperatures, this factor is likely to be important only in particular circumstances. One such circumstance concerns influent seepage into an aquifer from a surface water body whose temperature varies seasonally. Kazmann [76] found that, in these conditions, groundwater flow to pumped wells increased by about 50 per cent between $4.5°C$ and $24°C$.

Again, variations of salinity are not likely to be significant in normal groundwater conditions. Where saline infiltration occurs, however, hydraulic conductivities may be affected both by changes in the ionic concentration of the groundwater, and also by the chemical effect of the saline water on the aquifer material itself, particularly where this is of an argillaceous nature [63].

7.14 Flow nets

Since it is impossible to observe groundwater flow directly, it is frequently convenient to make use of the relationship between flow and the hydraulic or potential gradient and thereby to examine two-dimensional groundwater flow indirectly by reference to the subsurface distribution of groundwater potential, which can be quite easily measured. Lines joining points of equal potential (ϕ) are known as *equipotentials* and the potential surface may be contoured at regular intervals or increments of ϕ by a family of such lines (see Fig. 7.16). In accordance with Darcy's law, the gradient of potential, in the direction in which water is urged, is everywhere perpendicular to the equipotentials [24] so that the lines labelled S in Fig. 7.16 depict everywhere the direction of the force on the moving water and represent the path followed by a particle of water. Thus, the contours of equal S are known as *streamlines* and the network of meshes formed by the two families of equipotentials and streamlines is known as a *flow net* [24]. Since at any one point, the flow can have only one direction, it follows that streamlines never intersect [57].

The zone between any pair of neighbouring streamlines is known as a streamtube, and at every cross-section of a streamtube the total rate of flow (q) remains the same. If then the increments of potential between any pair of adjacent equipotentials is equal (i.e., $\Delta\phi$) and the streamlines have been selected to give the same rates of flow in all streamtubes (see Fig. 7.16), one can apply Darcy's law to any of the elements of the flow net having width W and length L so that

$$q = k\Delta\phi \ (W/L) \qquad (7.5)$$

As Childs [24] observed, with a porous medium having uniform hydraulic conductivity, the ratio of W to L, i.e., the shape of each of the rectangular elements of the flow net, is the same, since q and $\Delta\phi$ are the same for all.

As Fig. 7.17 indicates streamlines are refracted at the boundary between media

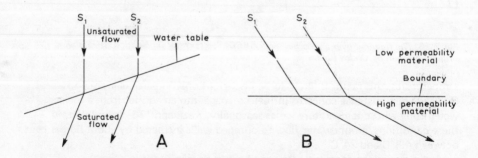

Fig. 7.17. *Refraction of streamlines; (A) across a water table and (B) across a permeability boundary. (From an original diagram by Todd [159].)*

210

of different permeabilities, or where unsaturated flow percolates to a water table. Streamline refraction permits the conservation of fluid mass when flow takes place across the permeability boundary. Thus, in accordance with Darcy's law, other factors being equal, the higher the permeability, the smaller the area required to pass a given volume of water in a given time [37]. The streamlines are thus widely spaced in the low permeability material and closely spaced in the high permeability material.

The use of flow nets, which are commonly constructed in either a vertical plane or a horizontal plane (i.e., potentiometric map), probably originated with Forchheimer in the nineteenth century [95] but was revived again with important modifications by Hubbert [61] and more recently, in connection with computer models of groundwater flow, by Toth [162]. With the growing interest in regional flow patterns in recent years, the use of flow nets has received a new impetus.

7.15 Two-dimensional groundwater flow

Although virtually all groundwater flow is three-dimensional, it is practically impossible to solve a three-dimensional flow problem with the Laplace equation. In most cases it is both necessary and reasonable to consider flow as taking place in two dimensions rather than three. This simplification is ideally suited to the use of

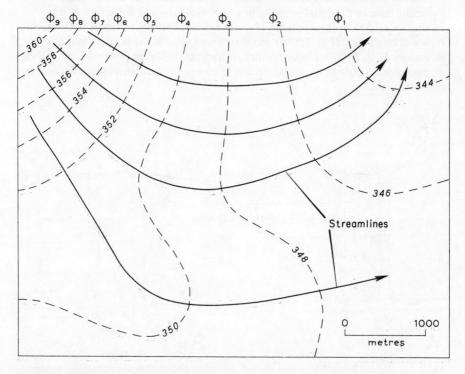

Fig. 7.18. *Water table contour map showing the direction of groundwater flow. Water table contours in metres.*

flow nets and also to the use of electrical analogues and parallel-plate Hele-Shaw models of groundwater flow.

In the horizontal plane, two-dimensional flow is represented by the potentiometric map. A simple example for unconfined groundwater flow in a non-homogeneous aquifer is shown in Fig. 7.18 where the equipotentials ($\phi_1 - \phi_9$) represent the water-table levels indicated in metres. Since the change in potential between adjacent pairs of equipotentials is equal and the hydraulic gradient varies inversely with the distance between equipotentials, then if inflow for any section is just balanced by outflow, the relative steepness of the hydraulic gradient reflects the hydraulic conductivity, as indicated in Darcy's law. Thus in Fig. 7.18 the hydraulic conductivity is greatest in the north-west of the area and decreases towards the east and south, as is evidenced by the wider spacing of the equipotentials.

If Fig. 7.18 represented potential distribution in a homogeneous, rather than a non-homogeneous, aquifer, the variable spacing of the equipotentials would reflect a variation in groundwater flow—in this case a decrease of flow in the direction of flow, as represented by the decrease in hydraulic gradient towards the east and south.

Because of the readier availability of the appropriate data, groundwater flow is normally considered as a two-dimensional problem in the vertical, rather than the horizontal, plane and in this case it is information on the rate of change of hydraulic head or potential with depth which facilitates the definition of the vertical groundwater flow pattern. If groundwater potential increases with depth, flow will be upward (Fig. 7.19A) and if it decreases with depth, flow will be downward (Fig. 7.19B). Under normal, or hydrostatic, pressure there is no change of groundwater potential with depth and therefore no groundwater movement in the vertical plane.

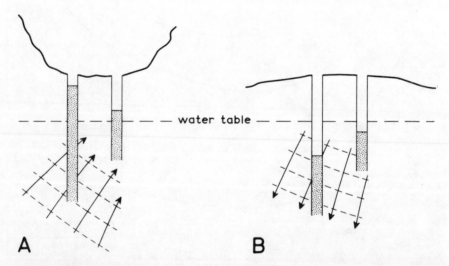

Fig. 7.19. *Vertical groundwater flow pattern where (A) potential increases with depth and (B) where potential decreases with depth. (From an original diagram by P. A. Domenico, Concepts and Models in Groundwater Hydrology, McGraw-Hill, New York, 1972.)*

However, a zero change of potential with depth is also characteristic of situations where the direction of unconfined groundwater flow is approximately horizontal, and where as a result the equipotentials are approximately vertical, and are labelled by the height at which they intersect the water table, and where the potential gradient is simply the slope of the water table immediately above the given point [24]. This might be the situation, for example, where an extensive thin permeable bed rests on an underlying horizontal impermeable bed and is an approximation proposed by Dupuit [39] and elaborated by Forchheimer [42] and subsequently known as the Dupuit-Forchheimer approximation.

7.16 Unconfined groundwater flow—classical models

Increasing emphasis on a hydrodynamic approach to the solution of groundwater problems including the application of flow net techniques which has been a notable feature of groundwater hydrology in the past decade or so stemmed largely from the classic work of Hubbert [61] who presented the groundwater system as a dynamic mechanism subject to all the requirements of the conservation equation within the context of accepted hydrodynamic principles [95]. Figure 7.20

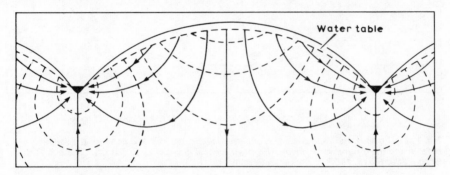

Fig. 7.20. *Approximate unconfined groundwater flow pattern in uniformly permeable material. (From an original diagram by M. K. Hubbert, J. Geology, XLVIII, University of Chicago, 1940. Used with permission of the University of Chicago Press.)*

illustrates the essential features of Hubbert's presentation. Under closed conditions, groundwater flow would ultimately result in the complete drainage of water from the topographic highs and the production of a flat surface of minimum potential energy (the hydrostatic condition) [37]. This tendency, however, is counteracted by continuous replenishment from precipitation. The result of this continuous movement and renewal is the flow pattern shown whose upper surface, the water table, is a subdued replica of the topography. Streamlines, indicating the groundwater flow paths, connect the source areas in which recharge is dominant with the sinks in which discharge is dominant. Equipotentials are shown as broken lines. The source areas are the topographic highs and in this diagram the sinks are shown as streams, and each groundwater flow cell is bounded by the lines of vertical flow beneath the groundwater divides and the sinks or by widely distributed impermeable beds or by both.

Toth [162] suggested that, whereas major streams may also be major groundwater sinks. as in the Hubbert model just discussed, for valleys having low-order streams only groundwater discharge is not concentrated at the stream but is broadly distributed on the down-gradient side of a mid-line between the valley bottom and the groundwater divide, as shown in Fig. 7.21. As can be seen, the mid-line is an approximately vertical equipotential about which the flow pattern is broadly symmetrical, giving a central area of lateral flow where groundwater

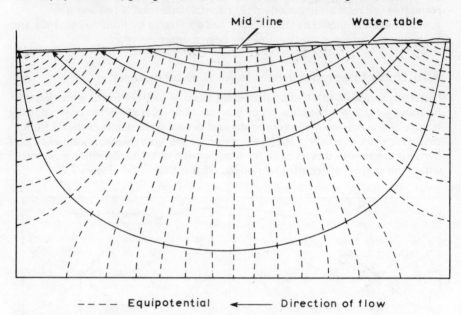

Fig. 7. 21 *Groundwater flow pattern symmetrical about the mid-line between the valley bottom and the groundwater divide. (From an original diagram by J. Toth, J. Geophys. Res., 67, p. 4380, Fig. 3, 1962. © American Geophysical Union).*

potential does not vary with depth, an up-gradient area of downward flow or recharge where groundwater potential decreases with depth and the down-gradient area of upward flow or discharge where groundwater potential increases with depth, as in Fig. 7.19. Toth's work stimulated other contributions and in particular the development of a similar model for arid and semi-arid areas having the same three major components but differing somewhat as a result of physiographic and topographic conditions [94] [96] . As Domenico [37] pointed out, however, the concepts of Toth and subsequent investigators are by no means new but were first introduced by Meinzer [99] to describe the flow system in Big Smokey Valley, Nevada.

The unconfined groundwater flow models which have been discussed here all assume that the porous medium is hydrologically isotropic and homogeneous. In many areas this may not be an unreasonable assumption, for as Maxey [95] observed, even in the Great Basin area in which he worked, the rocks have been broken up to such an extent by tectonic stresses that the whole mountain mass may be regarded as a singly hydrological unit, just as homogeneous and isotropic on a large scale as is a permeameter filled with sand on a small scale.

7.17 Confined groundwater flow

At the other extreme from the homogeneous isotropic unconfined groundwater situation is that shown in Fig. 7.2 where, with alternating beds of markedly different lithology and permeability, groundwater is confined beneath an impermeable layer and where the piezometric surface of the flow field is completely independent of surface topography and of the configuration of the water table in the upper, unconfined groundwater body. Domenico [37] suggested, however, that what is consistently found in actual conditions is neither a completely confined system such as that shown in Fig. 7.2, nor a completely unconfined system such as that assumed in Figs. 7.20 and 7.21 but a system of flow that possesses distinct characteristics of both extremes. For example, as has already been pointed out (page 187), confining beds rarely form an absolute barrier to water movement so that there is normally some degree of hydraulic continuity, which suggests that the potential distribution with depth in a confined groundwater body is partly affected by the potential distribution of the overlying water table. On the other hand, in an apparently unconfined situation, the flow field may possess characteristics of confinement whenever flow is refracted on emerging from a low-permeability bed so that it proceeds almost tangentially to the lower surface of that bed [37]. This is clearly illustrated in Fig. 7.22 which represents an unconfined system with hydraulic continuity in the vertical direction. Streamlines in the high-permeability bed are almost tangential to the lower surface of the low-permeability bed. It will be observed that there is very little difference in groundwater potential along an imaginary vertical line passing through the high-permeability bed but that if this line is extended upward, it crosses several

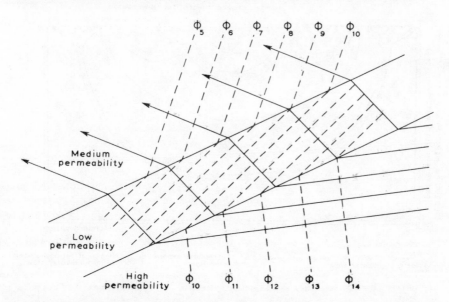

Fig. 7.22. *Streamline refraction in an unconfined system of layered permeability contrasts. (From an original diagram by M. K. Hubbert, J. Geology, XLVIII, University of Chicago, 1940. Used with permission of the University of Chicago Press.)*

equipotentials in the low-permeability bed and relatively few in the medium-permeability material. Thus, if a well was drilled along this line, there would be a large increase in piezometric head when it first entered the high-permeability bed because static levels can establish themselves more rapidly here than elsewhere in the system. This increase in head is often attributed to confinement of water under pressure, although in reality it results from the movement of water through the low-permeability bed, which suggests that the conditions implied by the term *confinement* will arise when a unit of low permeability overlies a unit of high permeability [37].

7.18 Regional groundwater flow

Recognition of the blurred distinction between confined and unconfined groundwater flow has facilitated the definition and study of regional groundwater flow systems which have been defined as large groundwater flow systems which encompass one or more topographic basins [111]. This work owes much to the earlier contributions of Meinzer [99] and especially of Hubbert [61] who was concerned with applying his theory of steady-state groundwater flow to large-scale regional problems. Unfortunately, as Freeze and Witherspoon [45] pointed out, this work was overshadowed for a quarter of a century up to 1960 by work on the application of mathematical solutions of transient flow to well and wellfield hydraulics. During and since the 'sixties, however, attention has once again been turned to the regional situation, with the groundwater basin as the unit of

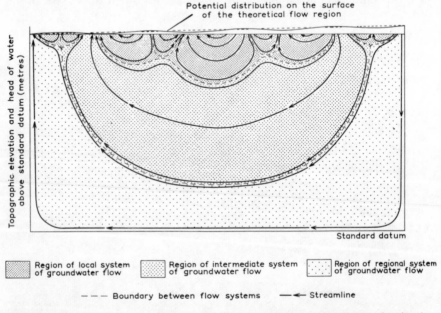

Fig. 7.23. *Theoretical flow pattern and boundaries between local, intermediate, and regional flow systems. (From an original diagram by J. Toth, J. Geophys. Res., 68, p. 4807, Fig. 3, 1963. © American Geophysical Union.)*

hydrological study. Especially prominent in this field have been a group of Canadian hydrologists including J. Toth, P. Meyboom, and R. A. Freeze.

Toth [162] [163] significantly extended Hubbert's work by introducing the concept that groundwater flow patterns can be obtained mathematically as solutions to formal boundary value problems. He assumed two-dimensional vertical flow in a homogeneous isotropic medium bounded below by a horizontal bed, and above by the water table which is a subdued replica of the topography. The lateral flow boundaries are the major groundwater divides. Toth [163] defined a flow system as a set of flow lines (streamlines) in which any two flow lines adjacent at one point remain adjacent through the whole region and which can be intersected anywhere by an uninterrupted surface across which flow takes place in one direction only. Toth recognized the three broad categories of flow system illustrated in Fig. 7.23. These are a *local* system which has its recharge area at a topographic high and its discharge area at an adjacent topographic low and is therefore identical to the classic Hubbert model shown in Fig. 7.20, an *intermediate* system having one or more topographic highs and lows located between its recharge and discharge areas, and a *regional* system, where the recharge area is at the main topographic high and the discharge area at the lowest part of the basin. Toth further suggested that if local relief variations are negligible but there is a general topographic slope, only regional systems will develop; that with pronounced local relief extensive unconfined regional systems are unlikely to develop across the valleys of large rivers or pronounced watersheds but rather that the greater the relief the deeper the local systems that will develop; and finally, that under extended flat areas neither regional nor local systems will develop and that such areas will be characterized by local waterlogging and groundwater mineralization from concentration of salts.

Meyboom [106] [109] described a general model of groundwater flow in a prairie environment which he designated the *prairie profile* (Fig. 7.24). By definition this consists of a central topographic high bounded on both sides by areas of major natural discharge. Geologically the profile consists of two layers of different permeability, the upper layer having the lower permeability, with a steady flow of groundwater toward the discharge areas. The ratio of permeabilities is such

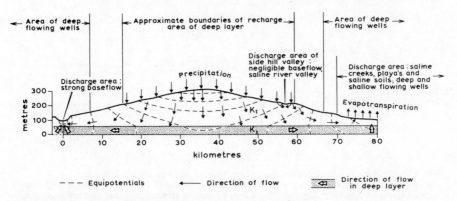

Fig. 7.24. *The prairie profile. (From an original diagram by Meyboom [106]. Reproduced by permission of Information Canada.)*

217

that groundwater flow is essentially downward through the low permeability material and lateral and upward through the more permeable underlying layer. Recharge and discharge areas are characterized by a decrease in head with depth and by an increase in head with depth, respectively, whilst those parts of the flow system in which there is virtually no change of head with depth, as for example in much of the lower layer, may be regarded as transmission zones.

Since most of the natural discharge occurs by means of evapotranspiration, Meyboom examined areas in which this component of the groundwater balance appeared to be important. He concentrated particularly on the occurrence of willow rings, areas of saline soil which occur where there is a net upward movement of mineralized groundwater, as in the major areas of regional groundwater discharge, and the occurrence of lakes and bogs and their relationship to groundwater flow [108] [109] [110].

The work of Toth and Meyboom was itself extended and generalized in a major contribution by Freeze and Witherspoon [45] [45A] [46] which introduced a more versatile mathematical modelling technique based on numerical solutions. Their model is capable of determining steady-state flow patterns in a

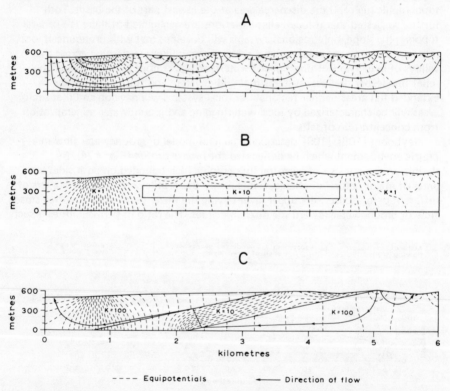

Fig. 7.25. *Potential field diagrams illustrating regional groundwater flow patterns in (A) a homogeneous, isotropic medium with a hummocky water table, (B) a high permeability lens with a hummocky water table, and (C) an area of sloping stratigraphy. See text for explanation. (From original diagrams by R. A. Freeze and P. A. Witherspoon,* Water Resources Res., **3**, *p. 625, Fig. 1, p. 629, Fig. 4, p. 630, Fig. 5, 1967. © American Geophysical Union.)*

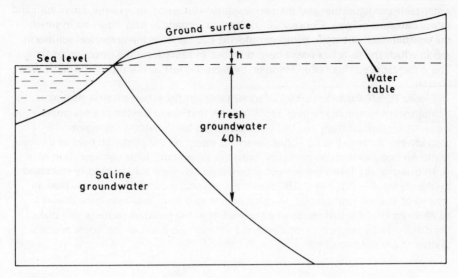

Fig. 7.26. *Diagrams illustrating, in a simplified manner, the Ghyben-Herzberg hydrostatic relationship in (A) a homogeneous coastal aquifer and (B) a layered coastal aquifer. ((B) is based on an original diagram by M. A. Collins and L. W. Gelhar,* Water Resources Res., 7, p. *972, Fig. 1, 1971. © American Geophysical Union.)*

three-dimensional, non-homogeneous anisotropic groundwater basin, with any water-table configuration given knowledge of the dimensions of the basin, the water-table configuration and the permeability configuration resulting from the subsurface stratigraphy. Figure 7.25 shows three potential field diagrams in which the equipotential net is the computer-plotted output from the numerical solution and in which the flow lines have been sketched in to indicate the direction of flow. The water-table configuration is shown by the solid line at the top of the contoured section.

Figure 7.25A shows the effect of water-table configuration on flow through a homogeneous isotropic medium and illustrates that the existence of a hummocky water-table configuration results in numerous sub-basins within the major groundwater system. Figure 7.25B shows the effect on the potential field of a lenticular body of high permeability, with the same water-table configuration as in A. In this case the flow lines are not shown although they may be readily envisaged by the reader. Finally, Fig. 7.25C shows the regional groundwater flow pattern in an area of sloping stratigraphy. In this case just two flow lines have been drawn in to illustrate that the difference of a few metres in the point of recharge will make the difference between the water entering a minor sub-basin or the major regional system of groundwater flow.

7.19 Groundwater flow in coastal aquifers

In natural conditions there is a hydraulic gradient towards the coast and the resulting seaward groundwater flow effectively limits the subsurface landward encroachment of saline water. Thus in normal unconfined groundwater conditions, with a water table sloping towards sea level at the coast, the groundwater body takes the form of a lens of fresh water 'floating' on more saline water beneath. The hydrostatic relationship between the fresh and saline water was investigated independently by Ghyben [6] and Herzberg [55], and now bears their names (i.e., the Ghyben-Herzberg relationship). As can be seen from Fig. 7.26A, saline groundwater is encountered, not at sea level, but at a depth below sea level equivalent to about forty times the height of water table above sea level. This relationship represents the condition of approximate hydrostatic equilibrium between the lighter fresh groundwater (density 1.00), and the heavier saline groundwater (density 1.03), and can thus be compared with the condition of isostatic equilibrium, whereby lighter sial 'floats' on heavier sima. In stratigraphically layered aquifers a number of investigators have indicated the presence of multiple saline wedges [26] [93] [136]. Figure 7.26B illustrates a simple situation in the presence of a limited semi-permeable layer and assuming steady flow and a sharp interface between fresh and saline groundwater. Collins and Gelhar [26] pointed out that whereas the saline wedge in the upper unconfined aquifer must intrude inland from the shoreline, the wedge in the lower aquifer may intrude even further inland, or as shown in the diagram may exist seaward of the shoreline when a sufficiently large freshwater head is available at the landward end of the semi-permeable layer.

The assumption of a sharp interface between saline and fresh groundwater is, in most cases, unrealistic. Tidal fluctuations as well as variations in recharge and discharge continually disturb the balance between the fresh water and the seawater

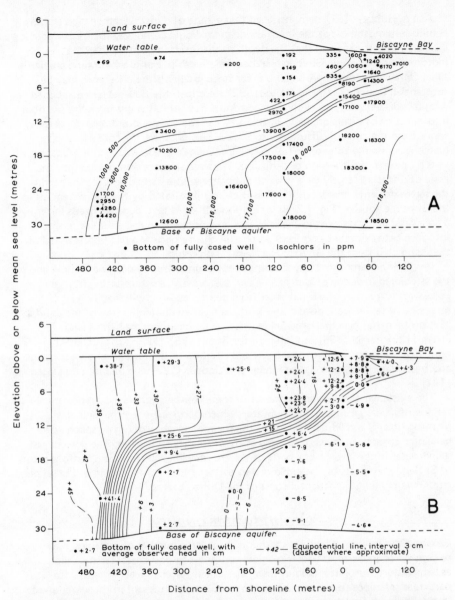

Fig. 7.27. *Hydrodynamic relationships between fresh and saline groundwater in the Biscayne aquifer, Florida. (A) Isochlors illustrate the zone of diffusion; (B) Equipotential lines in the zone of diffusion. (From original diagrams by Kohout and Klein [84] and by Stringfield and LeGrand [151].)*

and cause the interface to fluctuate. As a result of these fluctuations and the diffusion of salt water, the sharp interface is destroyed and a transitional zone of diffusion of brackish water is created [152]. This is well illustrated in Fig. 7.27A which shows the zone of diffusion in the Biscayne aquifer near Miami, Florida.

Again, Hubbert [61] demonstrated that a state of *dynamic* rather than static equilibrium must exist at the interface or there would be no way for fresh water to discharge into the sea and that the depth to the interface is more than that indicated by the simple Ghyben-Herzberg relationship. For low hydraulic gradients the differences in depth are small but for steep gradients the difference may be substantial. Cooper [27] and Kohout [83] described the flow pattern associated with this discrepancy, as involving a continual flow of salt water from the sea floor into the zone of diffusion. Only in this way can the continuous discharge of salty water from the zone of diffusion into the sea be explained, the concentration gradient across the zone of diffusion being too weak to account for the transport of salts by means of dispersion or diffusion. Thus the salts are transported into the zone of diffusion largely by a flow of salt water with a consequent loss of head in the salt-water environment [28]. This is clearly illustrated by the equipotentials in the Biscayne aquifer (Fig. 7.27B) which show that, in fact, negative heads in terms of seawater densities exist in the seawater environment. In this situation the interface between saline and fresh water will clearly occupy a position seaward of that estimated from the Ghyben-Herzberg relationship.

It will be apparent that any sustained reduction in fresh groundwater flow and the associated lowering of water table and piezometric levels resulting from the excessive abstraction of groundwater from coastal aquifers will readily lead to the incursion of saline groundwater and to the long-term contamination of the aquifer. This problem has been investigated in many areas including the Netherlands [9], Japan [139], Israel [136], and the United States [85] [122] [151]. Recent numerical solutions for the movement and position of the interface between saline and fresh water include those of Pinder and Cooper [123] and Shamir and Dagan [141].

The existence of saline groundwater in some coastal aquifers may represent the residue of an earlier invasion of seawater, which occurred when the land was relatively lower, and which now takes the form of a wedge of highly saline water beneath a superficial seaward flowing body of fresh groundwater. The presence of saline water in the chalk along parts of the Humberside, Norfolk, and Suffolk coasts of England may probably be explained in this way [16] as may some of the saline groundwater along the Atlantic coast of the United States [144].

Groundwater quality

Groundwater quality, which is determined by a large number of variables, may be considered in terms of bacteriological content, physical characteristics such as temperature, turbidity, colour, taste, and odour, or chemical content, with particular reference to the quantity of toxic substances present in the water and to its total salinity. For the most part, the quality of groundwater is of significance to the *applied* hydrologist concerned with the availability of water for supply purposes. But the physical hydrologist is also concerned with major departures from normal conditions, and with the reasons for these departures, especially where hydrological or hydrogeological factors have played an important part and particularly where groundwater quality itself can be used as an indicator of hydrological conditions. Apart from the bacteriological content of groundwater, which may be changed considerably by the accidental or deliberate disposal of waste material, especially in areas of high hydraulic conductivity or shallow groundwater, when the filtering action of the aquifer material is least effective, the two characteristics which are of most concern to the hydrologist are the *chemical composition* and the *temperature* of groundwater.

7.20 Chemical composition

In normal conditions, the composition of groundwater is different to that of surface water, by virtue of the larger proportion of minerals and salts in solution although, as Davis and De Wiest [34] observed, the number of *major* dissolved constituents in groundwater is quite limited. The concept of the biogeochemical

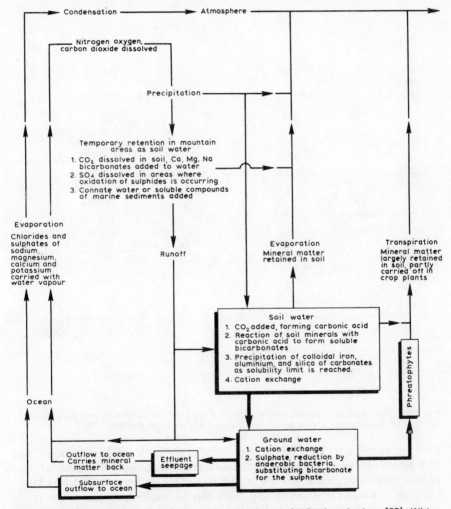

Fig. 7.28. *The geochemical cycle. (From an original diagram by Davis and others [32]. With acknowledgment to the US Geological Survey.)*

223

cycle, first proposed by Vernadskii in 1934 [172] and illustrated in Fig. 7.28, emphasizes the great complexity of the circulation of dissolved mineral matter within the hydrological cycle and also the great variety of sources for the mineral matter found in solution in groundwater. Thus some dissolved mineral matter is derived from ocean evaporation while some is transferred from the oceans in the liquid phase. On the windward coast of Barbados, for example, the infiltration of salt spray is believed to be at least partly responsible for extensive areas of shallow saline groundwater [140]. Again, some material is derived from the atmosphere

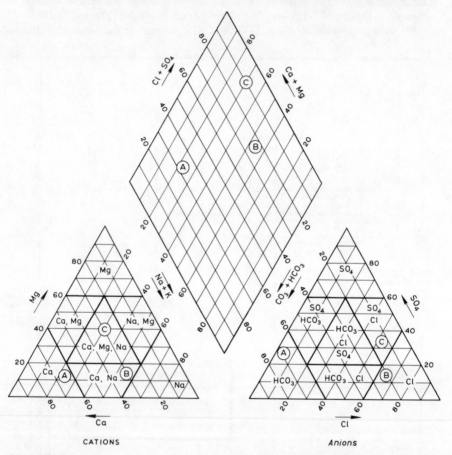

CATIONS   Anions

Fig. 7.29. *Piper trilinear diagram used to represent the chemical composition of groundwater. See text for explanation. (From an original diagram by Davis and DeWiest [34].)*

itself and some from biological decomposition at the earth's surface. However, one of the main sources of groundwater mineral content is normally the action of percolating rainfall on the first few centimetres of the soil and subsoil layers [15], although if rainfall percolates rapidly, there may be insufficient opportunity for its acid content to be neutralized and, in this case, chemical reactions will be delayed until deeper subsoil layers are reached, and these then play an important part in determining the chemical composition. Table 7.1 lists most of the major, secondary, minor, and trace constituents dissolved in groundwater and gives some information on the normal ranges of concentration.

Table 7.1. *Dissolved Constituents in Groundwater. (After Davis and DeWiest [34].*

Major constituents (range of concentration 1.0 to 1000 ppm)

Sodium	Bicarbonate
Calcium	Sulphate
Magnesium	Chloride
Silica	

Secondary constituents (range of concentration 0.01 to 10.0 ppm)

Iron	Carbonate
Strontium	Nitrate
Potassium	Fluoride
Boron	

Minor constituents (range of concentration 0.000 01 to 0.1 ppm)

Antimony	Lithium
Aluminium	Manganese
Arsenic	Molybdenum
Barium	Nickel
Bromide	Phosphate
Cadmium	Rubidium
Chromium	Selenium
Cobalt	Titanium
Copper	Uranium
Germanium	Vanadium
Iodine	Zinc
Lead	

Trace constituents (range of concentration generally less than 0.001 ppm)

Beryllium	Ruthenium
Bismuth	Scandium
Cerium	Silver
Cesium	Thallium
Gallium	Thorium
Gold	Tin
Indium	Tungsten
Lanthanum	Ytterbium
Niobium	Yttrium
Platinum	Zirconium
Radium	

The chemical composition of groundwater may be effectively represented on the trilinear diagram developed by Piper [124] who proposed that groundwater be treated as though it contained three cation constituents i.e., sodium, calcium, and magnesium and three anion constituents, i.e., bicarbonate, sulphate, and chloride. Silica, the other major constituent (see Table 7.1), is omitted from consideration and the less abundant cation and anion constituents are summed with the major constituents to which they are related, e.g., barium with calcium, potassium with sodium, carbonate with bicarbonate, etc. As can be seen in Fig. 7.29, the Piper trilinear diagram combines three plotting fields, two triangular fields and an intervening diamond-shaped field, all having scales of 0 to 100 representing per

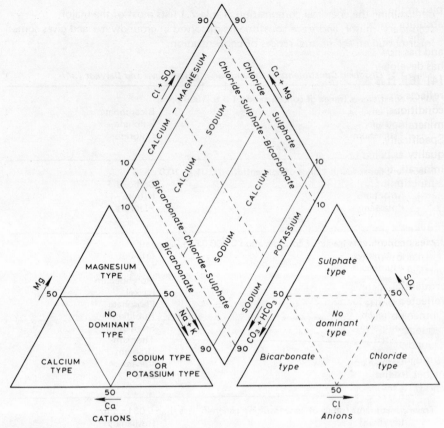

Fig. 7.30. *Trilinear diagram showing hydrochemical facies. (From original diagrams by Back [4], with acknowledgment to the U.S. Geological Survey, and by P. A. Domenico,* Concepts and Models in Groundwater Hydrology, *McGraw-Hill, New York, 1972.)*

centage of total equivalents per million.* The proportion of each cation constituent is plotted as a single point in the left-hand triangular field and of each anion constituent in the right-hand triangular field. The central diamond-shaped field is used to show the overall chemical character of the groundwater by a third single-point plotting which lies at the intersection of the rays projected from the cation and anion points. Thus in Fig. 7.29 the groundwater indicated by point A would represent a calcium bicarbonate water, by point B a calcium, sodium, chloride water, while point C would represent sodium, calcium, magnesium, chloride, and sulphate water [34]. It should be noted that relative and not absolute concentrations are indicated in the trilinear diagram so that it is often convenient to

---

*Equivalents per million (epm) are a convenient way of expressing the results of chemical analysis and are calculated by dividing parts per million by the equivalent weight of the ion under consideration, when equivalent weight is equal to the atomic weight divided by the valence [169]. For example, calcium has an atomic weight of 40.08 and a valence of 2 so that its equivalent weight is 20.04. Thus the concentration in groundwater containing 80 ppm is 4 epm [37].

plot in the central field, not a point, but a circle whose area is proportional to the absolute concentration of the water [169].

The type of single point nomenclature referred to above is of limited usefulness and has now been largely superseded by the concept of *hydrochemical facies* which has developed from the work of Chebotarev [22], Schoeller [137] and Back [3] [4] [5]. It is assumed that the chemical composition of groundwater at any point reflects a tendency toward equilibrium with the matrix rocks under the prevailing conditions. In general, the type of facies that develops is controlled mainly by the mineralogy of the rocks, and its distribution is controlled by the flow pattern [37]. Specifically, Toth [165] listed the main factors that determine point groundwater quality as being types of soluble minerals, amounts of soluble minerals, solubility of minerals, temperature of the rock, area of contact between rock and water, which is a function of porosity, antecedent composition of the water, antecedent temperature of the water, fluid pressure, and flow velocity and geometry both of which are a function of the groundwater potential. The Piper trilinear diagram provides a useful framework for illustrating the classification of hydrochemical facies proposed by Back [4] as can be seen in Fig. 7.30.

In addition, Chebotarev [22] and Schoeller [137] referred to the essentially vertical zonation of groundwater which may occur as a result of its progressive sequential chemical modification between recharge and discharge areas and which is reflected in the tendency of groundwater toward the composition of sea water. Domenico [37] observed that in terms of facies development the Chebotarev sequence may be regarded as three vertical zones: (a) an upper zone of bicarbonate groundwater with low mineralization associated with rapid groundwater circulation through well-leached rocks, (b) an intermediate zone of sulphate groundwater and higher total mineralization associated with a less intensive groundwater circulation, and (c) a lower zone of highly mineralized chloride groundwater associated with near stagnant groundwater circulation and virtually unleached rocks. The sequence provides a useful conceptual model but is rarely observed in its entirety, largely as a result of the dominant effects of the local physical environment.

Reviewing the concept of hydrochemical facies Domenico [37] suggested that, although no quantitative limits can be set that would apply to geochemical systems in general, three main points of agreement have emerged from the work done in this field, particularly by Chebotarev, Schoeller, and Back. First, the concentration of dissolved mineral matter is directly proportional to the length of the flow path and to the residence time of the groundwater. Certainly, groundwater which moves only very slowly, and which does not play a very active part in the hydrological cycle, tends to become highly mineralized and Buchan [16] quoted examples of such water occurring below the level of circulation and drainage in a number of areas where the movement of water is restricted by an overburden of clay or similar relatively impermeable material. Connate water also falls into this category and is normally found to be highly saline [70]. Second, within a given system, a major recharge area is characterized by groundwater which is lower in dissolved solids than that in a minor recharge area and the recharge areas themselves may be leached of all soluble mineral matter. These effects are clearly seen in arid and semi-arid areas, where evaporation is high and rainfall is low, and where, as a result, infiltration and percolation are not usually sufficient to flush soluble salts from the sediments, so that a highly mineralized groundwater tends to result [8]. Where,

227

however, surface runoff and infiltration are concentrated, as in wadis and playas, there is sufficient infiltration and percolation to produce mounds of fresh groundwater, which moves downward and outward in the direction of the hydraulic gradient. At the same time, the mineral content of the groundwater increases as a result of contact with the salts in the sediments [8]. Third, it is quite common for the ratio of sulphate to chloride to decrease in the direction of groundwater flow, as a result of chemical reduction. Thus, under the chalk outcrop of the London basin [16], the groundwater is hard, but down dip there is a change to softer water, rich in sodium salts, with a higher chloride and total solid content, which is maintained for some distance under the overlying layer of Tertiary sands and clays. At the eastern end of the synclinal basin, the groundwater has an even higher total solid content in which sodium chloride predominates. It is believed that this may be water which has not yet been forced out by the pressure of incoming groundwater from the chalk outcrop areas around the rim of the basin.

The foregoing discussion has illustrated that the chemical composition of groundwater tends towards equilibrium with the matrix rocks through which it is flowing. From this it may be inferred that, in some circumstances, it will be possible to reconstruct the flow path of groundwater from an analysis of its chemical composition. Work along these lines is still at a comparatively elementary stage [95], although as early as 1960 Back [2] developed a groundwater flow pattern from chemical data which closely resembled the pattern developed from data on groundwater energy potential. Subsequently Henningsen [54] distinguished groundwater from two separate recharge areas in Texas on the basis of chemical composition, while Toth [164] distinguished areas of groundwater recharge and discharge in Alberta using the same criterion. Maxey and Mifflin [98] correlated groundwater quality changes with distance from source in Nevada and suggested that the method showed considerable promise in the delineation of flow systems. Again, using chemical data Swenson [154] was able to shed new light on groundwater flow in the much-studied artesian basin of the Dakotas, while more recent work by Horowitz [59] indicated that an examination of palynological spectra combined with geochemical analysis might yield still more useful information on groundwater movement.

Finally, it should be noted that in some circumstances variations in the chemical composition of groundwater may themselves be responsible for groundwater movement since free energy decreases with an increase in the concentration of dissolved material [37]. Thus in the presence of a semi-permeable membrane *chemical osmosis* results in movement into the more saline solution. Domenico [37] discussed this phenomenon and referred to the situation where if the salinities of water in two aquifers separated by a thin layer of shale or clay are unequal, a net movement of water may occur from the low salinity aquifer to the high salinity aquifer. A number of pressure and salinity anomalies in sedimentary basins can only be explained in terms of chemical osmosis.

7.21 Temperature

Temperature is probably the most conservative property of groundwater, normally showing only a small seasonal and diurnal range in the case of shallow groundwater, and virtual constancy in the case of deeper groundwater below about 30 m [36].

228

Studies in the United States showed that the temperature of groundwater occurring at depths of 10 to 20 m will normally exceed the mean annual air temperature of that locality by some 1° to 2°C [159]. Accordingly, groundwater temperatures tend to be lower than air temperatures during the summer and higher than air temperatures during the winter months. At greater depths, groundwater temperatures reflect the geothermal gradient in the earth's crust: the average increase of water temperature with depth is approximately 3°C for every 100 m, but varies widely with local conditions, and in some places may be as high as 9°C for every 100 m [36].

It thus follows that marked spatial variations of groundwater temperature are largely the result of the groundwater flow pattern and particularly of the extent to which vertical movement occurs. As Domenico [37] observed, where groundwater temperature is uniform and approximates mean annual air temperature, the depth of circulation is probably small, as in local systems or in close proximity to recharge areas. Where temperature varies widely, both shallow and deep circulation may be present as in intermediate systems, while uniformly high temperature is believed to mark deeply circulating groundwater, such as that found in regional flow systems.

Although the role of groundwater as an agent of heat transfer in normal thermal areas has long been recognized by geophysicists [18], methods to determine the vertical component of groundwater velocity from thermal data are comparatively recent [13] [113] [119] [147]. Working in the Illinois basin, Cartwright [19] tested the hypothesis that areas of vertical groundwater movement should be reflected by temperature anomalies by comparing isothermal maps constructed from well temperatures with theoretical isothermal maps for the same depths constructed by projecting temperature gradients from the ground surface. Differences between the isothermal distributions were expressed as positive or negative anomalies which were assumed to be related to areas of upward movement or discharge and downward movement or recharge respectively. Combined computer modelling and field studies of groundwater temperatures in northern Ontario by Parsons [121] demonstrated that the effect of groundwater flow on the temperature at any point in a groundwater basin depends on the direction and magnitude of flow, the disposition of temperatures at the water table relative to areas of recharge and discharge, the disposition of flow systems within the basin, the depth of the basin and on whether groundwater is discharged directly through seepage and springs or indirectly by evapotranspiration.

Factors which result in *extremes* of temperature are of particular interest to the hydrologist. At the lower end of the scale, freezing is obviously important, especially in high latitude areas where permafrost occurs to considerable depths and where three groundwater zones may be delimited [159]. Below the permafrost which is believed to extend to depths of 600 m in parts of Siberia [114], *subpermafrost groundwater* occurs quite normally, apart from the fact that it is virtually isolated from the usual factors of recharge and discharge at the ground surface. Some groundwater may be trapped within the permafrost zone itself; this *intrapermafrost groundwater* occurs in areas of 'weakness' within the permafrost, for example, near rivers, lakes, faults, or fractures, and normally moves, like artesian water, as a result of differences of hydraulic head. The groundwater in this zone may give rise to the remarkable cryo-laccoliths or pingos [58] [77] found in Greenland and in northern Canada and Siberia. These are mounds or cones covered with deeply fissured layers of surface deposits, which have been upheaved by the intrusion of a thick lens of ice, and which are believed to grow as a result of the artesian 'intrusion', and subsequent freezing and expansion of summer melt-water in the permafrost zone. Third, groundwater exists in a shallow zone above the permafrost, which thaws each summer. During the warm season, therefore, this *suprapermafrost groundwater* behaves as a normal shallow groundwater body interacting with surface water and climatic factors but, during the long winter period, it is frozen solid and plays no active part in the hydrological cycle.

At the other extreme, the high temperatures associated with thermal springs have long attracted attention, the main problems being to explain both the source of water and the source of heat. It is now believed that most thermal springs result from the underground heating of meteoric groundwater [29] [35] [173] [174] [176], although some are almost certainly a mixture of meteoric water with juvenile and hot, connate water [59] often rising from deep zones of metamorphism and granitization [58]. Possible sources of heat, are, first, the normal increase in temperature with depth in the earth's crust: second, an underlying body of very hot or even molten rock, third, the effects of faulting, fourth, chemical reactions in the earth's crust and, finally, energy derived from the disintegration of radioactive elements, although of these, only the first two have been shown to be of any marked general significance [78], [175].

7.22 Selected groundwater measurements

It will be clear from the discussions in this chapter that a wide range of measured data is needed in order to understand and explain the occurrence, storage, and movement of groundwater. Unfortunately, space permits only a brief listing of the principal measurement techniques.

In the case of *porosity* and *specific yield* determinations standard procedures are involved which have been well documented elsewhere [105] [135]. The measurement of *piezometric head* involves the precise location of the water surface in a well or bore-hole penetrating the aquifer. In the unconfined situation the main problems concern the definition of the upper surface of the zone of saturation and measurement errors introduced by the presence of the observation bore itself [24]. With confined aquifers the problems vary depending on whether the piezometric surface is above or below ground level since in the former case the piezometer tube needs either to be extended above ground level or capped with a pressure indicating gauge.

The *hydraulic conductivity* of an aquifer is most reliably measured by means of well-point techniques. In the case of deep groundwater the conventional pumping test has been well documented [62] [64] [66] [67] [68], while in the case of shallow groundwater the piezometer [92], tube [47], and auger hole methods are used. An excellent review of the theory and practice of the auger hole method was given by Bonell [11] who compared results using the methods of Kirkham and van Bavel [7] [82] and Ernst [41]. Laboratory measurements of the hydraulic conductivity of small cores may also be made, using fixed- or falling-head permeameters.

Finally, the *speed and direction of groundwater movement* may be determined using tracers [12] [34] [48] [86] [133] [155] [159] [181], cartographic techniques such as flow nets and potentiometric maps, or laboratory equipment such as parallel plates and electrical analogues to model groundwater flow patterns [75] [34].

References

1. AMERICAN SOCIETY OF CIVIL ENGINEERS, *Hydrology Handbook,* New York, 1949.
2. BACK, W., Origin of hydrochemical facies of groundwater in the Atlantic coastal plain, paper presented at the 21st Congress of the International Geologic, Copenhagen, 1960. Quoted by Maxey [95].
3. BACK, W., Techniques for mapping of hydrochemical facies, *USGS Prof. Pap.,* 424-D: 380-382, 1961.
4. BACK, W., Hydrochemical facies and groundwater flow patterns in northern part of Atlantic Coastal Plain, *USGS Prof. Pap.,* 498-A, 1966.
5. BACK, W., and B. HANSHAW, Chemical geohydrology, in Chow, V. T. (Ed.), *Advances in Hydroscience,* **2**: 49-109, 1965.
6. BADON GHYBEN, W., Nota in verband met de voorgenomen putboring nabij Amsterdam, *Tijdschrift van het Koninklijk Inst. van Ingenieurs,* The Hague, 1888-89. Quoted by Todd [159].
7. BAVEL, C. H. M. van, and D. KIRKHAM, Field measurement of soil permeability using auger holes, *Proc. SSSA,* **13**: 90-96, 1949.
8. BERGSTROM, R. E., and R. E. ATEN, Natural recharge and localization of fresh ground water in Kuwait, *J. Hydrol.,* **2**: 213-231, 1965.
9. BIEMOND, C., Dune water flow and replenishment in the catchment of the Amsterdam water supply, *J. Inst. Water Eng.,* **11**: 195-213, 1957.
10. BONELL, M., An assessment of possible factors contributing to well level fluctuations in Holderness boulder clay, East Yorkshire, *J. Hydrol.,* **16**: 361-368, 1972.
11. BONELL, M., The application of the auger hole method in Holderness glacial drift, *J. Hydrol.,* **16**: 125-146, 1972.
12. BOROWCZYK, M., J. GRABSZAK, and A. ZUBER, Radio isotope measurement of groundwater flow direction by the single-well method, *Inst. Nuclear Res. Rept.* No. 502/VI, Warsaw, 1964. Quoted by Wurzel and Ward [181].
13. BREDEHOEFT, J. D., and I. S. PAPADOPULOS, Rates of vertical groundwater movement estimated from the earth's thermal profile, *WRR,* **1**: 325-328, 1965.
14. BROWN, R. H., A. A. KONOPLYANTSEV, J. INESON, and V. S. KOVALENVSKY (Eds.), *Groundwater Studies,* Studies and Reports in Hydrology, No. 7, UNESCO, Paris, 1972.
15. BUCHAN, S., Variations in the mineral content of some groundwaters, *Proc. Soc. Water Treatment and Examination,* **7**: 11-29, 1958.
16. BUCHAN, S., Geology in relation to groundwater, *J. Inst. Water Eng.,* **17**: 153-64, 1963.
17. BUCHAN, S., Hydrogeology and its part in the hydrological cycle, Informal discussion of the Hydrological Group, *Proc. Inst. Civil Eng.,* **31**: 428-431, 1965.
18. BULLARD, E. C., Heat flow in South Africa, *Proc. Roy. Soc. London,* **A173**: 474-502, 1939.
19. CARTWRIGHT, K., Groundwater discharge in the Illinois basin as suggested by temperature anomalies, *WRR,* **6**: 912-918, 1970.
20. CEDERSTROM, D. J., Groundwater resources of the Fairbanks area, Alaska, 1950, *USGS Water-Supply Pap.,* 1950, 84 pp., 1963.
21. CEDERSTROM, D. J., P. M. JOHNSTON, and S. SUBITZKY, Occurrence and development of groundwater in permafrost regions, *USGS Circ.* 275, 30 pp., 1953.
22. CHEBOTAREV, I. I., Metamorphism of natural waters in the crust of weathering, *Geochim. Cosmochim. Acta,* **8**: 22-48, 137-170 and 198-212, 1955.
23. CHILDS, E. C., The physics of land drainage, in Luthin, N. J. (Ed.), *Drainage of Agricultural Lands,* Amer. Soc. Agron., Madison, pp. 1-78, 1957.
24. CHILDS, E. C., *An Introduction to the Physical Basis of Soil Water Phenomena,* Wiley, London, 493 pp., 1969.

231

25. CHILDS, E. C., and N. COLLIS-GEORGE, The permeability of porous materials, *Proc. Roy. Soc. A,* **201**: 392-405, 1950.
26. COLLINS, M. A., and L. W. GELHAR, Seawater intrusion in layered aquifers, *WRR,* **7**: 971-979, 1971.
27. COOPER, H. H., A hypothesis concerning the dynamic balance of fresh water and salt water in a coastal aquifer, *JGR,* **64**: 461-467, 1959.
28. COOPER, H. H., F. A. KOHOUT, H. R. HENRY, and R. E. GLOVER, Sea water in coastal aquifers, relation of salt water to fresh groundwater, *USGS Water Supply Pap.,* 1613C, 84 pp., 1964.
29. CRAIG, H., G. BOATO, and D. E. WHITE, Isotopic geochemistry of thermal waters (abstr.), *Bull. Geol. Soc. Amer.,* **65**: 1243, 1954.
30. CUEVAS, J. A., Foundation conditions in Mexico City, *Proc. Intern. Conf. Soil Mech.,* 3, Cambridge, Mass., 1936.
31. DARCY, H., *Les Fontaines publiques de la ville de Dijon,* V. Dalmont, Paris, 1856.
32. DAVIS, G. H., J. H. GREEN, F. H. OLMSTED, and D. W. BROWN, Groundwater conditions and storage capacity in the San Joaquin Valley, California, *USGS Water Supply Pap.,* 1469, 1959.
33. DAVIS, G. H., J. B. SMALL, and H. B. COUNTS, Land subsidence related to decline of artesian pressure in the Ocala limestone at Savannah, Georgia, *Geol. Soc. Amer., Eng. Geol. Case Histories,* **4**: 1-8, 1963.
34. DAVIS, S. N., and R. J. M. DE WIEST, *Hydrogeology,* Wiley, New York, 463 pp., 1966.
35. DEGRYS, A., Some observations on the hot springs of central Chile, *WRR,* **1**: 415-428, 1965.
36. DIXEY, F., *A Practical Handbook of Water Supply* (2nd Ed.), Murby, London, 1950.
37. DOMENICO, P. A., *Concepts and Models in Groundwater Hydrology,* McGraw-Hill, New York, 405 pp., 1972.
38. DOMENICO, P. A., D. A. STEPHENSON, and G. B., MAXEY, Groundwater in Las Vegas Valley, *Desert Res. Inst. Tech. Rept.,* 7, Reno, 1964.
39. DUPUIT, J., *Etudes théoriques et pratiques sur le mouvement des eaux,* Edn. 2, Dunod, Paris, 1863.
40. ECKIS, R., South Coastal Basin investigation, geology and groundwater storage capacity of valley fill, *Bull.* 45, Calif. Div. Water Resources, Sacramento, 279 pp. 1934.
41. ERNST., L. F., *A new formula for the calculation of the permeability factor with the auger hole method,* TNO Groningen, 1950.
42. FORCHHEIMER, P., *Hydraulik,* Teubner, Leipzig and Berlin, 1914.
43. FREEZE, R. A., The mechanism of natural groundwater recharge and discharge I. One-dimensional, vertical, unsteady, unsaturated flow above a recharging or discharging groundwater flow system, *WRR,* **5**: 153-171, 1969.
44. FREEZE, R. A. & J. BANNER, The mechanism of natural groundwater recharge and discharge 2. Laboratory column experiments and field measurements, *WRR,* **6**: 138-155, 1970.
45. FREEZE, R. A., and P. A. WITHERSPOON, Theoretical analysis of regional groundwater flow, 1. Analytical and numerical solutions to the mathematical model. *WRR,* **2**: 641-656, 1966.
45A. FREEZE, R. A., and P. A. WITHERSPOON, Theoretical analysis of regional groundwater flow, 2. Effect of water-table configuration and subsurface permeability variation, *WRR,* **3**: 623-634, 1967.
46. FREEZE, R. A., and P. A. WITHERSPOON, Theoretical analysis of regional groundwater flow, 3. Quantitative interpretation, *WRR,* **4**: 581-590, 1968.
47. FREVERT, R. K., and D. KIRKHAM, A field method for measuring the permeability of soil below a water table, *Proc. Highway Res. Bd.,* **28**: 433-442, 1948.
48. GAT, J. R., Comments on the stable isotope method in regional groundwater investigations, *WRR,* **7**: 980-993, 1971.
49. GEORGE, W. O., and F. E. ROMBERG, Tide-producing forces and artesian pressures, *Trans. AGU,* **32**: 369-371, 1951.
50. GILLULY, J., and U. S. GRANT, Subsidence in Long Beach Harbor area, California, *Bull. Geol. Soc. Amer.* **60**: 461-521, 1949.
51. GODWIN, H., Studies in the ecology of Wicken Fen, I. The groundwater level of the Fen, *J. Ecol.* **19**: 449-473, 1931.

52. HANTUSH, M. S., Modification of the theory of leaky aquifers, *JGR,* **65**: 3713-25, 1960.
53. HEM, J. D., Study and interpretation of chemical characteristics of natural water, *USGS Wat. Sup. Pap.,* 1473, 1959.
54. HENNINGSEN, E. R., Water diagenesis in Lower Cretaceous Trinity aquifers of central Texas, *Baylor Geol. Studies Bull.,* **3**, 1962.
55. HERZBERG, B., Die Wasserversorgung einiger Nordseebäder, *J. Gasbeleuchtung und Wasserversorgung,* **44**: 815-819, 1901. Quoted by Todd [159].
56. HEWLETT, J. D., and A. R. HIBBERT, Factors affecting the response of small watersheds to precipitation in humid areas, in Sopper, W. E. and H. W. Lull (Eds.), *Forest Hydrology,* pp. 275-290, Pergamon, Oxford, 1967.
57. HILLEL, D., *Soil and Water,* Academic Press, New York, 288 pp., 1971.
58. HOLMES, A., *Principles of Physical Geology,* (2nd Ed.), Nelson, London, 1965.
59. HOROWITZ, A., Palynological tracing of saline water sources in Lake Kinneret region (Israel), *J. Hydrol.,* **10**: 177-184, 1970.
60. HORTON, J. H., and R. H. HAWKINS, Flow path of rain from the soil surface to the water table, *Soil Sci.,* **100**: 377-383, 1965.
61. HUBBERT, M. K., The theory of groundwater motion, *J. Geology,* **48**: 785-944, 1940.
62. INESON, J., Some observations on pumping tests carried out on Chalk wells, *J. Inst. Water Eng.* **7**: 215-225, 1953.
63. INESON, J., Darcy's law and the evaluation of 'Permeability', *IASH Symposia Darcy,* **2**: 165-172, 1956.
64. INESON, J., Yield-depression curves of discharging wells, with particular reference to Chalk wells, and their relationship to variations in transmissibility, *J. Inst. Water Eng.* **13**: 119-163, 1959.
65. INESON, J., Fluctuations of groundwater levels due to atmospheric pressure changes from nuclear explosions, *Nature,* **195**: 1082, 1962.
66. INESON, J., Some Aspects of groundwater hydrology and hydrogeology, *Water and Wat. Eng.,* **66**: 333-338, 1962.
67. INESON, J., The techniques used in pumping-test observations, *Wat. Res. Assoc., Special Rept.,* **2**: 141-145, 1962.
68. INESON, J., Applications and limitations of pumping tests: (b) Hydrogeological significance, *J. Inst. Water Eng.,* **17**: 200-215, 1963.
69. INESON, J., The assessment of aquifer yield, *Inst. Civil Eng., Hydrological Group,* Cyclostyled notes for informal discussion, 1966.
70. INESON, J., and D. A. GRAY, Electrical investigations of borehole fluids, *J. Hydrol.,* **1**: 204-218, 1963.
71. INSTITUTE OF GEOLOGICAL SCIENCES, Private communication from R. Kitching, 1973.
72. JACOB, C. E., Fluctuations in artesian pressure produced by passing railroad-trains as shown in a well on Long Island, New York, *Trans. AGU,* **20**: 666-674, 1939.
73. JACOB, C. E., On the flow of water in an elastic artesian aquifer, *Trans. AGU,* **21**: 574-586, 1940.
74. JACOB, C. E., Correlation of groundwater levels and precipitation on Long Island, New York, *Trans. AGU,* **25**: 928-939, 1944.
75. KARPLUS, W. J., *Analog Simulation,* McGraw-Hill, New York, 427 pp., 1958.
76. KAZMANN, R. G., Contribution to the Report of the Committee on Groundwater 1943-44, *Trans. AGU,* **25**: 712-721, 1944.
77. KENNEDY, B. A., Periglacial morphometry, in Chorley, R. J. (Ed.), *Water, Earth and Man,* pp. 381-388, Methuen, London, 1969.
78. KENT, L. E., The thermal water of the Union of South Africa and South West Africa, *IASH Gen. Ass. Oslo,* **3**: 201-228, 1948.
79. KENT, L. E., and J. F. ENSLIN, Groundwater prospecting methods used in the Republic of South Africa, *Republic of South Africa, Dept. of Water Affairs Tech. Rept.* No. 32, 1962.
80. KIDDER, E. H., and W. F. LYTLE, Drainage investigations in the plastic till soils of northeastern Illinois, *Agric. Eng.,* **30**: 384-389, 1949.
81. KING, F. H., Principles and conditions of the movements of groundwater, *USGS 19th Ann. Rept. 1897-98,* 1899.

82. KIRKHAM, D., and C. H. M. van BAVEL, Theory of seepage into auger holes, *Proc. SSSA,* **13**: 75-89, 1948.

83. KOHOUT, F. A., Cyclic flow of salt water in the Biscayne aquifer of south-eastern Florida, *JGR,* **65**: 2133-2141, 1960.

84. KOHOUT, F. A., and H. KLEIN, Effect of pulse recharge on the zone of diffusion in the Biscayne aquifer, *IASH Sympos. Haifa, Publn.,* **72**: 252-270, 1967.

85. LEGGETTE, R. M., Salt water encroachment in the Lloyd Sand on Long Island, New York, *Water Works Eng.,* **100**: 1076-1079 and 1107-1109, 1947.

86. LIBBY, W. F., Tritium in nature, *Sci. Amer.,* **190**: 38-42, 1954.

87. LINDENBERGH, P. C., Movement of underground water below and above phreatic level, *IASH Gen. Ass. Toronto,* **2**: 193-198, 1958.

88. LINSLEY, R. K., M. A. KOHLER, and J. L. H. PAULHUS, *Applied Hydrology,* McGraw-Hill, New York, 1949.

89. LOEHNBERG, A., Aspects of the sinking of Mexico City and proposed countermeasures, *J. Amer. Water Works Assoc.,* **50**: 432-440, 1958.

90. LOHMAN, S. W., Compression of elastic artesian aquifers, *USGS Prof. Papers,* 424-B: 47-48, 1961.

91. LONG, R. R., *Mechanics of Solids and Fluids,* Prentice-Hall, Englewood Cliffs, 156 pp., 1961.

92. LUTHIN, J. N., and D. KIRKHAM, A piezometer method for measuring permeability of soil *in situ* below a water table, *Soil Sci.,* **68**: 349-358, 1949.

93. McCOLLUM, M. J., and H. B. COUNTS, Relation of salt-water encroachment to the major aquifer zones, Savannah area, Georgia and South Carolina, *USGS Water Supply Pap.,* 1613D, 26 pp., 1964.

94. MAXEY, G. B., Hydrogeology of desert basins, *Groundwater,* **6**: 10-22, 1968.

95. MAXEY, G. B., Subsurface water—groundwater, in *The Progress of Hydrology,* Univ. of Illinois, Urbana, Illinois, **2**: 787-815, 1969.

96. MAXEY, G. B., and R. N. FARVOLDEN, Hydrogeologic factors in problems of contamination in arid lands, *Groundwater,* **3**: 29-32, 1965.

97. MAXEY, G. B., and J. E. HACKETT, Application of geohydrologic concepts in geology, *J. Hydrol.,* **1**: 35-45, 1963.

98. MAXEY, G. B., and M. D. MIFFLIN, Occurrence and movement of groundwater in carbonate rocks of Nevada, *Natl. Speliol. Soc. Bull.,* **28**: 141-57, 1966.

99. MEINZER, O. E., Geology and water resources of Big Smokey, Clayton, and Alkali Spring Valleys, Nevada, *USGS Water Supply Pap.* 423, 1917.

100. MEINZER. O. E., Outline of groundwater hydrology: *USGS Wat. Sup. Pap.* 494, 1923.

101. MEINZER, O. E., Occurrence of groundwater in the United States, *USGS Water Supply Pap.,* 489, 1-321, 1923.

102. MEINZER, O. E., Compressibility and elasticity of artesian aquifers, *Econ. Geol.,* **23**: 263-291, 1928.

103. MEINZER, O. E., Outline of methods for estimating groundwater supplies, *USGS Water Supply Pap.* 638: 99-144, 1932.

104. MEINZER, O. E., Occurrence, origin, and discharge of groundwater, in Meinzer, O. E. (Ed.), *Hydrology,* McGraw-Hill, New York, 1942.

105. MEINZER, O. E., The occurrence of groundwater in the United States, *USGS Water Supply Pap.,* 489, 1959.

106. MEYBOOM, P., Patterns of groundwater flow in the prairie profile, in *Groundwater, Proc. Hydrology Sympos.* No.3: 5-20, National Research Council of Canada, Ottawa, 1963.

107. MEYBOOM, P., Three observations on streamflow depletion by phreatophytes, *J. Hydrol.,* **2**: 248-261, 1965.

108. MEYBOOM, P., Unsteady groundwater flow near a willow ring in hummocky moraine, *J. Hydrol.,* **4**: 38-62, 1966.

109. MEYBOOM, P., Groundwater studies in the Assiniboine River drainage basin, *Geol. Surv. Can. Bull.,* 139, 64 pp., 1967.

110. MEYBOOM, P., Mass transfer studies to determine the groundwater regime of permanent lakes in hummocky moraine of western Canada, *J. Hydrol.,* **5**: 117-142, 1967.

111. MIFFLIN, M. D., Delineation of groundwater flow systems in Nevada, *Desert Res. Inst., Tech. Rept. Ser.* H-W, no. 4, Reno, 1968.

112. MILLER, R. J., and P. F. LOW, Threshold gradient for water flow in clay systems, *Proc. SSSA,* **27**: 605-609, 1963.
113. MISENER, A. D., and A. E. BECK, The measurement of heat flow over land, in Runcorn, S. K. (Ed.), *Methods and Techniques in Geophysics,* pp. 10-61, Interscience, New York, 1960.
114. MONKHOUSE, F. J., *A Dictionary of Geography,* Arnold, London, 1965.
115. MULLER, S. W., *Permafrost or permanently frozen ground and related engineering problems,* Edwards, Ann Arbor, 231 pp., 1947.
116. NACE, R. L., Human use of groundwater, in R. J. Chorley (Ed.), *Water, Earth and Man,* Methuen, London, pp. 285-294, 1969.
117. NERPIN, S., S. PASHKINA, and N. BONDARENKO, Evaporation from bare soil and the ways of reducing it, *Proc. Wageningen Sympos. on Water in the Unsaturated Zone,* **2**: 595-602, IASH/UNESCO, 1969.
118. NEWSON, M. D., A model of subterranean limestone erosion in the British Isles based on hydrology, *Trans. IBG,* **54**: 55-70, 1971.
119. NEWSTEAD, G. N., and J. C. JAEGER, The determination of underground water movements from measurements in drill holes, *The Engineer, (London),* **202**: 76-78, 1956.
120. PARKER, G. G., and V. T. STRINGFIELD, Effects of earthquakes, trains, tides, winds, and atmospheric pressure changes on water in the geologic formations of Southern Florida, *Econ. Geol.,* **45**: 441-460, 1950.
121. PARSONS, M. L., Groundwater thermal regime in a glacial complex, *WRR,* **6**: 1701-1720, 1970.
122. PERLMUTTER, N. M., and J. J. GERAGHTY, Geology and groundwater conditions in southern Nassau and southeastern Queens counties, Long Island, New York, *USGS Water Supply Pap.,* 1613A, 205 pp., 1963.
123. PINDER, G. F., and H. H. COOPER, A numerical technique for calculating the transient position of the saltwater front, *WRR,* **6**: 875-882, 1970.
124. PIPER, A. M., A graphic procedure in the geochemical interpretation of water analyses, *Trans. AGU,* **25**: 914-923, 1944.
125. PITTY, A. F., *An Approach to the study of karst water,* Occasional Papers in Geog., No. 5, Univ. of Hull, 70 pp., 1966.
126. POISEUILLE, J. L. M., Recherches expérimentales sur le mouvement des liquides dans les tubes de très petit diamètre, *Roy. Acad. Sci. Inst. France Math. Phys. Sci. Mem.,* **9**: 433-543, 1846.
127. POLAND, J. F., The coefficient of storage in a region of major subsidence caused by compaction of an aquifer system, *USGS Prof. Papers,* 424-B: 52-54, 1961.
128. POLAND, J. F., and G. H. DAVIS, Subsidence of the land surface in the Tulare-Wasco (Delano) and Los Banos-Kettleman City area, San Joaquin Valley, California, *Trans. AGU,* **37**: 287-296, 1956.
129. RICHARDS, L. A. (Chairman), Report of the subcommittee of permeability and infiltration, committee on terminology, Soil Science Society of America, *Proc. SSSA,* **16**: 85-88, 1952.
130. RICHARDSON, R. M., Tidal fluctuations of water level observed in wells in East Tennessee, *Trans. AGU,* **37**: 461-462, 1956.
131. ROBERTS, I., On the attractive influence of the sun and moon causing tides . . . in the underground water in porous strata, *Rept. Brit. Assn.,* p. 405, 1883. Quoted by Richardson [130].
132. ROBINSON, T. W., Earth-tides shown by fluctuations of water-levels in wells in New Mexico and Iowa, *Trans. AGU,* **20**: 656-666, 1939.
133. SALEEM, M., A simple method of groundwater direction measurement in a single bore hole, *J. Hydrol.,* **12**: 387-410, 1971.
134. SCHEIDEGGER, A. E., Directional permeability of porous media to homogenous fluids, *Geofis. Pura Appl.,* **30**: 17-26, 1954.
135. SCHEIDEGGER, A. E., *The Physics of Flow through Porous Media* (Revised edition), Univ. of Toronto Press, 313 pp., 1960. (First edition, 1957).
136. SCHMORAK, S., and A. MERCADO, Upconing of fresh water—sea water interface below pumping wells, field study, *WRR,* **5**: 1290-1311, 1969.

137. SCHOELLER, H., Arid zone hydrology, recent developments, *UNESCO Arid Zone Res.* **12**, 125 pp. 1959.
138. SCOTT, J. S., and F. W. RENDER, Effect of an Alaskan Earthquake on water levels in wells at Winnipeg and Ottawa, Canada, *J. Hydrol.,* **2**: 262-268, 1964.
139. SENIO, K., On the groundwater near the seashore, *IASH Gen. Ass. Bruxelles,* **2**: 175-177, 1951.
140. SENN, A., Geological investigations of the groundwater resources of Barbados, B.W.I., *Report of the British Union Oil. Co. Ltd.,* 1946.
141. SHAMIR, U., and G. DAGAN, Motion of the seawater interface in coastal aquifers: A numerical solution, *WRR,* **7**: 644-657, 1971.
142. SHUSTER, E. T., and W. B. WHITE, Seasonal fluctuations in the chemistry of limestone springs: a possible means for characterizing carbonate aquifers, *J. Hydrol.,* **14**: 93-128, 1971.
143. SILINE-BEKCHOURINE, A., *Hydrogeology of Irrigated Lands,* Foreign Languages Publishing House, Moscow, 1962.
144. SIPLE, G. E., Salt-water encroachment in coastal South Carolina, *Proc. Conf. on Hydrologic Activities in the South Carolina Region,* pp. 18-33, Clemson Univ., Clemson, 1965.
145. SLICHTER, C. S., The motions of underground waters, *USGS Wat. Sup. Pap.,* 67, 1902
146. SNOW, D. T., Anisotropic permeability of fractured media, *WRR,* **5**: 1273-1289, 1969.
147. STALLMAN, R. W., Steady one-dimensional fluid flow in a semi-infinite porous medium with sinusoidal surface temperature, *JGR,* **70**: 2821-2829, 1965.
148. STEARNS, N. D., Laboratory tests on physical properties of water-bearing materials, *USGS Water Supply Pap.,* 596f, 1928.
149. STRINGFIELD, V. T., and H. E. LEGRAND, Hydrology of carbonate rock terranes—A review, *J. Hydrol.,* **8**: 349-376, 1969.
150. STRINGFIELD, V. T., and H. E. LEGRAND, Hydrology of carbonate rock terranes—A review, *J. Hydrol.,* **8**: 377-417, 1969.
151. STRINGFIELD, V. T., and H. E. LEGRAND, Relation of sea water to fresh water in carbonate rocks in coastal areas, with special reference to Florida, U.S.A., and Cephalonia (Kephallinia), Greece, *J. Hydrol.,* **9**: 387-404, 1969.
152. STRINGFIELD, V. T., and H. E. LEGRAND, Effects of karst features on circulation of water in carbonate rocks in coastal areas, *J. Hydrol.,* **14**: 139-157, 1971.
153. SWARTZENDRUBER, D., Non Darcy behaviour in liquid saturated porous media, *JGR,* **67**: 5205-5213, 1962.
154. SWENSON, F. A., New theory of recharge to the artesian basin of the Dakotas, *Bull. Geol. Soc. Amer.,* **79**: 163-182, 1968.
155. TAMERS, M. A., Groundwater recharge of aquifers as revealed by naturally occurring radiocarbon in Venezuela, *Nature,* **212**: 489-492, 1966.
156. THEIS, C. V., Relation between the lowering of the piezometric surface and the rate and duration of discharge of a well using groundwater storage, *Trans. AGU,* **16**: 519-524, 1935.
157. THEIS, C. V., Earth tides expressed as fluctuations of the water level in artesian wells in New Mexico, Internat. Union Geodesy and Geophysics, Washington, D.C., 11 pp., September 1939 (processed). Quoted by Richardson [130].
158. THOMAS, H. E., Underground sources of our water, in *Water,* USDA Yearbook, USGPO, Washington, 1955.
159. TODD, D. K., *Groundwater Hydrology,* Wiley, New York, 1959.
160. TODD, D. K., Groundwater, in Chow, V. T. (Ed.), *Handbook of Applied Hydrology,* Section 13, McGraw-Hill, New York, 1964.
161. TOLMAN, C. F., *Groundwater,* McGraw-Hill, New York, 1937.
162. TOTH, J., A theory of groundwater motion in small drainage basins in central Alberta, Canada, *JGR,* **67**: 4375-4387, 1962.
163. TOTH, J., A theoretical analysis of groundwater flow in small drainage basins, *JGR,* **68**: 4795-4812, 1963.
164. TOTH, J., Groundwater geology, movement chemistry and resources near Olds, Alberta, *Res. Council Alberta, Geol. Div., Bull.* 17, 1966.

236

165. TOTH, J., A conceptual model of the groundwater regime and the hydrogeologic environment, *J. Hydrol.,* **10**: 164-176, 1970.
166. TROXELL, H. C., The diurnal fluctuation in the groundwater and flow of the Santa Ana River and its meaning, *Trans. AGU,* **17**: 496-504, 1936.
167. TUINZAAD, H., Influence of the atmospheric pressure on the head of artesian water and phreatic water, *IASH Gen. Ass. Rome,* **2**: 32-37, 1954.
168. VREEDENBURG, C. G. F., On the steady flow of water percolating through soils with homogeneous-anisotropic permeability, *Proc. Int. Conf. Soil Mech. Found. Eng.,* pp. 222-225, 1936.
169. WALTON, W. C., *Groundwater Resource Evaluation,* McGraw-Hill, New York, 664 pp., 1970.
170. WATER RESEARCH FOUNDATION OF AUSTRALIA, Nuclear chemical study of the age and renewal rate of the Great Artesian Basin, *Water Research Foundation 9th Annual Report,* pp. 12-13, 1964.
171. WATER RESOURCES BOARD, *Artificial Recharge of the London Basin, I Hydrogeology,* Water Resources Board, Reading, 1972.
172. WATTS, D., *Principles of Biogeography,* McGraw-Hill, London, 1971.
173. WHITE, D. E., Thermal waters of volcanic origin, *Bull. Geol. Soc. Amer.,* **68**: 1637-1657, 1957.
174. WHITE, D. E., Magmatic, connate, and metamorphic waters, *Bull. Geol. Soc. Amer.,* **68**: 1659-1682, 1957.
175. WHITE, D. E., and W. W. BRANNOCK, Sources of heat, water supply, and mineral content of Steamboat Springs, Nevada, *IASH Gen. Ass. Oslo,* **3**: 168-176, 1948.
176. WHITE, D. E., J. D. HEM, and G. A. WARING, Chemical composition of subsurface waters, *USGS Prof. Pap.,* 440-F, 1963.
177. WHITE, W. M., A method of estimating groundwater supplies based on discharge by plants and evaporation from soil, *USGS Water Supply Pap.,* 659-A, 1932.
178. WILSON, G., and H. GRACE, The settlement of London due to underdrainage of the London clay, *J. Inst. Civ. Eng.,* **19**: 100-127, 1942.
179. WINSLOW, A. G., and L. A. WOOD, Relation of land subsidence to groundwater withdrawals in the upper Gulf coastal region of Texas, *AIME, Mining Div., Mining Eng.,* pp. 1030-1034, 1959.
180. WISLER, C. O., and E. F. BRATER, *Hydrology,* 2nd Ed., Wiley, New York, 1959.
181. WURZEL, P., and P. R. B. WARD, A simplified method of groundwater direction measurement in a single borehole, *J. Hydrol.,* **3**: 97-105, 1965.

8. Runoff

8.1 Introduction

Runoff or streamflow comprises the gravity movement of water in channels which may vary in size from the one containing the smallest ill-defined trickle to the ones containing the large rivers such as the Amazon, the Congo, and the Mississippi. In a general sense, this water, flowing in small rivulets, in ditches and dykes, in streams and rivers, is a residual item, representing the excess of rainfall over evapo-transpiration, when allowance is made for storage on and under the ground surface. But, whereas rainfall on the land surface is spasmodic and irregular in space, time, and amount, runoff from the land surface is a comparatively constant factor. This contrast between runoff and the rainfall from which it is directly or indirectly derived, results mainly from the great storage capacity of the surface layers of the earth, by means of which much of the excess rainfall is held back and only gradually released into the streams. It has, in fact, been estimated that normally, 30 cm of soil holds more water than the entire overlying atmosphere [109]. Almost inevitably, therefore, much of the following discussion of runoff will be concerned with the relationships between precipitation and the surface on which it falls, and the effects of those relationships upon the character, amount, and distribution of runoff.

Runoff, which may be variously referred to as *streamflow, stream* or *river discharge,* or *catchment yield,* is normally expressed as a volume per unit of time. The *cumec,* i.e., one cubic metre per second, and *cumecs per square kilometre* are commonly used units. Runoff may also be expressed as a depth equivalent over a catchment, i.e., millimetres per day or month or year. This is a particularly useful unit for comparing precipitation and runoff rates and totals since precipitation is almost invariably expressed in this way. Alternative runoff expressions still found in the literature include millions of gallons per day (m.g.d.) and, particularly in American irrigation literature, acre feet per day, i.e., the volume of water which would cover one acre to a depth of one foot in one day.

8.2 Sources and components of runoff

The persistent misuse of runoff terminology has resulted in much confusion and ambiguity about the sources and components of runoff and this problem has been recently intensified by our revised concepts of the runoff process. Freeze [47] provided a consistent and unambiguous terminology which has been adopted with only slight modification in this chapter. The total runoff from a typically heterogeneous catchment area may be conveniently divided into four component parts: channel precipitation, overland flow, interflow, and groundwater flow (see Fig. 8.1).

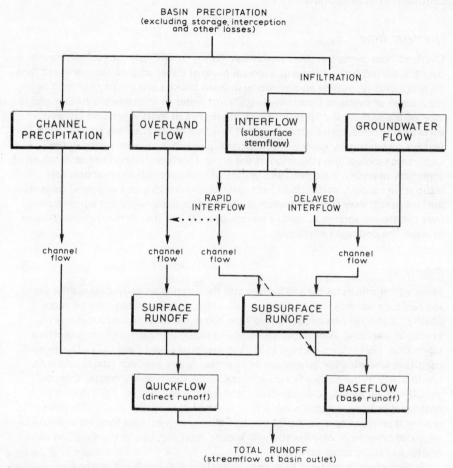

Fig. 8.1. *Diagrammatic representation of the runoff process.*

Channel precipitation

Direct precipitation onto the water surfaces of streams, lakes, and reservoirs makes an immediate contribution to streamflow. In relation to the other components, however, this amount is normally small in view of the small percentage of catchment area normally covered by water surfaces. Thus the perennial channel system occupies only about one per cent of the total area of the Eastern United States [66] and although during a prolonged storm this area will increase as a result of the extension of the stream into intermittent and ephemeral channels, the water surface for most catchments does not exceed 5 per cent of the total area even at high water levels [97]. However, Rawitz and others [122], in an analysis of ten storms over a small Pennsylvania watershed, estimated that channel precipitation accounted for between 3 and 61 per cent of total runoff. In catchments containing a large area of lakes or swamps channel precipitation may make a substantial contribution to streamflow.

Overland flow

Overland flow comprises the water which, failing to infiltrate the surface, travels over the ground surface towards a stream channel either as quasi-laminar sheet flow or, more usually, as flow anastomosing in small trickles and minor rivulets. The main cause of overland flow is the inability of water to infiltrate the surface and in view of the high value of infiltration characteristic of most vegetation-covered surfaces it is not surprising that overland flow is a rarely observed phenomenon (except on laboratory models!). Conditions in which it assumes considerable importance include the saturation of the ground surface, the hydrophobic nature of some very dry soils, the deleterious effects of many agricultural practices on infiltration capacity as discussed in chapter 6, and freezing of the ground surface. Surface runoff may then be defined as that part of the total runoff which travels over the ground surface to reach a stream channel and thence through the channel to reach the drainage basin outlet.

Interflow

Water which infiltrates the soil surface and then moves laterally through the upper soil horizons towards the stream channels, either as unsaturated flow or, more usually, as shallow perched saturated flow above the main groundwater level is known as interflow. Alternative terms found in the literature include subsurface storm flow [79], storm-seepage [13], and secondary base flow [12]. The general condition favouring the generation of interflow is one in which lateral hydraulic conductivity in the surface horizons of the soil is substantially greater than the overall vertical hydraulic conductivity through the soil profile. Then during prolonged or heavy rainfall water will enter the upper part of the profile more rapidly than it can pass vertically through the lower part, thus forming a perched saturated layer from which water will 'escape' laterally, i.e., in the direction of greater hydraulic conductivity.

Except where conditions have been artificially disturbed, e.g., surface compaction, the situation described above is the one most commonly found. Even in a deep relatively homogeneous soil profile hydraulic conductivity will tend to be greater in the surface layers than deeper down in the profile, thereby encouraging the generation of interflow, although still more favourable conditions exist where a thin permeable soil overlies impermeable bedrock, with a markedly stratified soil profile, or where an iron-pan occurs a short distance below the surface. There may thus be several levels of interflow below the surface corresponding to textural changes between horizons and to the junction between weathered mantle and bedrock. In addition, some hydrologists argue that water may travel downslope through old root holes and animal burrows and other subsurface pipes. In view of the variety of possible interflow routes it is to be expected that some will result in a more rapid movement of water to the stream channels than will others, so that it is sometimes helpful to distinguish between rapid and delayed interflow (see Fig. 8.1). It is almost certain, however, that apart from pipe flow the very rapid arrival of interflow at the stream channels, which has been observed by some investigators must result from the translatory flow or 'push-through' mechanism which was described in section 7.8. Further complications result from the fact that some interflow does not discharge directly into the stream channel but comes to the surface at some point between the stream and catchment divide, and may then continue to move over the soil surface. Freeze [47] suggested that interflow discharging close to the stream channel should be considered as part of the total subsurface runoff, as also should interflow discharging at greater distances from the main channel wherever (as is often the case) it feeds an intermittent tributary channel. Otherwise this outflow might be considered as a contribution to overland flow and surface runoff and this assumption is indicated by the dotted line in Fig. 8.1.

The role of interflow in total runoff will be discussed in more detail in subsequent sections of this chapter although it is interesting to note at this stage that experimental evidence has long indicated that it may account for up to 85 per cent of total runoff [60].

Groundwater flow

Most of the rainfall which percolates through the soil layer to the underlying groundwater will eventually reach the main stream channels as groundwater flow through the zone of saturation. Since water can move only very slowly through the ground, the outflow of groundwater into the stream channels may lag behind the occurrence of precipitation by several days, weeks, or often years. Groundwater flow also tends to be very regular, representing as it does, the overflow from the slowly changing reservoir of moisture in the soil and rock layers. It must not be inferred from this that groundwater may not show a rapid response to precipitation. Indeed the push-through mechanism of translatory flow (section 7.8) frequently results in a rapid response of groundwater flow to precipitation during individual storm periods and especially on a seasonal basis. Since translatory flow can only operate in moist soil and subsoil conditions, however, the replenishment of large moisture deficits created particularly during summer conditions, may result in a considerable lag of groundwater outflow after precipitation during and immediately following prolonged dry periods. In general, groundwater flow represents the main long-term component of total runoff and is particularly important during dry spells when surface runoff is absent.

Snowmelt

In some areas, particularly at high altitudes or in high latitudes, a large proportion of streamflow may be derived from the melting of snows and glaciers. Where this melting occurs gradually, over a long period of time, the resulting contribution to streamflow will resemble that of groundwater flow. Where, however, it occurs suddenly as a result of a föhn or chinook type wind, for example, a large volume of water will enter the streams during a short period of time, giving a runoff peak which closely resembles that derived from storm rainfall.

Although, in terms of the phase relationship between precipitation and runoff, snow accumulation and melt pose particular problems, in terms of the present discussion snowmelt does not represent a special case or merit consideration as a fifth component of runoff. Snow falling directly on to the stream surface has already been discussed under the heading of channel precipitation, while water generated by the process of snowmelt will either flow over the ground surface as overland flow or will infiltrate to become interflow and groundwater flow depending on whether the sub-snowpack surface is saturated and/or frozen.

Quickflow and baseflow

The foregoing discussion should have clarified the definition and role of four other runoff terms, illustrated in Fig. 8.1, which are used somewhat indiscriminately in the literature, i.e., surface and subsurface runoff and quickflow and baseflow. Surface runoff, as has been shown, is that part of total runoff which reaches the drainage basin outlet via overland flow and the stream channels, although it may in some circumstances also include interflow which has discharged at the ground surface at some distance from the stream channel. Subsurface runoff is the sum of interflow and groundwater flow and is normally equal to the total flow of water arriving at the stream as saturated flow into the stream bed itself, and as percolation from the seepage faces on the stream bank [47].

Quickflow, or direct runoff, is the sum of channel precipitation, surface runoff and rapid interflow and will clearly represent the major runoff contribution during storm periods and is also the major contributor to most floods. It will be observed that quickflow and surface runoff as defined above can *not* be used synonymously. Baseflow or base runoff may be defined as the sustained or fair-weather runoff [30] and is the sum of groundwater runoff and delayed interflow, although some hydrologists prefer to include the total interflow as illustrated by the broken line in Fig. 8.1. Again it will be observed that baseflow and groundwater flow, as defined above, can not be used synonymously; indeed Hewlett [62] [64] demonstrated that baseflow from steep mountain drainage basins may consist almost entirely of unsaturated lateral flow from the soil profile. Hewlett and Nutter [66] suggested that in upland forested catchments about 85 per cent of total runoff may consist of baseflow and that in the eastern United States as a whole about 80 per cent of total runoff is baseflow and only 20 per cent is quickflow.

Types of stream

The long-term relationship between baseflow and quickflow determines the main characteristics of a stream or river; whether, for example, it will flow steadily or flashily through the year, whether, indeed, it will flow throughout the year or only for part of the time; and this provides a basis for classifying streams into three main types, i.e., *ephemeral, intermittent,* or *perennial* [160].

Ephemeral streams are those which comprise quickflow only and which, therefore, flow only during and immediately after rainfall or snowmelt. Normally there are no permanent or well-defined channels, and the water table is always below the bed of the stream, so that groundwater flow can make no contribution to runoff.

242

In the case of intermittent streams, which flow during the wet season, but which dry up during the season of drought, streamflow consists mainly of quickflow, but baseflow makes some contribution during the wet season, when the water table rises above the bed of the stream. A particular example of an intermittent stream occurs when groundwater becomes frozen during the winter months, as, for example, in high latitude areas.

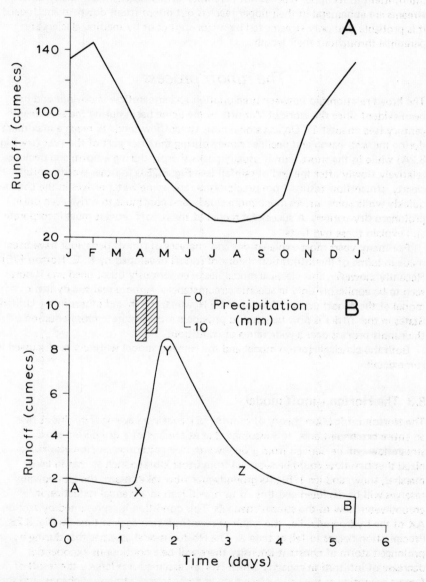

Fig. 8.2. *(A) Long-period hydrograph of the River Thames showing seasonal variation of flow; (B) Short-period storm hydrograph for a hypothetical catchment.*

Finally, as the term suggests, perennial streams flow throughout the year, because even during the most prolonged dry spell, the water table is always above the bed of the stream, so that groundwater flow can make a continuous and significant contribution to total runoff at all times.

In only a few cases will it be possible to classify the entire length of a stream under only one of these three headings. A chalk bourne, for example, is normally intermittent in its upper reaches but perennial farther downstream; many other streams are ephemeral in their upper reaches but intermittent downstream. Indeed, it is probable that only streams fed by major springs or by melting glaciers are perennial throughout their length.

The runoff process

The broad relationship between precipitation and streamflow is obvious and has been evident since the work of Mariotte in the Seine basin during the seventeenth century (see chapter 1). On a seasonal basis streamflow tends to reach a maximum during the wet season and declines slowly during the drier part of the year (see Fig. 8.2A) while in the short term it usually peaks sharply during a storm and declines relatively slowly after the end of rainfall (see Fig. 8.2B). In other words, quite clearly, streamflow results from precipitation and some water arrives in the channel quickly while some arrives much more slowly and continues to arrive even during prolonged dry periods. A successful model of the runoff process must incorporate and explain these two facts.

For many years most explanations and analyses of runoff behaviour have been made in terms of the infiltration theory of runoff developed by R. E. Horton [76]. Recently, however, this classical model has been seriously questioned and is now seen to be applicable only in specific circumstances. A more realistic dynamic model of the runoff process developed by Hewlett, Hursh and others in the United States in the 'fifties is now believed to provide a more accurate representation of the runoff process over a wide range of conditions.

Both the classical Horton model and the Hewlett model will now be discussed in some detail.

8.3 The Horton runoff model

The Horton infiltration theory of runoff [76] considers average conditions over an entire catchment area. It is assumed that at the end of a dry period most streamflow will be derived from groundwater flow, although Horton also recognized that baseflow could be supplied from other storage such as that in lakes, marshes, snow, and ice [75]. As groundwater flow takes place, the groundwater reserves will be depleted and this, in turn, will lead to a gradual reduction in groundwater flow to the stream channels. This condition is represented by section AX of the hydrograph, i.e., the graph of runoff plotted against time, in Fig. 8.2B. Precipitation begins to fall at time X. The Horton model assumes that during a prolonged storm of constant intensity there will be a continuous exponential decrease of infiltration capacity which Horton assumed was largely the result of factors operating at the soil surface such as compaction, structural change, and the inwashing of fine particles (see section 6.15 *et seq*). Eventually a constant low value

244

of infiltration capacity is reached over the entire catchment area. If this value falls below that of rainfall intensity, so that rain is falling at a faster rate than the soil can absorb it, overland flow and subsequently surface runoff will occur. Overland flow, having built up to a sufficient depth, will reach the stream channels quite quickly, giving rise to the marked increase in streamflow represented by the line XY in Fig. 8.2B. Soon after the end of rainfall, the surface runoff will begin to diminish, rapidly in the initial stages, and later more slowly, as first the minor channels, and then the larger ones begin to drain dry (see section YZ).

Any water which is left on the surface as depression storage will soak into the soil and may perhaps percolate to the groundwater. Finally, once all the surface runoff has been disposed of, the flow in the streams will again consist almost entirely of groundwater flow (see section ZB), which will gradually decrease as the reserves are used up.

The hydrograph of storm runoff described here, thus consists of three main sections; a steep *rising limb* from X to Y, a *peak* or crest at Y, and a shallower

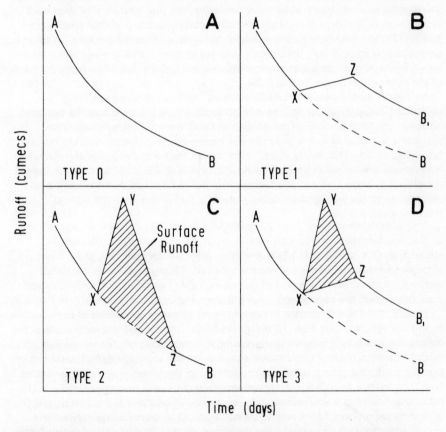

Fig. 8.3. *Stream hydrographs showing the main types of runoff increase according to R. E. Horton. (From an original diagram by R. E. Horton,* Trans. Amer. Geophys. Union, **14:** *446-460, 1933.)*

recession limb from Y to B. It will be apparent that most of the large increase in streamflow, which is represented by the triangle XYZ, is the result of surface runoff. Some part of that increase, however, will be the direct result of an increase in groundwater flow and the exact relationship between these two components will depend largely on how much of the rainfall is absorbed by the soil—it will, in other words, depend upon the relationship between rainfall intensity and infiltration capacity.

On this basis Horton [76] described the four main types of rainfall-induced increase in runoff shown in Fig. 8.3. Type 0 was so called by Horton because, according to him, nothing happens as far as the stream hydrograph is concerned. The rain which falls is light, and does not exceed the infiltration capacity of the soil, so that no surface runoff occurs; neither is it sufficient to make good the soil moisture deficit, so that no accretion to the groundwater and, therefore, no increase in groundwater flow occurs either. Thus, the stream hydrograph follows the form of the normal dry weather depletion curve, with groundwater-derived baseflow gradually decreasing from A to B as the groundwater reserves are used up. The significance of Type 0 conditions lies in the fact that a number of successive occurrences of this type may bring the soil to field capacity and thus enable any further rain to contribute to the groundwater reserves. Strictly speaking, of course, since some of the rain will fall directly into the stream channels, there will inevitably be *some* increase in streamflow, causing a slight departure from the form of the hydrograph shown in Fig. 8.3A.

In Type 1 conditions, the rainfall intensity, again, does not exceed the infiltration capacity of the soil, so that no surface runoff is produced. In this case, however, the total amount of infiltration exceeds the soil moisture deficit so that some percolation to the groundwater reserves takes place, causing an increase in groundwater flow (XZ in Fig. 8.3B), after which the normal groundwater depletion hydrograph is resumed at a higher level (ZB_1 in Fig. 8.3B). It is possible for Type 1 conditions to occur without producing a rise in the hydrograph if the additional percolation to the groundwater takes place at a slower rate than the normal groundwater depletion.

The rainfall intensity in Type 2 conditions exceeds the infiltration capacity so that overland flow and surface runoff occur, causing a marked increase in streamflow (XY in Fig. 8.3C), but any infiltration which does take place is not sufficient to make good the soil moisture deficit. Thus groundwater depletion carries on throughout the period of increasing runoff and once the surface runoff peak has passed, the hydrograph rejoins the normal depletion curve (ZB in Fig. 8.3C).

Finally, the Type 3 increase in runoff results from the generation of both surface runoff and groundwater flow. In these conditions, the rainfall intensity exceeds the infiltration capacity, giving the surface runoff increase represented by the line XY while the total amount of infiltration exceeds the soil moisture deficit, resulting in percolation to the groundwater and an increase in the groundwater component of base flow from X to Z. Thus, after the surface runoff peak has passed, the hydrograph follows a new groundwater-based depletion curve from Z to B_1. Again, as in the case of Type 1, the rate of percolation to the groundwater may be less than the normal rate of groundwater depletion, so that groundwater flow, in fact, decreases during the period of the overall increase in runoff represented by line XYZ in Fig. 8.3D.

246

The Horton runoff model is clearly most applicable in areas having low infiltration capacity where only a limited length of rainstorm is necessary to saturate the entire catchment surface and to initiate overland flow on all the slopes. Badlands are an obvious example, although others include areas of frozen ground and areas where the natural soil and vegetation characteristics have been severely modified by agricultural and other human activity resulting in compacted surfaces with sparse vegetation cover.

8.4 The Hewlett runoff model

One of the most important contributions to modern hydrology arose from work at Coweeta, in the southern Appalachians, by C. R. Hursh and particularly J. D. Hewlett. For various reasons, including editorial resistance to the refutation of existing concepts, Hewlett's work was not published until 1961 and then only in abbreviated form in the Annual Report of the Southeastern Forest Experiment Station of the USDA Forest Service [61]. Hewlett eventually presented a full account of his ideas at a symposium on forest hydrology in 1965 although the proceedings of this symposium were not finally published until 1967 [65]. By that time these ideas had been borrowed by other workers, particularly Betson in the TVA to whom the original concepts are often wrongly ascribed. Thus Freeze [47] quoted Betson [21] when referring to the . . . original espousal of the partial area concept. . .'

The Hewlett runoff model, embodying the variable source area concept makes the basic assumption that infiltration is seldom a limiting factor, i.e., that only in special conditions does rainfall intensity exceed infiltration capacity. Hewlett derived this basic premise from a number of fundamental considerations. Firstly, there was considerable field evidence that most precipitation infiltrated the ground surface, particularly in well-vegetated areas. Even during a 100-year storm which delivered more than 500 mm of rain in five days on one Coweeta watershed no overland flow was detected [67]. Secondly, only about 10 per cent of the annual precipitation in humid areas such as the eastern United States appears as quickflow in headwater streams [163]. Thirdly, most of this quickflow leaves the drainage basins hours after the cessation of rainfall. In view of the high speed of overland flow, e.g., up to 270 m per hour [67] and the comparatively short distance between divide and channel in the eastern United States of 180 m [67] this quickflow '. . . cannot possibly be classified as simple overland flow. . .' [67]. Fourthly, Hewlett [61] described as the most intriguing of all the findings at Coweeta the evidence that unsaturated soil moisture movement may account entirely for baseflow from some mountain watersheds.

Essentially the core of the Hewlett runoff model, which is described in detail by Hewlett and Hibbert [65] and Hewlett and Nutter [67], is the response of the channel system to precipitation. Overland flow is treated as an extension of the perennial channel system into zones of low storage capacity which quickly become saturated as a result of infiltration and more particularly interflow into the lower valley sides, minor depressions, swampy spots, and intermittent channels. Thus interflow is regarded as feeding the expanding channel from below while rainfall feeds it from above. In this way the channel system may grow to many times its perennial width and length as was demonstrated by Gregory and Walling [52] and it may also continue to grow, as a result of interflow, for some time after the end of rainfall [67]. Hewlett and Hibbert [65] described this phase in the following way

... when the subsurface flow of water from upslope exceeds the capacity of the soil profile to transmit it, the water will come to the surface and channel length will grow. This in essence is the variable source area concept (Reprinted with permission from *Forest Hydrology* © 1967 Pergamon Press Ltd.)

thereby elaborating the earlier statement of Hewlett [61]:

Under prolonged and heavy rainfall, the storm-flow-contributing area contiguous to stream channels may grow wider and wider, depending on the nature and depth of the earth mantle.

The main significance of this channel-system expansion is that the very rapid contribution of channel precipitation to quickflow is effectively increased in a directly proportional way.

The fact that interflow, which moves so slowly in comparison with overland flow (e.g., 5 m per day), should be able to make such a substantial contribution to quickflow is explained partly by the fact that the expanding channel network 'reaches out' to tap the subsurface flow systems [67] but particularly by the push-through mechanism of translatory flow (sections 8.2 and 7.8) [65] [74], whose role is illustrated in Fig. 8.4. As one proceeds upslope in Fig.8.4 each unit of

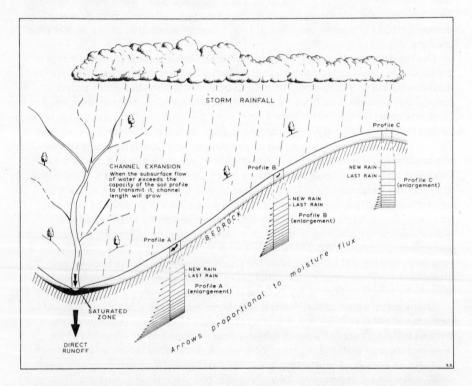

Fig. 8.4. *Diagram showing the source of direct runoff from a forested upland catchment with a uniform soil mantle. This illustrates the basic concepts of variable source areas proposed by J. D. Hewlett. (From original diagrams by Hewlett [61] and by J. D. Hewlett and A. R. Hibbert in* Forest Hydrology, *© 1967, Pergamon Press Ltd. [65])*

rainfall contributes more to temporary storage and less to direct runoff. The contribution to direct runoff will in part comprise some of the rain falling during the present storm, but the remainder will consist of flow produced by a system of displacement (translatory flow) and involving water already stored in the soil mantle before the present rainfall began. Translatory flow may thus result in an immediate discharge of interflow into the expanding stream channel as a direct consequence of infiltration-induced interflow higher up the slope although, clearly, the water discharged is not the actual water infiltrating upslope. Translatory flow will be quantitatively important only when the soil moisture content is high, e.g., at or near field capacity, and its contribution to quickflow will tend to increase downslope (see Fig. 8.4). On the upper valley slopes translatory flow may be regarded as a pulse in soil moisture which will migrate slowly downhill but which will not make a direct contribution to quickflow.

A further proposition of the Hewlett runoff model is that the shallow subsurface movement of water, resulting from the very large percentage of total precipitation infiltrating the surface, makes a vital contribution to the recession limb of the hydrograph as well as to the rising limb and peak. In some circumstances the contribution of interflow may entirely account for recession flows which could previously have been explained (somewhat improbably) as the result of groundwater flow. Thus Hewlett [62] and Hewlett and Hibbert [64] reported detailed measurements of unsaturated flow from a sloping soil block which showed that the slow drainage of water from unsaturated soil profiles is sufficient to sustain and account for the baseflow of small headwater streams during dry spells, a conclusion subsequently confirmed by field observation in other upland catchments [85]. Hewlett [61] suggested that, in any case, this was a much more probable mechanism than the traditional 'groundwater' explanation of such recession flows in small mountain catchment areas having a range of elevation of several hundreds of metres and where groundwater would have to be held for many weeks at hundreds of metres of hydraulic potential along steeply sloping stream channels. It seems reasonable to assume that groundwater flow will tend to make an increasingly important contribution to baseflow as catchment slope and the depth to groundwater decrease and as subsoil permeability increases.

Referring again to the hydrograph illustrated in Fig. 8.2B it will be appreciated that this may be readily explained, in terms of the Hewlett runoff model, in the following manner: the rapid increase of discharge between X and Y results from the generation of quickflow from an expanding channel network which is fed by interflow; the early part of the recession (YZ) will be sustained largely by saturated interflow whilst the later stages of recession (ZB and AX) will be largely sustained by unsaturated interflow in small, steep, impervious catchments, by groundwater flow in flat, permeable catchments, or by a combination of both processes in intermediate situations.

The main advantages of the Hewlett model over the Horton model are that it accommodates a broad diversity of field observations of runoff, that it more realistically incorporates the more important dynamic aspects of the runoff process, e.g., the inherently non-linear effect of a variable source area, and that it accommodates the entire range of runoff formation from Horton overland flow, including the extreme case of the car-park hydrograph, at one end of the range to the deep porous basin, with stable channel length, in which total runoff is derived almost entirely from subsurface components at the other end of the range. It will be immediately apparent that acceptance of the Hewlett model has a number of important implications, some of which will be discussed in more detail in later sections. For the present it will suffice to note that traditional techniques of hydrograph analysis are rendered meaningless while techniques of runoff-prediction must be carefully re-examined. Thus, for example, in relation to the unit hydrograph method, the time base of a given magnitude of hydrograph is not constant but depends on the routes taken by the subsurface components of runoff, which in turn depend on a number of antecedent and precipitation conditions and on the rate of expansion or shrinkage of the channel system [67]. Again, in relation to statistical prediction techniques, the relationship between runoff and precipitation may be non-linear simply because the more common, smaller runoff peaks are generated from small contributing areas while the comparatively rare large events are generated from substantially larger contributing areas. Because the prediction model is fitted to the typical event it is difficult to predict large events from a knowledge of small ones [21]. In more general terms, as Freeze [46] observed, there seems to be no physical reason why catchments should respond in a linear fashion or indeed in any consistent non-linear fashion.

The Hewlett model has now established itself in hydrological theory although there are still many hydrologists to whom it is at least partially unacceptable. The two main areas of controversy concern the evidence for variable source areas and for the role of interflow, particularly in the generation of quickflow. Both aspects will now be discussed in more detail.

8.5 Variable source areas

It is now widely accepted that widespread Hortonian overland flow resulting from saturation from above when rainfall intensity exceeds the infiltration capacity of the ground surface is rare. A number of authors have discussed the relationships between rainfall intensity and infiltration capacity and have demonstrated that over a wide range of conditions typical rainfall intensities are less than the infiltration capacities of many soils and that in any case normal storm durations tend to be shorter than the time required for most soils to become saturated at the surface, c.f. [47] [87]. Some areas, e.g., the semi-arid south-west of the USA, are still regarded by some investigators as being conducive to the development of Hortonian overland flow, although it should be noted that in more than 30 years of hydrological investigations in some steeply sloping Arizona drainage basins widespread overland flow has never been observed. In general, slope, slope material, and slope vegetation are in such an equilibrium that all precipitation is able to infiltrate. Only where one or more of these factors has been drastically modified, usually by man, or during the course of 'catastrophic' meteorological events, is widespread overland flow generated. Were this not so the entire land surface would be scarred by gullies. The concept of variable (or partial) source areas is an attempt to reconcile the absence of widespread overland flow with the rapid response of most streams to precipitation by postulating that over-the-surface movement of water is restricted to limited areas of a drainage basin.

A number of definitions of variable source areas have been implied, although rarely specified, in the work of the principal investigators. The variable source area was conceived of by Hewlett in terms of the expanding and shrinking channel network, as has already been discussed in section 8.4, and was subsequently confirmed by independent field observations elsewhere. Thus in two small drainage basins in Vermont, Ragan [121] found that the areas contributing to quickflow were generally located in the floodplain near the stream channels. Hewlett proposed that the principal mechanism for expanding the channel network source area is saturation from below resulting from the downward and downslope movement of infiltrating water and interflow which leads to the saturation of valley-side soil profiles from the lower slopes upwards (see Fig. 8.4). In addition, channel expansion may reach quickly into areas with shallow soils and bare rock surfaces, or where overland flow does occur due to the existence of roads, trails, and compacted soil [66].

Dunne and Black [38] found that apart from channel precipitation on the stream surface itself, significant amounts of quickflow were only generated from small areas where the water table rose to the ground surface. Others have suggested alternative hydrologically responsive areas. Thus Kirkby and Chorley [88] enumerated four localities, at the base of slopes, in hollows, in slope-profile concavities and in areas of thin or less permeable soils. Betson [21], postulating the Hortonian mechanism of saturation from above, found that in small drainage basins in the Tennessee Valley runoff usually originates from a small, but relatively consistent, part of the basin. He later [22] suggested that saturated soil conditions frequently occurred where the A horizon was thin. Still others have suggested a random variable disposition of source areas. Thus Amerman [5], discussing runoff from small basins in Ohio, suggested that runoff-producing areas were located '... in seemingly random fashion on ridge-tops, valley-slopes, and valley bottoms' and Zavodchikov [165] considered runoff production to be controlled by an 'effective area' (source area) which was a function of drainage basin topography and soil moisture content. However, much of the overland flow on upslope areas reinfiltrates into downslope soils before reaching the stream channels [5] [22] [121] and cannot, therefore, be counted as a direct source of runoff. This means that only those wet patches which are connected to the stream channels and adjacent floodplain areas will contribute directly to quickflow and therefore confirms Hewlett's equation of variable source area with the expanding and shrinking channel net.

Dunne and Black [39] confirmed the applicability of the variable source area concept to snowmelt runoff production where concrete frost and sub-snowpack saturated soils may provide the mechanism for the generation of overland flow from a continuously varying area.

Various methods have been suggested for quantifying the variable source area within a given drainage basin. Inevitably, early attempts were sometimes speculative. Thus Fig. 8.5 illustrates tentative suggestions about the relationship between varying source area and (A) the percentage of precipitation contributing to quickflow [61] and (B) initial moisture conditions and accumulated storm rainfall [144]. A method of estimating the varying contributing area was used by Betson [21] and Ragan [120] and subsequent workers, whereby channel precipitation is subtracted from total runoff volume and a runoff coefficient of 100 per cent (i.e., an infiltration capacity of zero) is assumed for the contributing area. Unfortunately this method raises problems of defining channel precipitation (e.g., precipitation on

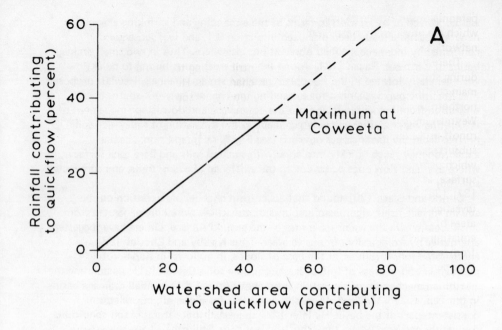

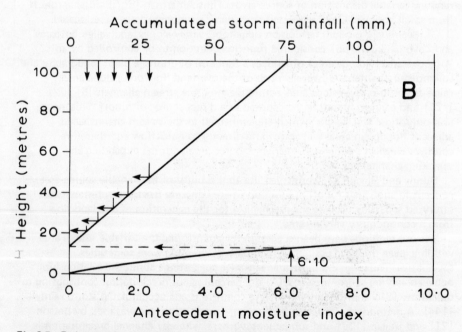

Fig. 8.5. *Early tentative attempts to quantify variable source areas according to (A) the percentage of rainfall contributing to quickflow and (B) initial moisture conditions and accumulated storm rainfall. (From original diagrams (A) by Hewlett [61] and (B) by Tennessee Valley Authority [144].)*

perennial/intermittent/ephemeral/other minor channels? at bankfull stage? etc.) which are obviated by Hewlett's view of the source area as being the channel network.

There have, as yet, been few rigorous field measurements of varying source areas during storm events. Apart from the work of Hewlett and Hibbert at Coweeta, the main contributions in this respect have come from Dunne and Ragan working in northern Vermont, Betson and coworkers in North Carolina, and Kirkby and Weyman in Britain. Dunne's work was part of a more general investigation of the runoff process in which the contribution to total streamflow of surface and subsurface water movement was examined with the aid of an interceptor trench which permitted the separate measurement of water movement over the ground surface, at the base of the root zone and in the zone of perennial groundwater flow [37] [38] [39]. Ragan [120] had earlier used similar instrumentation in a similar investigation. Betson and Marius [22] made a specific attempt to identify source areas of quickflow within a catchment by detecting surface soil saturation and equating surface saturation with source area, finding that storm runoff originates from 'a small portion of the total drainage area'. Similar work by Kirkby and Weyman [83] [89] used a dense network of hillside instruments in two Somerset

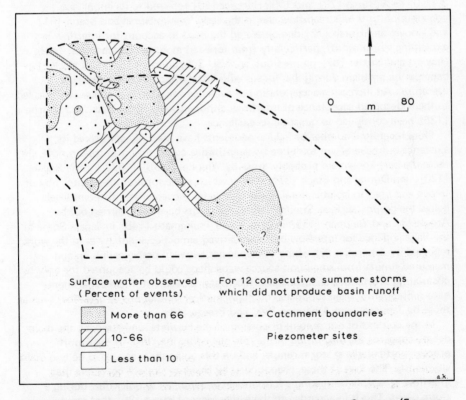

0   m   80

Surface water observed
(Percent of events)

For 12 consecutive summer storms
which did not produce basin runoff

More than 66   — — Catchment boundaries

10-66   • Piezometer sites

Less than 10

Fig. 8.6. *Surface water frequency map for East Twin Brook, Burrington, Somerset. (From an original map prepared by M. J. Kirkby and D. R. Weyman in Institute of Hydrology [83].)*

253

drainage basins to record the presence of surface water and subsurface saturation. Regular site visits provided a series of maximum surface-water area maps, each related to a recorded storm event. Preliminary data analysis indicated that measured areas of surface water were consistently higher than the theoretical runoff-contributing area calculated from peak rainfall and peak discharge, that measured surface water areas increased with peak discharge and that despite contrasts in land use, soil type, and topography, both basins showed similar internal growth patterns of contributing area in which low angle slopes near the channel and on the divides tended to saturate easily but yielded a true storm hydrograph only when connected by channelled flow along topographic hollows. Figure 8.6 shows the relative frequency of surface water for one of the drainage basins in which high frequency areas are connected by lower frequency areas crossing the hillslopes.

It is evident from this discussion that such work as has been done tends to confirm Hewlett's basic concepts, although it is to be hoped that further fieldwork will be initiated in order to provide additional evidence about this problem.

8.6 Interflow

Interflow has long been a recognized component of the runoff process (see section 8.2). In 1936 Hursh [79] and 1939 Hertzler [60] referred to its important contribution to total streamflow and in the early 'forties Hursh and Brater [81] and Hoover and Hursh [72] demonstrated the need to account for interflow in explaining storm runoff, particularly from forested mountain catchments. Then, as Hewlett and Nutter [67] pointed out, an AGU Subcommittee on Subsurface Flow debated the problem during the 'forties with 'little effect on the hydrologic literature'. An increasing accumulation of evidence since that time, however, has led to the widespread acceptance of interflow, although some hydrologists, c.f. Roche [126] have continued to question its existence.

More recently a number of field experiments have not only confirmed its existence but have begun to define its quantitative contribution to total runoff. Of these the best-known are probably those by Whipkey [156] [157] [158], Ragan [120], and Dunne and Black [37] [38] in which some form of interceptor box or trench was used to isolate lateral subsurface water movement at different levels below the ground surface. Similar field experiments have been carried out by Arnett [8] and Weyman [155] in Britain and Tsukamato [148] in Japan. Some of the field evidence for interflow has been derived almost accidentally as in the work reported by Amerman [5] where discrepancies between precipitation on and measured runoff from small unit-source watersheds could be accounted for only by postulating substantial, shallow subsurface water movement. Field experiments have subsequently been supported by mathematical simulations of interflow such as those by Jamieson and Amerman [84] and Freeze [47].

In the context of our present discussion of the Hewlett runoff model, the main debate concerns the *magnitude* of the role played by interflow in the runoff process, particularly in storm runoff, and on this point there appear to be two main viewpoints. The first of these, propounded by Hewlett himself, considers that interflow is capable of making a substantial contribution to quickflow during a storm event. This is in accord with the earlier view of Hursh [80] that rapid interflow is a primary source of stormflow from forested catchments.

Corroborating field evidence was provided by Whipkey [156] [157] [158] and was accepted by Kirkby and Chorley [88] as proof that interflow '. . . is capable of producing runoff peaks in river hydrographs'. Whipkey [158] noted that with deep, coarse-textured soils interflow was generally a function of hydraulic conductivity but that in other conditions it was largely dependent on the existence of subsurface 'biological and structural channels'. Subsequently, the role of pipe flow in subsurface water movement was considered by Jones [86] and illustrated by work in the Institute of Hydrology's experimental catchments in central Wales [82]. Thus Fig. 8.7 shows rainfall and discharge data for a number of ephemeral flowing

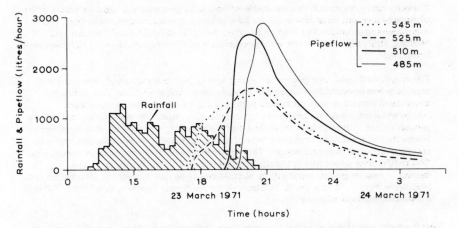

Fig. 8.7. *Rainfall and discharge data for four ephemeral flowing pipes at various elevations in the Nant Gerig basin, Wales. (From an original diagram in Institute of Hydrology [82].)*

pipes in the Nant Gerig basin. The Institute of Hydrology investigations showed that ephemeral pipes carry large quantities of water during and after rainfall. They do not respond immediately to rainfall but begin to flow only after a 'storage deficit has been satisfied' [82] (water could enter a macro-cavity of this type only if the soil water pressure in the surrounding soil was positive (see section 6.3)) and carry about 20 per cent of the total rainfall measured in the contributing catchment area. It was concluded that pipe flow represents a dominant storm runoff process. Further confirmation of Hewlett's view was provided by two years' observations from a small agricultural watershed in Pennsylvania which indicated that less than 5 per cent of the rainfall could be assumed to have reached the stream without 'flowing through the soil' [122].

A second viewpoint regards interflow as only a minor contribution to storm flow and stresses the overwhelming role of overland flow from limited areas of the drainage basin. Thus Dunne and Black [37] found that although interflow occurred during large storms it was not an important contributor to total storm runoff and that it was the ability of an area 'to generate overland flow' which determined its importance as a producer of storm runoff. Again, theoretical simulations of runoff generation by Freeze [47] showed that '. . . there are stringent limitations on the occurrence of subsurface storm flow as a quantitatively significant runoff-generating mechanism'.

To a large extent the debate outlined above is a matter of semantics. We have seen that Hewlett regarded as channel flow virtually any water movement other than interflow so that interflow is, by definition, a major component in the storm hydrograph. On the other hand, Dunne and Black and others treated as overland flow much of what Hewlett regarded as the extension of the channel net during precipitation. The close identity of what have previously been interpreted as differing viewpoints is clearly illustrated by the following quotations:

As soil water moves downward and concentrates, it must finally saturate soil and then surface to make its contribution to streamflow. . . (Hewlett, 1961) [61]

It would not be impossible to treat overland flow as an exceptionally rapid extension of the channel system into areas where the soil cannot transmit water as subsurface flow. This seems more appropriate than treating all direct runoff as overland flow. (Hewlett and Hibbert, 1967) [65] (Reprinted with permission from *Forest Hydrology* © 1967 Pergamon Press Ltd.)

The only runoff-producing mechanism which approached channel runoff in amount and sensitivity was overland flow on a small concave portion of the hillside. . . Significant amounts of stormflow were produced on the hillside by return flow and by direct precipitation onto the saturated area that developed . . . where the water table reached the ground surface. Although the return flow had the same origin as the flow that remained below the soil surface, its emergence from the ground caused it to have features that were lacking in its subsurface counterpart. The first of these was its greater velocity, which was 100 to 500 times greater than the calculated velocities of subsurface flow. . . The second . . . was its greater sensitivity to fluctuations of rainfall intensity. . . (Reprinted with permission from T. Dunne and R. D. Black, *Water Resources Research*, 6, 478-490, 1970, © American Geophysical Union).

Two further comments may be added by way of emphasis. In terms of the runoff process both infiltration and interflow act in a substantially similar way as agents of surface saturation from below except that the former is likely to dominate on flat areas and the latter on steep slopes. In other words, Dunne and Black's water table rise and Hewlett's basal slope saturation by interflow are similar responses to essentially similar mechanisms neither of which is likely to be rapid enough to make a significant contribution to storm runoff except through the medium of translatory flow or through pipe flow in saturated conditions.

8.7 Hydrograph analysis

A standard hydrological technique which has, in the past, been used in the attempt to solve a wide range of hydrological problems is that of hydrograph analysis. It has become customary to attempt either a simple or a more detailed genetic separation of the components of the flood hydrograph. Thus in the simple case the hydrograph is separated into direct runoff and base runoff while in the more detailed case the various components of runoff, i.e., channel precipitation, overland flow, interflow, and groundwater flow, may be 'identified' and their changing magnitude plotted against time.

Some of the assumptions made in hydrograph analysis are quite elaborate. Thus Brater [26] suggested that a rapid increase in stream water levels could cause water to flow from the stream channel into the surrounding flood plain materials, thereby creating a period of negative groundwater flow as is illustrated in Fig. 8.8. At the

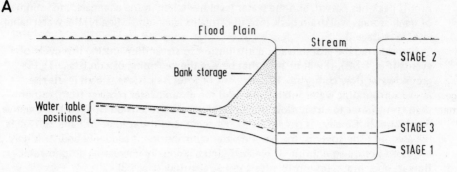

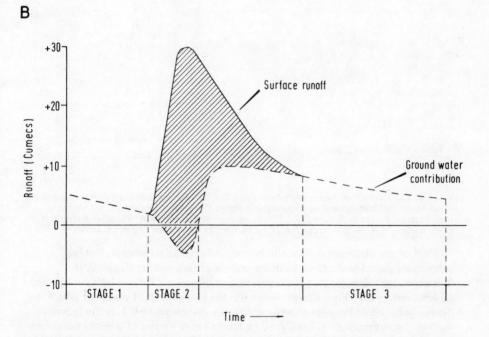

Fig. 8.8. *(A) Stream-groundwater relationships and (B) the stream hydrograph before, during, and immediately after storm rainfall.*

end of a dry period when groundwater flow is the only component of streamflow, the water table will slope towards the main streams (stage 1 in Fig. 8.8A). Subsequent rainfall and infiltration may result in an increase in water table height but, at the same time, the concentration surface runoff into the comparatively limited stream channels may result, initially, in a much larger increase in stage there. There is now, therefore, a reversed water table gradient close to the main streams as a result of seepage from the rivers into the surrounding floodplain areas (stage 2). In this way, an often significant proportion of total runoff is stored in the river banks, and is thus, appropriately known as *bank storage*. After the surface runoff peak has passed, and the water level has fallen in the channels, this volume of bank storage will drain back into the streams again until, finally, the water table level is at stage 3.

Considering the groundwater contribution to streamflow during this cycle of events (Fig. 8.8B), it will be seen that before the beginning of rain (stage 1), groundwater flow comprises 100 per cent of total runoff. As stream levels rise above surrounding water table levels, and the groundwater receives from, rather than contributes to streamflow, the groundwater contribution is, in fact, a negative one (stage 2). Finally, as stream levels fall again, and the volume of bank storage is released into the streams, the groundwater contribution is suddenly and strikingly increased before equilibrium is reached, and a gradually decreasing groundwater flow is, once more, responsible for total streamflow (stage 3).

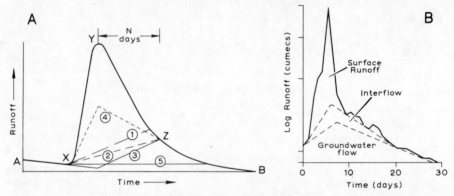

Fig. 8.9. *(A) Methods of hydrograph separation. See text for explanation. (B) Hydrograph of River Stour, Suffolk, plotted semi-logarithmically to illustrate the method of hydrograph separation suggested by Barnes [13]. (From an original diagram by A. Bleasdale and others, in* Conservation of Water Resources, *pp. 121-136, Institution of Civil Engineers, London, 1963.)*

Most of the assumptions made in hydrograph analysis, however, involve simplifying generalizations of an often quite arbitrary nature. Figure 8.9A illustrates, for example, that the separation of direct runoff from base runoff may be achieved by drawing a straight line from the sharp break of slope (X) where discharge begins to increase to some arbitrarily chosen point (Z) on the recession limb of the hydrograph. Point Z may be located at the point of greatest curvature near the lower end of the recession limb (line 1) [81] or at a given time interval after the occurrence of peak flow (line 2). The time interval (N) may be determined from hydrograph inspection or from a simple empirical equation

$$N = A^{0.2} \qquad (8.1)$$

where N is in days and A is the drainage area in square miles [97A]. Alternatively, the pre-storm base flow recession curve (AX) may be projected forward in time to a point beneath the peak of the hydrograph and then connected by another straight line to the arbitrarily chosen point Z (line 3) or the average base flow recession curve, determined from a number of recession limbs, may be extended backward in time to a point beneath the peak and then joined by a straight line to point X (line 4). Finally, the simplest and probably the most logical approximation is that of a horizontal line drawn from point X to its intersection with the recession limb (line 5).

An apparently more sophisticated method of hydrograph separation was first suggested by Barnes [13] and is simply illustrated in Fig. 8.9B. In this method it is assumed that after the hydrograph peak the decline of each component of total runoff, i.e., surface runoff, interflow, and groundwater flow may be represented by a recession curve, having a particular recession constant, which will plot on semi-log paper as a straight line. The individual contribution of each component to total runoff may then be determined by extending each recession line back to an appropriate point under the peak of the hydrograph, and then drawing a further straight line to the initial point of rise. A number of the assumptions implicit in this method have been questioned and in addition neither Barnes nor subsequent workers have clearly defined the location of the peaks of interflow or groundwater flow components [92]. Each of the foregoing techniques, to which only brief reference has been made in this context, are discussed in some detail in the standard engineering texts on hydrology.

The growing awareness in recent years that the generation of runoff cannot be explained simply in terms of the Horton infiltration theory of runoff at first cast doubt upon traditional methods of hydrograph analysis and finally discredited them altogether. Thus in the early 'forties Hoover and Hursh [72] and Hursh and Brater [81] emphasized the need to account for interflow when explaining storm hydrographs and W. M. Snyder has consistently questioned traditional hydrograph separation procedures [139] [140]. More recently Freeze [46] referred to hydrograph separation as appearing to be '. . . little more than a convenient fiction' while Nash and Sutcliffe [115] observed that with all the traditional techniques of hydrograph separation the observed volume of direct runoff for a given storm event is not the total volume of water contributed to streamflow in that event but some arbitrary part of it, i.e., the part above the line of separation. They further called for a rejection of 'the *a priori* division of hydrographs into ill defined components', arguing that there is 'a continuum of different paths by which runoff reaches the streams'.

The last point is fundamental and has been developed in detail in the earlier part of this chapter. It is now recognized that it is the speed of arrival of the water in the stream channel which is the most important factor determining the shape of the hydrograph rather than the precise route followed by the water in reaching the channel. Furthermore, although it is hoped that continuing research into physical processes will eventually yield a solution, it is at present impossible to quantify the contribution of the major hydrograph components to total streamflow. For these reasons modern techniques of hydrograph analysis are based almost entirely on the speed of response of a catchment to precipitation and in this connection it is interesting to note that Rogers [129] demonstrated a relationship between channel length frequency distribution and the shape of the runoff.

One possible approach, developed by Snyder [140] enables the *total* response of streamflow to all storms on a given watershed to be analysed. Only the baseflow of the recession antecedent to the storm is separated but no attempt is made to isolate groundwater flow or interflow resulting from the storm. In this method the hydrograph is 'open ended', so that there is no need to specify a time at which a

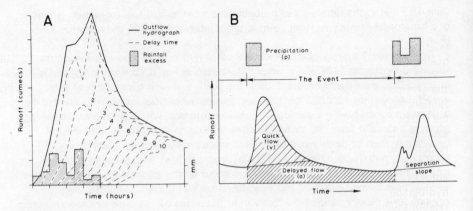

Fig. 8.10. *(A) Outflow hydrograph developed from the Snyder (1968) method, showing one-hour time separation lines; (B) Hydrograph separation into quick flow and delayed flow using a constant separation slope of 0.05 cusec/sq. mile hour, i.e., 1.13 mm/day. (From original diagrams by (A) Snyder [140] and (B) J. D. Hewlett and A. R. Hibbert, in* Forest Hydrology, © *1967, Pergamon Press Ltd. [65])*

certain component of flow ceases, and is separated into a large number of time-of-flow categories. Thus in Fig. 8.10A the solid line represents total stream response to the indicated rainfall excess, i.e., the increase in streamflow above antecedent flow irrespective of whether this is surface runoff, interflow, or groundwater flow. The broken time-separation lines are constructed for delay-time increments of 1 hour and areas on the graph between indicated delay times represent volumes of flow between these delay times.

A second, now widely used, approach is that proposed by Hewlett and Hibbert [65] in which 'quick flow' (*not* as defined in Fig. 8.1) is separated from 'delayed flow' by a line of constant slope (0.05 cubic feet per second per square mile per hour) projected from the beginning of a stream rise to the point where it intersects the falling limb of the hydrograph (Fig. 8.10B). The value of 0.05 cusec/mi^2 (1.13 mm/day) was chosen because it was greater than the normal diurnal fluctuation of flow, gave a relatively short time base to the largest single-peaked hydrographs in the study area and permitted large storms separated by a period of about three days to be calculated as separate events. The same value has, in fact, been used by subsequent workers in entirely different areas, c.f. Walling [150] in south-west England. Hornbeck [73] used a value of 1.25 mm/day in New Hampshire and Woodruff and Hewlett [163] slightly modified the technique for use in the wider area of the eastern USA.

The separation of quick flow and delayed flow may be achieved graphically or, more usually, by means of computer programs such as those originally referred to by Hewlett and Hibbert [65], and Hibbert and Cunningham [69] and by Woodruff and Hewlett [163]. It will be appreciated that this technique is no less arbitrary than most of the early methods of hydrograph separation, although at least the arbitrariness is logical and the method is consistent wherever it is applied and avoids the illusion of precision created by many of the outmoded techniques where the separate components of runoff were named, albeit without being defined. In this connection it should again be emphasized that the terms 'quick flow' and 'delayed

260

flow' which were proposed by Hewlett and Hibbert are in no sense genetic terms, i.e., quick flow may comprise surface runoff, interflow, and groundwater flow—the essential feature being that it is the water which gets into the stream channel quickly.

The separation of quick flow and delayed flow, so defined, has contributed to the developing analysis of hydrologic response, which was defined by Hewlett and Hibbert [65] as either the fraction of total precipitation yielded as quick flow or the fraction of total runoff which is quick flow, although the former is the one which has been most frequently calculated. For individual storms, hydrologic response varies between about 1 and 75 per cent, while the average for all storms in the eastern USA is about 10 per cent [66]. This figure may seem surprisingly low although, as Hewlett and Nutter [66] pointed out, precipitation and the storm hydrograph are often plotted together using different units which mask the real relationship between them. Plotted in the same units the average storm hydrograph

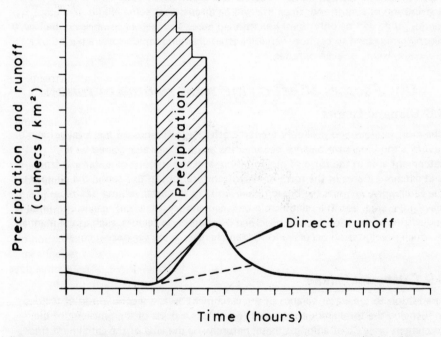

Fig. 8.11. *Average precipitation and storm hydrograph for the eastern United States. (From an original diagram by J. D. Hewlett and W. L. Nutter,* An Outline of Forest Hydrology, *University of Georgia Press, 1969.)*

for the eastern USA would resemble that in Fig. 8.11. Regional variations of hydrologic response in the eastern USA were calculated and mapped by Hewlett and Nutter [66] and by Woodruff and Hewlett [163] and were shown to provide a useful indication of spatial and temporal changes of flood potential.

It should be noted, in conclusion, that although the genetic separation of the storm hydrograph has been shown, in this section, to be unrealistic at the present time, it may yet prove feasible, with further research developments, to identify the phase relationships of the various components of the hydrograph, either by measurements of water temperature [134] or by the analysis of water chemistry. Some work on the chemical analysis of runoff has been done by Russian workers [149] and a useful method was proposed by Pinder and Jones [119] in which a chemical mass-balance is used to calculate the groundwater flow component of total streamflow. Kunkle [93], Toler [147], and Nakamura [113] used conductivity to estimate the groundwater contribution to base flow.

Factors affecting runoff

It will be convenient at the outset of this discussion of the factors affecting runoff to differentiate between those factors which combine to influence the *total volume of runoff* over, say, a period of several years, and those other factors which combine to influence the *distribution of runoff in time,* say, over a period of one year or less. The former group of factors will significantly affect the areal distribution of runoff and, since this will be discussed at some length in a later section, they will be only briefly mentioned here. The factors comprising the second group tend to be more variable, often changing rapidly over a period of a few days, hours, or even minutes.

(A) *Factors affecting the total volume of runoff*

8.8 Climatic factors

The most obvious and probably the most effective influence on the total volume of runoff is the long-term balance between the amount of water gained by a catchment area in the form of precipitation, and the amount of water lost from that catchment area in the form of evapotranspiration. In this sense, the climate of the catchment area sets the broad upper limits to the total volume of streamflow leaving the area, but this relationship between annual totals and means of rainfall and evapotranspiration may be modified by short-term factors, such as the manner in which precipitation occurs, and sudden changes in the vegetation cover.

8.9 Catchment factors

In addition to the water balance of the catchment area, a second group of factors influencing the total amount of runoff comprises aspects of the physique of the catchment area. Chief amongst these, naturally, is the *area* of the catchment since, other factors being equal, this determines the total amount of precipitation caught. It should be noted, however, that the effect of area may depend upon the prevailing climatic regime. Thus, in a region where PE exceeds $P,$ a larger catchment is just as likely as a smaller one to have a zero or very low runoff whereas, in a region in which P exceeds $PE,$ the larger the catchment area is, the larger will be the total amount of runoff.

   Slope, soil, and *rock type* may indirectly influence the total runoff from a catchment through their effects in delaying water movement after precipitation thereby possibly affecting the amount of evapotranspiration. In general, the highest annual runoff would be expected from steeply sloping areas having thin soils and impermeable rocks. Finally, the average *height* of the catchment may affect total runoff, again indirectly, through its direct orographic influence on precipitation amounts.

(B) *Factors affecting the distribution of runoff in time*

From the hydrologist's point of view, climate, which it has been suggested sets the broad upper limits to total runoff, is a comparatively stable environmental factor; even more so are the catchment factors which have been discussed. It is, then, the second group of factors which influence the distribution of runoff in time, and which, themselves, tend sometimes to be more variable and unpredictable, which have attracted the most attention. For convenience, this second group may be further subdivided into *meteorological* and *catchment factors.*

8.10 Meteorological factors

In the sense that precipitation forms the raw material of streamflow, meteorological factors are obviously of great importance and their variation with time tends to be closely related to similar variations of runoff.

Type of precipitation

For the purpose of this discussion, it may be considered that precipitation occurs either as rainfall or as snowfall. Other forms, such as hail and sleet may be conveniently grouped with rainfall; hail, for example, often occurs in conditions which favour rapid melting at the ground surface, and the snow content of sleet will normally tend to liquefy soon after contact with the ground.

The most important feature of a snow blanket is its storage capacity, and the resulting time interval between the occurrence of precipitation and the occurrence of runoff. Thus, in a simple and rather obvious example, precipitation falling as snow during the winter months between December and February will not contribute to runoff until melting occurs during the spring. But a further factor is that the snow blanket itself may have a storage capacity for liquid water, comparable with that of a highly porous soil, so that, in the early stages of spring melting, large quantities of meltwater and rain that may have fallen on the snow surface will be accumulated in the snow layer [162]. Then, with further melting, this blanket may collapse and suddenly add a considerable volume of water to the streams. Flood investigations in Scotland [161] showed that a blanket of snow up to 30 cm thick may have completely absorbed a heavy rainfall, and stored it for several hours, before collapsing to cause a runoff peak some 35 per cent in excess of that to be expected.

Only in high latitude and high altitude areas is the effect of the accumulation and melt of snow of long-term significance, although comparatively little work has been done in these areas. Ward [152] listed the principal Canadian contributions and additional findings were presented by McCann and others [103]. Dingman [34] reported measurements of summer runoff from a small drainage basin in Alaska. Figure 2.20 suggests that in the British Isles runoff will be materially affected by snow only in the highland areas of Wales, Scotland, and the Pennines.

Some valuable work has been done in the Pennines by Smith [135] and other investigations by the Institute of Hydrology [83] are currently in progress in Wales.

Precipitation falling as rain may, of course, contribute directly to runoff, but the extent to which it does so will depend upon the interaction of numerous meteorological and other factors, the most important of which will be discussed in the following paragraphs.

Rainfall intensity and duration

It has already been shown that, in some circumstances, the intensity with which rain falls may be an important factor in determining the proportions of the rainfall which go to overland flow, interflow, and groundwater flow and, therefore, the speed with which water may reach the stream channel. The duration of rainfall is important, particularly in relation to the hydrologically responsive areas of a catchment referred to on pages 250–254 and particularly in flat, low-lying areas. Here, if rainfall is sufficiently prolonged, infiltration may raise the surface of saturation to the ground surface itself, thereby reducing the infiltration capacity to zero and causing a sudden increase of surface runoff.

The duration of rainfall also becomes significant when considered in relation to the *mean travel time* of a drop of water from its point of impact on the catchment area as rainfall, to its exit from the catchment area as streamflow. If the rainfall duration is equal to or greater than this mean travel time, then the whole of the catchment area is likely to be contributing to runoff during the later stages of the storm, so that the potential runoff is at a maximum. If, on the other hand, the duration of rainfall is less than the mean travel time, then the potential runoff will be lower than the maximum because only part of the catchment will be contributing to runoff before rainfall ceases. In this context, it is apparent that the importance of rainfall duration will tend to vary with the size and nature of the catchment. In a small catchment, with steep slopes, maximum potential runoff is likely to be caused by a rainfall of much shorter duration than would be required in a large, gently undulating catchment.

Rainfall distribution

The time relationship between rainfall and runoff may be greatly affected by the distribution of rainfall over the catchment area. A given volume of rainfall, which is uniformly distributed over the whole of a catchment, will have lower intensities and is, therefore, less likely to produce quickflow than is the same volume of rain falling on a small, localized part of the catchment. The first type of rainfall distribution will tend to result in an increase in baseflow, and consequently a long-term increase in streamflow, while the second sort of distribution will tend to give larger volumes of quickflow and thus, a more sudden, short-lived increase in streamflow.

When discussing rainfall distribution, one is rarely concerned with a stationary pattern. Much of the rainfall in the British Isles, for example, is related to series of depressions moving in from the Atlantic, causing belts of frontal rain to cross the country. In this way, a rainstorm may begin in one part of a catchment area, and end in another, and the direction of movement of the storm may be very important. Figure 8.12, for example, illustrates one effect of the direction of

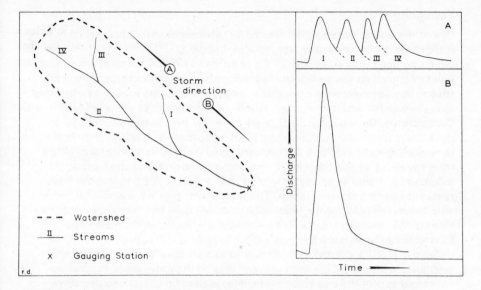

Fig. 8.12. *Diagram showing the probable runoff hydrographs resulting at gauging station X from two storms of equal rainfall amount moving in opposite directions over a hypothetical catchment.*

movement of a storm in a long, narrow catchment area. When the storm is moving in an upstream direction, the flood peaks from the lower tributaries will pass the gauging station at X before those from the middle and upper tributaries. When, on the other hand, the storm moves downstream, the flood peaks from the individual tributary streams are more likely to arrive at the gauging station at approximately the same time, with the result that the maximum runoff at this point will be considerably higher than in the case of the storm moving upstream.

8.11 Catchment factors

The various meteorological factors which have been discussed can all be measured with reasonable accuracy thus facilitating their correlation with runoff characteristics. It is, however, much more difficult to apply such precise determinations to many catchment factors. Furthermore, some of these factors, such as shape, topography, and soil type, remain fairly constant over long periods while others, such as those associated with land use, may change very rapidly. The main problems, however, arise partly from the fact that the various components of runoff may be differently affected by each of the catchment factors, so that the net effect of any one catchment factor on variations of total runoff with time are difficult to establish, and partly from the fact that only limited, hydrologically responsive, areas of a catchment make a substantial contribution to quickflow, so that it is the influence of catchment factors in these limited areas, and not over the whole of the catchment, which must be determined. The fact that the variable source areas within a catchment are difficult to define means that the present discussion must be largely qualitative in nature.

265

Topography

One of the several factors which may be included under this rather general heading is the *shape of the catchment area,* which is known to influence runoff through its effects on flood intensities, and on the mean travel time of a drop of water from its point of impact on the surface of the catchment to its point of exit in the main stream. In a generally square or circular catchment area, the tributaries often tend to come together and join the main stream near the centre of the area [118]. Consequently, the separate runoff peaks generated by a heavy fall are likely to reach the main stream in approximately the same locality at approximately the same time, thereby resulting in a large and rapid increase in the discharge of the main stream. If, on the other hand, the catchment area is long and narrow, the tributaries will tend to be relatively short, and are more likely to join the main stream at intervals along its length. This means that, after a heavy rainfall over the area, the runoff peaks of the lower tributaries will have left the catchment before those of the upstream tributaries have moved very far down the main stream. Elongated catchments are thus less subject to high runoff peaks.

Snyder [138] suggested that one way to express the *effective shape* of a catchment was to draw isopleths of travel time of the water above the drainage outlet and to plot the area between isopleths against time. The resulting curve would then express the shape of the catchment by giving the increment of the area that is at any particular travel-time distant from the outlet.

A second pertinent topographical factor is the *slope of the catchment area* which, as has already been suggested, may affect the relative importance of predominantly vertical movement of water by means of infiltration and the predominantly lateral movement of water by means of interflow and overland flow, the former tending to be more important in flat areas, the latter in steeply sloping areas. Furthermore, because the speed of water movement will tend to increase with slope, runoff in steeply sloping areas will reach stream channels quickly.

The shape of the runoff hydrograph depends not only on the speed with which water gets into the stream channel but also on the speed with which it moves down the channel to the outlet of the catchment. *Channel slope* may therefore be as important as catchment slope and has frequently yielded more significant correlations with runoff characteristics.

Geology

By its very nature, the geology of a catchment area will exert a fundamental influence on runoff which will be felt in many ways. Two aspects are, however, of chief importance—the type of rock in which the catchment area has been eroded, and the main structural features of the area.

The influence of rock type on runoff may be seen in the close relationships which often exist between geology and the texture of drainage. This is a problem which will be discussed in a subsequent section but, at this stage, it is relevant to emphasize the fundamental contrast between, say, the drainage networks developed upon chalk and clay terrain. The chalk will probably have a few large streams with wide interfluves, in contrast to the fine, vein-like pattern dissecting the clay into small interstream spaces, no part of which is very far from a drainage channel, and thus encouraging the rapid disposal of overland flow and interflow. The extreme

case which can be quoted in this context, is, of course, that of the very coarse
drainage network normally found in areas of massive limestone. This is emphasized
in Fig. 8.13, which shows that, on the limestone outcrop in the Ingleborough region
of the Pennines, surface drainage virtually disappears, and that the greatest

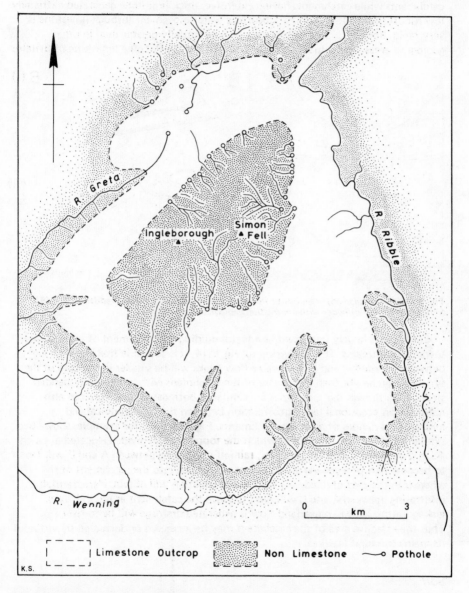

Fig. 8.13. *Simplified map of the Ingleborough district showing the relationship between
drainage network and geology. (From an original diagram by A. A. Miller,* The Skin of the
Earth, *Methuen, London, 1953.)*

proportion of water movement takes place *below* and not over, the surface. Where the dominance of subsurface water movement is associated with an aquifer having high storage and retention properties water resurgence will take place slowly during the period after a storm and this will result in reduced runoff peaks. Thus studies in Pennsylvania showed that high runoff peaks occurred in shale and sandstone catchments while catchments having extensive, thick limestone generated extremely low runoff peaks [159]. Where subsurface water movement through limestone is very rapid, however, runoff peaks may be substantially greater than in other geological areas. This situation is found in parts of the Pennine limestone of Britain.

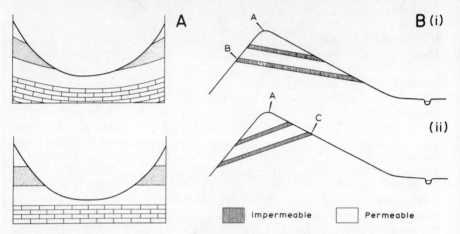

Fig. 8.14. *(A) Similar valley sections through synclinal and horizontally bedded strata; (B) Similar topographic divides with contrasting geological structures.*

Structure is largely important as a factor guiding the movement of groundwater towards the streams. Thus, referring to Fig. 8.14, it is probable that the time lag between the rainfall and groundwater flow peaks will be smaller in the case of the synclinal catchment than in the case of the catchment with horizontally bedded strata, even though the rock types are similar in both cases. Structure can also influence an occasional lack of correlation between the topographical and hydrological divides of adjacent catchments. Figure 8.14B, for example, shows two similar relief sections, in both of which the topographical divide is located at point A. In the case of divide (i), however, rainfall infiltrating between A and B will tend to become part of a general groundwater movement into the catchment of river (i), whereas in the case of divide (ii) the reverse situation will obtain. Here, rainfall infiltrating between A and C will tend to leave the catchment as ground water leakage, although when overland flow is generated drainage will be to river (ii). Thus the effective area of the catchment may be increased or decreased by virtue of its main structural features.

Soil

The influence of *soil type* on infiltration characteristics, and its consequent effect upon the disposition of rainfall as either overland flow, interflow or groundwater flow, has already been discussed and it must suffice at this stage merely to remark that open-textured sandy soils will tend to be associated with higher infiltration values than fine-grained, closely compacted clay soils and will, therefore, tend to generate smaller volumes of quickflow.

Soil profile characteristics are important in relation to their effects upon infiltration and the generation of interflow. In particular, marked reductions of hydraulic conductivity with depth, especially in the upper horizons, facilitate the formation of interflow and, during prolonged rainfall, the saturation of the soil surface and the generation of overland flow. On the other hand, deep uniformly permeable soils tend to encourage continued vertical infiltration and the dominance of baseflow over quickflow.

Hydrologists have for a long time recognized the relationship between the amount of runoff (especially quickflow) produced by a given rainfall and the *moisture content* of the soil, expressed either as a direct measurement or indirectly as an antecedent precipitation index. Earlier workers, using Hortonian concepts, tended to ascribe this relationship to the reduction of infiltration capacity which accompanies the increase of soil moisture content (see section 6.18). In the light of modern views on runoff formation, however, it will be appreciated that, in general, the source areas within a catchment will tend to expand as the catchment becomes wetter and that this is a much more likely explanation of observed relationships between runoff and catchment moisture indices.

Freezing conditions may change the runoff characteristics of soils to a greater or lesser extent, depending upon a number of considerations. The overall effect of freezing is to produce an impermeable surface, and this will be intensified both by the moisture content of the soil and by the degree of frost. A moist soil, for example, will freeze into a more impermeable mass than a relatively dry soil, but will take longer to do so. The depth of frost penetration is also an important factor since, if this is great, daytime thawing of the surface layers will have little effect upon the infiltration capacity or the runoff characteristics of the soil, whereas a thin layer of frozen soil may thaw completely during the daytime, and thus allow the downward movement of accumulated water.

Vegetation

The effect of vegetation on the distribution of runoff with time has probably received more attention from hydrologists than the effect of any other catchment factor and yet because of the complexity of the interactions involved there is still much confusion. The complexity of the problem becomes evident when it is considered that the total effect of vegetation on runoff is comprised of its individual effects on interception, evapotranspiration, soil moisture movement (particularly in terms of infiltration and interflow), and also the pattern of snow accumulation and melt. This type of complex inter-relationship, which will be touched upon again in the concluding chapter, may be briefly illustrated by comparing a forested and a non-forested area on which all precipitation falls as rain and which are identical, apart from vegetation differences. In the non-forested area transpiration and interception losses will probably be lower, thus producing wetter soils with a reduced capacity for additional water storage, and thereby resulting in higher volumes of dry season quickflow and higher instantaneous peak flows. In the wet season the storage capacity of the soils under both vegetation types will be at a minimum so that the response of runoff to rainfall will be similar from both areas.

However, the partitioning of precipitation between overland flow, interflow, and groundwater flow and also the speed of entry of water into the stream channels may depend more on soil, slope, precipitation, and other factors than on the simple contrast between vegetation types. Also the situation in many areas is further complicated by the fact that vegetation cover has been and is still being extensively altered by man with resulting short-term and long-term hydrological implications. This aspect of the problem will be further discussed in the next section.

In a less direct way, vegetation may influence runoff through its effect on soil type. Many types of zonal soil appear to evolve only beneath specific types of vegetation [42] so that in Eurasia, for example, brown forest soils are found over much of the area formerly covered by deciduous woodland, chernozems are found beneath the sub-humid grasslands, and chestnut-brown soils beneath the drier steppes. Since the hydrological characteristics of the major soil types differ, albeit only slightly, these will tend to effect large-scale contrasts between, say, infiltration characteristics and, thus, between the proportions of baseflow and quickflow. Lvovitch [101], for example, suggested that if the flood peaks from solonetses and solonchaks were assigned the value of 100, those from podsols and clayey soils would be 80-85, from chestnut soils 65-70, from highly fertile clayey chernozems 40-50, and from sandy soils 20-35.

Drainage network

The character of the drainage pattern is important because of the extent to which it may reflect many of the physical characteristics of catchment areas which have already been discussed under preceding headings. Attention was earlier drawn, for example, to the contrast in the texture of drainage on clays and chalk and, in Fig. 8.13, to the contrast between limestone drainage and that developed on other rocks, both instances emphasizing the control which hydrogeology exerts on the drainage pattern. Thus in a catchment where subsurface storage and hydraulic conductivity are low the ground surface will be prone to saturation resulting both from the downslope movement of interflow and the increase in elevation of shallow water tables during and immediately after precipitation. In this situation the channel network will expand rapidly during precipitation, or, stated in alternative terms, a large proportion of the precipitation will be evacuated as overland flow. On the other hand, in a catchment where subsurface storage and hydraulic conductivity are high, deep infiltration will be encouraged, channel network expansion during precipitation will be small, and only a small proportion of the precipitation will be evacuated as overland flow. Consequently, the surface, topographic expression of overland flow, i.e., interconnected depressions, rills and valleys, will be correspondingly smaller.

In an area of given subsurface storage and hydraulic conductivity the relative proportions of quickflow and baseflow will depend on the amount of precipitation. In a dry area all precipitation may be evacuated as baseflow whereas in a wet area the input of precipitation may exceed the subsurface storage and transmission capacity of the catchment and result in over-the-surface movement of water during precipitation. Thus, climate is also reflected in the drainage network. Figure 8.15

270

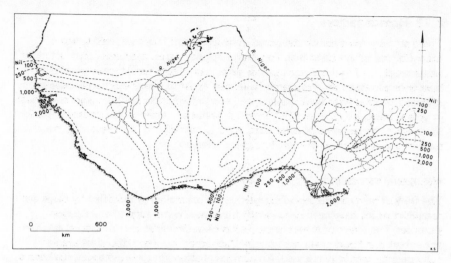

Fig. 8.15. *Map showing the River Niger and its main tributaries and isolines of mean annual water surplus (mm). (From an original diagram by Garnier [48].)*

shows the Niger and its main tributaries, together with an estimate of the average annual water surplus; the evident correlation between areas of high surplus and a dense tributary network, and areas of low surplus and a sparse tributary network, needs no further emphasis. Similar relationships are apparent for other large rivers, such as the Nile, which flow through a series of distinct climatic zones.

A final point which can be made in connection with the influence of the drainage network on runoff concerns the presence of lakes and swamps in a catchment area. Where these occur they tend to 'absorb' high runoff peaks and thus to exert a moderating influence on the hydrograph, which is particularly beneficial in catchments which generate large volumes of quickflow. Examples of the influence of lakes in this respect are numerous and normally well known. The effect of swamps, however, has been less well documented, and it is, thus, of interest to note the role of the swampy Barotse Plain in regulating the flow of the Zambesi. Reeve and Edmonds [123] suggested that this natural reservoir is capable of storing over 17.2×10^9 cubic metres at the peak of high floods, and of delaying the flood peak by some three to five weeks. In this particular case, a large part of the moderating effect of the swamp results from substantial evaporative losses which would probably be less important in, say, higher latitude areas.

Although the effects on runoff of each of the catchment factors which have been discussed have been individually investigated for a wide range of conditions, there have been comparatively few studies of the effects of many variables over a large area. Valuable work was done, however, by Lull and Sopper [100] and Reich [125] in the north-east of the USA. Lull and Sopper examined the effects on average annual and seasonal runoff of a wide range of factors and found that precipitation, percentage of forest cover, and mean maximum July temperatures had the most influence. Reich investigated a number of ways of stratifying runoff characteristics by using physiographic criteria and found no way of improving upon the basic physiographic divisions first suggested by Fenneman.

271

8.12 Human factors

There are very few areas of the world in which runoff is not affected to some extent by the influence of man. In remote uplands dams have been constructed for water supply and hydro-electric power generation. Elsewhere, former grasslands have been ploughed up, moorlands have been forested, semi-desert areas have been irrigated, swamps have been drained, and everywhere there has been a great increase in urbanization, and the resulting spread of artificial, impermeable surfaces. In all these ways, the response of catchment areas to rainfall and, consequently, the pattern and distribution of runoff has been changed.

Hydraulic structures

The flow of many of the world's large rivers is controlled or modified by dams and reservoirs which have been constructed for power, water supply, or irrigation purposes. The effect of these structures has been similar to that of natural lakes to the extent that flood peaks are normally 'absorbed' by the artificial lakes and subsequently gradually released. An additional effect, in some instances, has been a marked diminution of flow, particularly where multi-purpose schemes are in operation in which the impounded water is used not only for water supply and power but also for irrigation. Lvovitch [101] estimated that the average amount of water withdrawn each year from the world's rivers represented approximately 2 per cent of their combined annual discharge but that, in the case of some rivers, such as the Nile and the Syr Darya, this figure rose to almost 50 per cent, although much of this water is not consumed, and re-enters the main stream below irrigation projects.

Streamflow may also be affected by artificial modifications to stream channels, especially in areas which are prone to flooding. Such modifications commonly include channel straightening and enlargement and the construction of relief and by-pass channels in order to reduce both discharge and water levels at critical points.

Structures located away from the stream channels may also influence runoff. For example, the use of snow fences to control the spatial accumulation of snowfall and thereby, also, the timing and yields of water from subsequent melting, were described by Berndt [20] and Martinelli [106].

Agricultural techniques

A second aspect of human influence on runoff may result from the application of specific agricultural techniques and practices, particularly where these cause a sudden change in catchment characteristics, e.g., vegetation cover. Some of these changes have been brought about deliberately, as *conservation measures,* and in this context the important work carried out over a long period of time by the Tennessee Valley Authority [143] affords an obvious example. Other important long-term work on the influence of land use and treatment on hydrology has been done at Coshocton, Ohio and was presented by Harrold and others [56]. Again, it has been estimated that more than 4000 km^2 of marginal cropland in the Great Plains of the USA have already been converted to permanent grass, as a result of federal conservation programmes, with a corresponding significant reduction in quickflow. Three years after conversion, runoff was representative of that from native meadow [36].

Other changes have brought about 'accidental' and sometimes initially surprising hydrological results. In Russia, for example, experiments showed that autumn *ploughing* may decrease runoff by as much as 45 per cent, and a 20 per cent reduction in runoff resulted in some areas from the cultivation of virgin lands [101]. Again, Ayers [9] reported that in southern Ontario ploughed soils in good management yielded much less runoff during the winter months than grass-covered areas. It was suggested that this was because the rough texture of the ploughland results in extensive narrow ridges of dark soil being exposed to solar energy. Heat is then absorbed and transmitted readily by these patches to melt the snow and in addition, a large potential for depression storage occurs on ploughed land.

Some of the most dramatic land use changes are those associated with *afforestation* and *deforestation*, the hydrological effects of which continue to engender considerable controversy. Numerous experiments, many of which were reviewed by Hibbert [68] and Ward [152], have demonstrated that the volume and timing of runoff may be substantially modified by forest cutting removal practices, e.g., clear cutting, block or strip cutting, and selective thinning. A major objective, particularly in important water supply catchments, is to manage the forests so as to permit optimum yields of both timber and water. Work done to date has emphasized the clear relationships between forest cutting and increased runoff, and the possibility of developing sound forest management practices which are also sound hydrological practices. Other possibilities include the replacement of one type of forest cover by another or the replacement of forest by grassland or other agricultural crops. In general it has been shown that runoff is reduced when deciduous trees are replaced by conifers and increased when forest is replaced by lower growing vegetation such as grass or crops. Understandably, little work has been done to demonstrate the hydrological effects of reforestation and such evidence as does exist is inconclusive.

A major problem is that although it is comparatively easy to identify the main ways in which forest and other vegetation covers are hydrologically different, e.g., in terms of soil stabilization and infiltration characteristics, improved soil moisture storage occurring beneath vegetation covers having high interception and evapotranspiration losses, and spatial and temporal effects on snow accumulation and melt, most of the experimental work which has been done has been of the 'black box' variety which does not permit these various differences to be individually quantified. Thus vegetation changes within a catchment area have been related to the modified output of runoff from the catchment but few attempts have been made to detail the hydrological effects or to relate the effects of land use manipulation to modern ideas on the runoff process although, significantly, Hewlett himself has attempted to do this [63].

In low-lying areas, where the water table is close to the ground surface, crops can often be introduced in place of rough marshy grazings, or the yields of existing crops can be improved, as the result of *artificial drainage.* Both open ditches and buried tile drains are extensively used, for example, in the low flat areas bordering much of the southern part of the North Sea, in the eastern counties of England, and in the Netherlands. Here, the aims of agricultural management are to maintain the water table at that height which permits the optimum growth of crops but, although groundwater control is the objective, the secondary effects of artificial drainage upon runoff are often quite considerable.

Closely spaced open drains behave in a similar way to a high density natural drainage network. In other words, lateral flow distances are comparatively short so that water reaches the channels very quickly. There is, therefore, normally a rapid increase in quickflow after rainfall. In addition, since the water table is lowered evapotranspiration is reduced and so dry period base flow is normally increased. Tile drains are usually buried below the water table and behave in a similar way. In that tile drainage encourages the lateral movement of water at some small distance below the ground surface, its effects on the runoff hydrograph often closely resemble those of interflow [160]. In addition, since the water table level is lowered, evapotranspiration is reduced and therefore dry period runoff tends to be increased.

Urbanization

The effects on runoff of the spread of settlement and ancillary features such as roads, pavements, and airfields has been discussed in detail [14] [132] and needs little emphasis. Over large areas, infiltration capacity is considerably reduced; falling precipitation is caught by rooftops and roads, and is passed through drainage systems which have been designed to dispose of it into nearby streams as rapidly as possible. The result is that, immediately below large urban areas, there tends to be a marked and rapid build-up of surface runoff which will be accentuated where slopes are steep. Increases in the magnitude of peak flows are thus a result partly of an increase in the volume of quickflow and partly of the more rapid movement of runoff which is possible in an urbanized area. Anderson [6] found that these two factors combined to increase flood peaks in northern Virginia by between 2 and nearly 8 times while, in Texas, Espey and others [41] found that urban development resulted in peak discharges which were from 100 per cent to 300 per cent greater than those from undeveloped areas.

Apart from peak flows, urbanization also affects water quality and a wide range of other hydrological variables. In addition, although urban areas account for only a small percentage of the total land area they accommodate a large percentage of the population. In the USA, for example it was estimated [4] that by the year 2000 urban land area would represent about 2.4 per cent of the national land area but accommodate 80 per cent of the population and the bulk of the nation's economic wealth. Recognition of these factors had led to the comparatively recent growth of Urban Hydrology as an identifiable topic, with its own specialized literature, concerned largely with the application and adaptation of basic hydrologic principles in urban areas.

Severity of flooding

In conclusion, it is of interest to note that it is largely the human factors which have been discussed above which are believed to be responsible for the apparent increased severity of floods during recent times. Floods, which may be defined as unusually high rates of discharge often leading to the inundation of land adjacent to the streams, are nearly always the result of quickflow, rather than baseflow, and are thus usually caused by intense or prolonged rainfall, snowmelt, or a combination of these factors.

274

Any increase in the severity of floods is, therefore, likely to be caused by increased rainfall intensity, or duration, reduced infiltration capacity, or the changed efficiency of the drainage network. There is some evidence to suggest that storms are increasing in intensity; the effects of urbanization in reducing infiltration capacities have already been noted and, in addition, such factors as forest clearance and the burning, accidentally or otherwise, of large areas of peat moorland must also be taken into account. Finally, the efficiency of drainage channels is likely to be impeded by bridges, levees, flood walls, and similar structures, and although the individual effect of each may be small, their combined effect in large built-up areas may be surprisingly significant [160].

Variations of runoff

Preceding discussions have shown that many factors combine to influence runoff. So numerous are these factors, indeed, that all variations of runoff are unique since no combination of all the variables is ever likely to be repeated. However, certain general similarities and patterns may be observed both in the spatial variation of annual runoff totals and in the variation of runoff with time at one particular place.

8.13 Areal variations

Figure 8.16A shows average runoff in Great Britain in cumecs per square kilometre for the period 1953-63 and was compiled from annual runoff data published in the *Surface Water Yearbooks* of Great Britain. An outstanding feature of this distribution is the very low values representative of the English Plain and the eastern coastal areas, where rainfall is low and evaporative losses are generally high. At the other extreme, the western fringes of Wales and Scotland have high values in excess of 0.3 cumec/km^2, and there is a fairly rapid transition between the areas of high and low runoff. Another interesting feature is the area of low values extending across the Cheshire plain in the rain-shadow of the Welsh mountains. A similar pattern was noted by Linton [98] in relation to the 1955-6 figures. Linton also showed that this pattern was further emphasized when the discharge ratio, i.e., streamflow as a percentage of rainfall, is plotted for the same area (see Fig. 8.16B). In the wetter areas of the north and west, a higher percentage of the rainfall reaches the streams than in the drier south and east.

Similar maps showing the geographical variation of runoff have been produced for those areas for which sufficient data are available. Although these are too numerous to be reproduced here, the basic data were collected by Lvovitch [102] who compiled the world map of average annual runoff shown in Fig. 8.17. The reliability of this map varies from one part of the world to another but, according to the author, is good for the USSR, most of Europe, the USA, Venezuela, and Australia and varies elsewhere according to the availability and accuracy of published runoff data.

8.14 Seasonal variations

Most rivers show a seasonal variation in flow which, although influenced by many factors, is largely a reflection of climatic variations and, in particular, of the balance

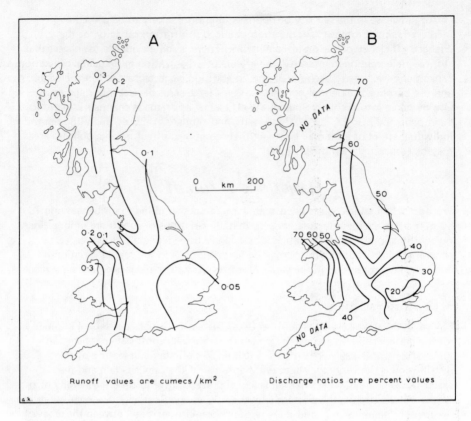

A

B

0.3 0.2

0.1

0.2 0.1

0.3

0.05

0   200
        km

70

NO DATA

60

50

70 60 50

40

30

20

40

NO DATA

Runoff values are cumecs/km²

Discharge ratios are percent values

a.k.

Fig. 8.16. *Maps of runoff in Great Britain. See text for explanation. (From an original diagram by Ward [151].)*

between rainfall and evaporation. The pattern of seasonal variations which tends to be repeated year after year is often known as the *regime* of the river or stream. Thus, equatorial rivers tend to have a fairly regular regime, tropical rivers show a marked contrast between runoff in the rainy and dry seasons, while in other climatic areas complications may arise from the fact that precipitation falls as snow and does not, therefore, contribute directly to runoff until melting occurs. In this way river regimes may be considered in relation to the climatic zones from which they principally derive [15] [53] [101]. Beckinsale [15] presented a useful climatic classification of river regimes based directly upon the Köppen climatic classification. This is illustrated and very briefly summarized in Fig. 8.18, in which it will be seen that Beckinsale made some allowance for the difficult cases of rivers which originate in mountain snow and ice environments over a wide range of climatic types.

Inevitably, however, the larger and therefore the more important rivers cross one or perhaps several significant climatic boundaries and are difficult to analyse in this way. This problem was largely overcome by Pardé, who was one of the earliest geographers to attempt a systematic classification of river regimes based upon the

276

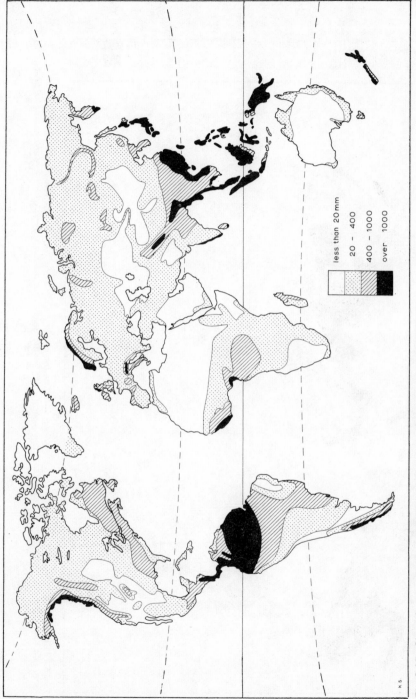

Fig. 8.17. *Simplified world map of mean annual runoff. (From an original map by M. I. Lvovitch, Trans. Amer. Geophys. Union,* **54***, p. 34, Fig. 1, 1973. © American Geophysical Union.)*

	less than 20 mm
	20 – 400
	400 – 1000
	over 1000

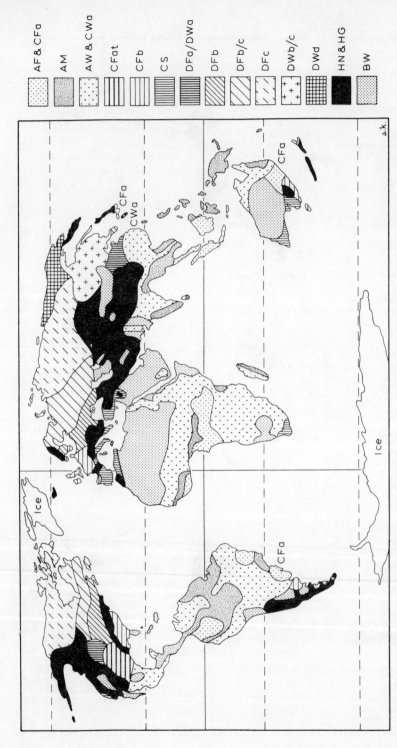

Fig. 8.18. *World distribution of characteristic river regimes based on the climatic terminology of Köppen. NH and HG refer to mountain river regimes which are shown where the scale permits; BW denotes desert and other dry areas where streams cannot originate. (From an original diagram by R. P. Beckinsale in R. J. Chorley (Ed.), Water, Earth and Man, Methuen, London, 1969.)*

Legend:
- AF & CFa
- AM
- AW & CWa
- CFat
- CFb
- CS
- DFa/DWa
- DFb
- DFb/c
- DFc
- DWb/c
- DWd
- HN & HG
- BW

278

nature, rather than upon the causes of the seasonal variations in flow. Pardé, in fact, distinguished three main types of river regime: simple, complex I, and complex II, in his book *Fleuves et Rivières* first published in 1933 [118].

Simple regimes

Simple regimes are those variations of river flow throughout the year in which a simple distinction may be made between one period of high water levels and runoff,

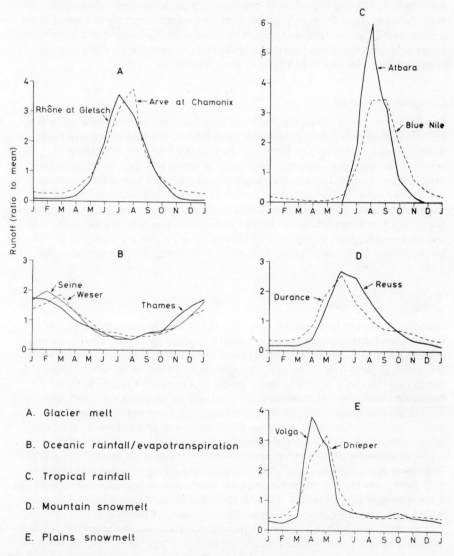

A. Glacier melt

B. Oceanic rainfall/evapotranspiration

C. Tropical rainfall

D. Mountain snowmelt

E. Plains snowmelt

Fig. 8.19. *Examples of simple regimes. (Drawn from data in Pardé [118].)*

and one period of low water levels and runoff. Such regimes may result from one of several contrasting factors (see Fig. 8.19): thus, many European mountain rivers have a high water period in July and August when the glaciers feeding them melt most rapidly, and a very low or zero flow during the winter months when temperatures are low and ice melt is negligible. Again, in many of the oceanic areas of Europe, rainfall is fairly evenly distributed throughout the year, but the peak of evapotranspiration during the summer months results in low runoff during this season, in contrast to high runoff values during the winter months when evapotranspiration is small. In tropical areas, on the other hand, evapotranspiration tends to be uniformly high through the year, so that the rainfall distribution is the main determinant of the river regimes, with high runoff occurring as a result of the summer rains. Finally, simple regimes may result from the melting of a snow cover, either in mountainous areas during early summer, or over the great plains areas, such as those of Eurasia or North America in late spring.

Complex I regimes

Complex I regimes are characterized by at least four, and sometimes as many as six, hydrological phases, although normally there are two low runoff and two high runoff periods. In the case of European streams, the first high runoff period, resulting perhaps from snowmelt, may occur in spring and then be followed by a period of low runoff. Later in the year, a second period of high water levels and runoff may occur in the summer as a result of, say, convectional rainfall over a 'continental' area, or in autumn as a result of Mediterranean storms, or in winter as the result of an excess of rainfall over evapotranspiration in an oceanic area. This sequence results, then, in two periods of peak runoff which are separated by two periods of lower discharge, giving four distinct hydrological phases through the year.

Complex II regimes

Complex II regimes form the third and probably the most important group in Pardé's classification and are found on most of the world's large rivers. Since these normally flow through several distinct relief and climatic regions, and may receive the waters of large tributaries which themselves flow over varied terrain, rivers comprising this group normally have a simple or a complex I regime in their headwater reaches but, downstream, are gradually influenced by a variety of factors such as snow or glacier melt, rainfall, and evaporation regimes which may emphasize the trends found in the headwater regime or which, because they work in opposite ways, may cancel each other out.

   Three examples of the main types of complex II regime are shown in Fig. 8.20. The Rhine typifies those rivers which are definitely glacier or snowmelt streams in their headwater reaches, but which become increasingly influenced downstream by a single type of rainfall regime. Thus, at Kehl, the Rhine has a simple meltwater regime with a summer maximum; after the confluence of the Neckar at Frankenthal, the slightest signs of a winter rainfall peak become apparent, and this feature is further strengthened after the confluence of the Main at Mainz. After the entry of yet another large tributary, the Moselle, the regime is again altered

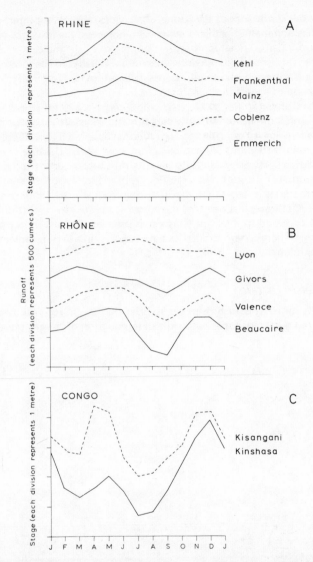

Fig. 8.20. *Regimes of (A) the Rhine, (B) The Rhone, and (C) the Congo. (From original diagrams by M. Pardé,* Fleuves et Rivières *(3rd Ed.), Armand Colin, Paris, 1955.)*

considerably, and at Coblenz, the runoff peak attributable to the excess of winter rainfall over evapotranspiration almost equals the summer meltwater peak. Finally, at Emmerich, on the Dutch border, the winter oceanic rainfall regime has become the dominant factor and the contribution of meltwater now gives rise only to a secondary peak in the summer months.

The Rhône is an interesting example of those rivers which are influenced initially by meltwater and then, farther downstream, by at least two types of rainfall regime. At Lyon, a glacier melt summer maximum is apparent, but at

Givors, below the confluence of the Saône, an oceanic rainfall-evapotranspiration regime becomes dominant. Farther downstream, however, the influx of water from the Isère, above Valence, re-establishes the dominance of the meltwater peak, although there is still a secondary winter maximum. Finally, in the lower reaches, as at Beaucaire, a Mediterranean rainfall regime is superimposed upon all this, resulting in a rapid increase of runoff in the autumn.

Thirdly, the Congo typifies those rivers whose regimes are influenced only by rainfall—but rainfall of two or more distinct climatic zones. Throughout its length, the Congo has a regime of double high waters corresponding to the double rainfall maxima of equatorial regions; but complications arise, partly because the Congo basin covers such a vast area, and partly because the greater part of it lies south of the equator, so that the runoff peak resulting from the inflow of the southern hemisphere tributaries is larger than that resulting from the northern hemisphere tributaries. At Kisangani, just north of the equator, the regime is most typical of the equatorial type but, by the time Kinshasa is reached, the vast inflow of waters from the southern hemisphere tributaries have resulted in a very prominent December maximum.

River regimes in Britain

Runoff records for most British rivers are so brief that the thirty-year average values often considered necessary for a reasonable determination of the regime are only

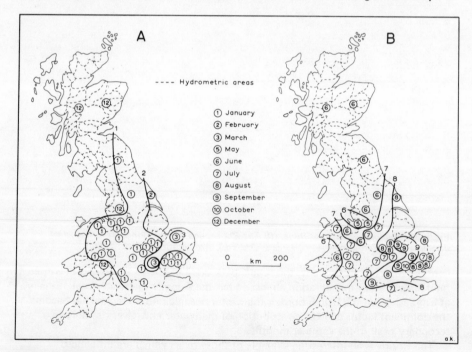

Fig. 8.21. *Maps showing the month of occurrence of (A) maximum and (B) minimum runoff in Britain, 1953-63. (From an original diagram by Ward [151].)*

282

rarely available. Consequently, apart from papers by Pardé [117] and Ward [151], little has been written about British regimes. Analysis of 37 regimes, calculated from a 10-year run of data, showed that 29 fell clearly within Pardé's 'simple' classification of the oceanic rainfall-evapotranspiration type with summer minima and winter maxima [151]. This is to be expected since even the largest rivers, such as the Shannon and the Thames, are short by world or even by European standards so that there is little possibility of the regimes being complicated by a succession of contrasting influences. The remaining 8 regimes showed certain characteristics of Pardé's 'complex I' category, having either a definite period of high water in both summer and winter, or an extended period of winter high water together with a tendency towards a period of high water in summer, or finally, two periods of autumn and winter high water. These complex regimes occur on rivers in the north and west and can probably be accounted for by a combination of geological and climatological factors, including spring snowmelt and reduced summer evapotranspiration.

There is also clear evidence that the time of occurrence of runoff extremes becomes successively later towards the south and east of Great Britain. This is illustrated in Fig. 8.21 which shows that mean monthly maximum runoff in north-west Scotland and the western extremities of England and Wales occurs in December but is as late as March in parts of East Anglia and Hertfordshire. Similarly, the mean minimum monthly runoff occurs in June over Scotland, northern England, and northern Wales, and even in May in the case of some of the Cheshire rivers, but is as late as September or even October in parts of eastern and central southern England. These runoff trends seem to reflect trends in the pattern of evapotranspiration but particularly in greater water holding capacity and the consequently later release of accumulated water from the porous rocks of the English Plain.

The estimation, prediction, and forecasting of runoff

Ideally, all hydrological problems would be solved by the use of measured data, thus obviating the necessity for estimation, prediction, and forecasting. There are many circumstances, however, in which the use of these techniques becomes necessary. Thus, for example, there may be a deficiency of measured data for a particular area, but there may be the possibility of extrapolating future runoff trends either from existing runoff data relating to adjacent or nearby areas or from precipitation data. Alternatively, measured data may be collected too late to be of any use. Such is the case in areas where peaks of quickflow constitute a flood problem which must be viewed and solved in the light, not only of hydrological factors, but also of factors of settlement and communications, agriculture, and economics. Inevitably, the relevant measured data cannot become available until the flood peaks themselves have occurred and so, in these circumstances, the need is for techniques for accurately forecasting the volume and timing of quickflow peaks. Again, in areas where water supplies for agriculture, industry, or domestic uses are likely to be limited at times of low flow, the need is for accurate forecasts of the magnitude of dry-weather flows, and the time of occurrence of minimum flow.

The main requirements, therefore, are for techniques to forecast, for a given point within a drainage basin, both the total volume of runoff and the magnitude of the instantaneous peaks normally associated with sudden increases of quickflow and also to forecast the timing and magnitude of the minimum flows which are likely to be associated with decreasing volumes of baseflow, particularly groundwater flow. Most of the techniques currently in use were developed before the newer concepts of runoff formation, which have been presented in this chapter, had been accepted or, in some cases, had even emerged. Interestingly, however, many of these methods yield reasonable results despite being conceptually weak or even erroneous. Such successes may be fortuitous but more often reflect the fact that the techniques are either highly empirical, and are often applicable only to restricted areas, or else are based upon factors which, although not directly cause-related to the patterns of runoff under consideration, are themselves directly affected by the real runoff-forming factors.

Although, in normal English usage, the terms forecasting and prediction are clearly synonymous, they are sometimes used in a more restricted sense by hydrologists. Thus, as Smith [136] observed, prediction, in this context, refers to the application of statistical concepts to long periods of data, usually relating to extreme events, with a view to defining the statistical probability or return period of a given magnitude of flow. In other words, there is no indication of when this particular flow will occur. Forecasting, on the other hand, refers to specific runoff events, whether floods or low flows, and to the use of current hydro-meteorological data in order to provide a forecast of the magnitude of the runoff event and also, in many cases, of its timing. As far as possible this distinction will be preserved in the ensuing discussions.

There are many techniques of runoff prediction and forecasting. Some of these are in widespread use, either because they work reasonably well over a wide range of conditions or else are easy to apply. The use of other techniques may be restricted to specific areas or to specific users, such as a particular Government agency. Most methods have little merit and yield poor results. It would clearly be impossible to deal with all methods or even a representative selection of the better ones. Indeed, in the present context this would not, in any case, be appropriate. Instead, the main lines of approach to the problem of runoff prediction and forecasting will be briefly reviewed in general terms and will be illustrated, where appropriate, by specific examples.

(A) *Long-period runoff*

8.15 Annual and seasonal runoff

Reasonably accurate estimates of annual and seasonal runoff totals for given catchment areas may, in some cases, be made from annual or seasonal rainfall totals, using a simple straight line regression between the two variables. Normally, however, such simple correlations between rainfall and runoff may be expected to yield forecasts of only token accuracy and certainly for shorter periods of hours, days, or even weeks, more refined techniques must be used.

Some improvement on the simple rainfall-runoff relationship may be derived from the use of a water-balance accounting procedure such as that proposed by Thornthwaite [146]. It was shown that a reasonably accurate forecast of monthly and even shorter period runoff totals for a small drainage basin in eastern England could be obtained from a simple consideration of rainfall and potential evapotranspiration values [153].

Finally, it may in some cases be appropriate to consider vegetation type as an indicator of potential runoff. Satterlund [131] investigated the use of forest vegetation in this respect and concluded that, in relation to the USA, the technique would be more useful where water supplies are scanty, as in the south-west, and less useful where a considerable excess of water over plant needs is available, as in southern Appalachia.

8.16 Snowmelt runoff

Some of the techniques of forecasting runoff in areas where a substantial part of the winter precipitation falls as snow were briefly discussed in section 2.13. Linsley [96] observed that in such situations forecasts of seasonal or annual runoff have proved particularly useful and that the delay between snowfall and subsequent melt is often sufficiently long for a forecast to indicate expected amounts of water likely to become available during the growing season. Virtually all the methods currently in use rely upon an accurate assessment of the total quantity of water stored in the snowpack (see section 2.12). Baker [11] achieved a high degree of sucess in forecasting spring runoff in Minnesota using only soil temperature and cold period precipitation, i.e., precipitation falling on frozen soils.

(B) *Peak flows*

It has earlier been shown that increases in runoff after rainfall or snowmelt tend to be rapid, leading quickly to a short-lived or instantaneous peak, which is followed by a rather longer period of declining runoff. Since it is likely to be the peak flow which causes damage to structures, or flooding of agricultural land adjacent to streams, the main problems of forecasting and prediction concern the magnitude and timing of this peak, and the frequency with which it is likely to occur, although it may also be of interest to know in advance, say, the total volume of runoff, and the length of the period between the beginning and the end of quickflow. Several techniques may be used, and in the ensuing discussion these will be considered under the headings of empirical methods, statistical methods, analytical methods of river forecasting, and catchment models. The choice of method will normally be determined by the purpose for which it is required, by the available data, and by the area and characteristics of the drainage basin. In relation to the last point, for example, there is clearly a major difference between a drainage basin of 1 km^2, where the runoff-producing storm may cover the entire area, and a basin of several thousands of square kilometres, where a flood peak generated on a small headwater stream by an isolated thunderstorm will move down the drainage system and where the main problem is to forecast its change in shape and magnitude as it does so.

8.17 Empirical methods

To the extent that virtually all methods of runoff prediction and forecasting contain an element of approximation and empiricism, the heading of this section may be considered misleading. However, as with equality, so with empiricism, some methods are more empirical than others. The main weaknesses of such methods are normally their non-generality and the difficulty of knowing the exact conditions under which they may be used.

One of the earliest and best-known attempts to estimate peak flows was the so-called rational formula, whose origin is obscure. As Chow [30] pointed out, it was first mentioned in American literature by Kuichling in 1889 [91], although some authors believe that the principles may be ascribed to the earlier work of Mulvaney [112], while in England the rational formula has often been referred to as the Lloyd-Davis method [99]. The method assumes that the maximum rate of runoff from a catchment area will occur when the entire area is contributing to runoff and may be expressed in the form:

$$Q = CIA \qquad (8.2)$$

where Q is the rate of runoff, C is a runoff coefficient indicating the percentage of rainfall which appears as surface runoff, I is the rainfall intensity, and A is the area of the catchment. As Wisler and Brater [160] pointed out, this simple relationship between rainfall intensity and runoff could only possibly apply to very small areas. Even then, the use of the coefficient C ignores the fact that, whether one applies the runoff-producing concepts of Horton or Hewlett, the relationship between rainfall and runoff may change dramatically during the progress of a storm.

Similar criticisms may be applied to another early formula developed by Meyer [108]. This may be expressed in the form:

$$Q = CC_1 A^{0.6} \qquad (8.3)$$

where Q is the maximum rate of runoff, C is a coefficient related to various catchment characteristics such as soil and vegetation types, relief and drainage, C_1 is a frequency coefficient, and A is the area of the catchment. Although this formula seems to take account of more of the relevant factors, it does so in the form of static coefficients rather than in terms of the changing relationships between the numerous variables during the course of the runoff-producing storm.

Gray and Wigham [50] tabulated more than thirty empirical formulae which are still in general use for estimating flood flows and which rely solely on the relationship between peak flow and drainage basin area—a value which can be readily obtained from topographic maps. More sophisticated empirical formulae in which various measures of drainage basin morphometry and climate have been combined with an explicit term for the frequency of occurrence of a given class of runoff peak, were discussed by Rodda [127] [128] and by Gray and Wigham [50]. Rodda [127] noted, however, that even the inclusion of three or more independent variables covering a wide range of climate and basin characteristics leaves considerable differences between observed and estimated runoff peaks.

8.18 Statistical methods

An estimate of the frequency with which a given magnitude of runoff may be exceeded in the future is based upon consideration of the recorded historical

Table 8.1. *Probability and recurrence interval calculated from a 24-year annual maximum series for a hypothetical stream.*

Runoff Peak (cumecs)	Order number m	Probability* $\dfrac{m}{n+1}$	Recurrence Interval* $\dfrac{n+1}{m}$
		per cent	years
150 000	1	4	25.00
135 000	2	8	12.50
126 000	3	12	8.33
117 000	4	16	6.25
114 000	5	20	5.00
108 000	6	24	4.17
105 000	7	28	3.57
102 000	8	32	3.13
100 500	9	36	2.78
96 000	10	40	2.50
93 000	11	44	2.27
91 500	12	48	2.08
87 000	13	52	1.92
84 000	14	56	1.79
81 000	15	60	1.67
79 500	16	64	1.56
78 000	17	68	1.47
75 000	18	72	1.39
72 000	19	76	1.32
70 500	20	80	1.25
66 000	21	84	1.19
63 000	22	88	1.14
57 000	23	92	1.09
51 000	24	96	1.04

*Where n is the number of years of record and m is the rank of the item in the array

pattern of runoff events. Unfortunately the duration of runoff records is normally so short that it is difficult to extrapolate them with a high degree of reliability. In essence the main problem is that flood frequency estimation involves the estimation of the tail of a probability distribution curve from a sample of values which usually does not include values within this tail [107]. However, statistical methods of extreme value frequency analysis have been designed to extract the maximum information from the recorded data and to evaluate the likely nature of the distribution of the parent population from which the sample has been drawn [28].

Various methods of plotting extreme value data and of drawing frequency curves through the plotted points have been devised, including several types of special plotting paper on which the data will plot as a straight line and are therefore, easier to extrapolate [28]. An excellent review of current techniques was presented by

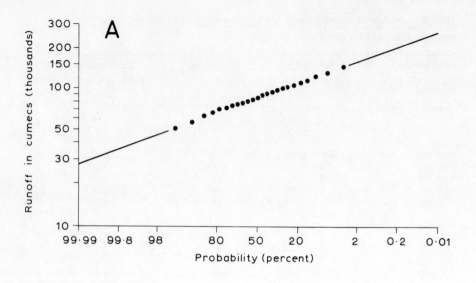

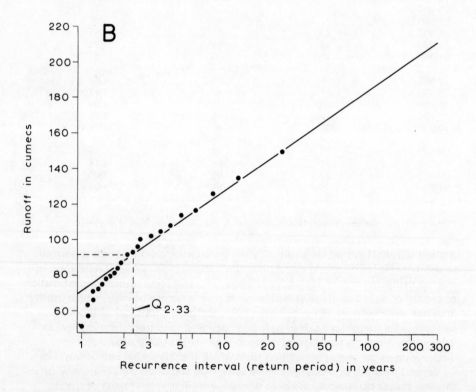

Fig. 8.22. (A) Log-probability and (B) recurrence interval graphs of the peak flows listed in Table 8.1.

Benson [18], the most commonly used distributions being two-parameter gamma, Gumbel, log-Gumbel, log-normal, log-Pearson Type III, and Hazen. Comparison of these distributions applied to selected USA runoff data led to the conclusion that no one of these methods could be considered as the best on statistical grounds, although on administrative grounds the log-Pearson Type III was selected pending further study and research.

The two main ways of extracting flood data for frequency analysis are in the form of the *annual maximum series* or the *partial duration series*. In the former case only the highest instantaneous peak discharge in each year is considered, which means that many high runoff peaks may be ignored and that the total number of peak flows considered is comparatively small. The partial duration series, on the other hand, includes all peak flows having a value greater than some chosen base and thus appears more useful for theoretical analysis than the annual maximum series [166]. The main weakness of the partial duration series is that the separate peak flows may not be fully independent although there are ways of overcoming this problem.

One of the simplest and most practical methods of computing flood frequency is illustrated in Table 8.1 and Fig. 8.22. In this hypothetical example the problem is to predict the likelihood of a certain sized runoff peak occurring during a specified time, on a stream whose flow has been measured for, say, 24 years. Using the annual maximum series, the peak flows recorded during that period are arranged in descending order of magnitude and the recurrence interval and probability are determined as shown in Table 8.1. If the size of the peak flow and the probability are then plotted on log-probability paper, as shown in Fig. 8.22A, it will be seen that an approximately straight line is described, which it is possible to extend in order to predict floods with probability factors, within or outside the range represented by the sample data. Alternatively, runoff may be plotted against recurrence interval on semi-logarithmic coordinates as in Fig. 8.22B. Mathematical analyses have demonstrated that the mean annual flood has a recurrence interval of 2.33 years, i.e., on average, the annual maximum flow will exceed the mean maximum flow once every 2.33 years, and this recurrence interval is indicated by the broken line.

This type of technique is quite successful provided that a long sequence of recorded data is available. Where, however, the records are brief or intermittent there is a considerable danger that the sample data will be inadequate, and that extrapolations based upon them will be grossly inaccurate. One frequently used method of reducing sampling error is *regional analysis*, whereby data from within a homogeneous area, or from adjacent similar areas, are combined to increase the sample size, and thus to improve the quality of the result [1], [131]. A homogeneity test was developed by Langbein [94] to define a homogeneous region and various methods of regional analysis have been developed, including the construction of regional flood frequency curves [33]. This involves plotting graphs in which the ordinate scale is the ratio of any given flood to the mean annual flood (Fig. 8.23). By reducing flood magnitude to this dimensionless ratio it is thus possible to group together the flood data from different areas and different periods. Figure 8.23 represents the preliminary analysis by the Institute of Hydrology of 5470 station years data from 440 flow-gauging stations in the United Kingdom and the curve shown is the average flood frequency curve for the whole of the country.

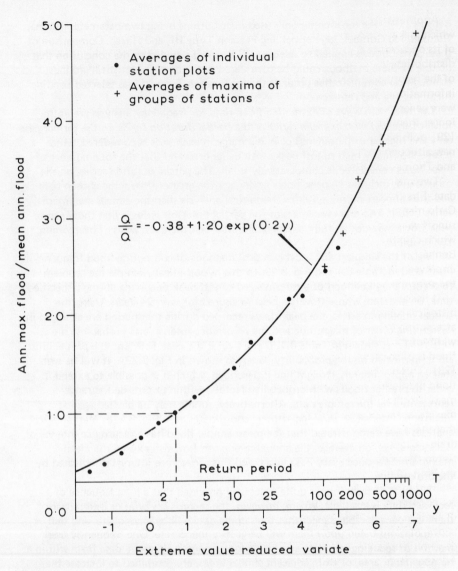

Fig. 8.23. *Average flood frequency curve for the United Kingdom based on 5470 station years of data from 440 flow-gauging stations. (From an original diagram by Institute of Hydrology [83].)*

In fact the country was divided into ten regions and the curve for each of these differs in position from the average curve. In order to determine the recurrence interval of a particular magnitude of peak flow at a given gauging station, the mean annual flood at that station is determined and the ratio of that flood magnitude to the mean annual flood is then entered on the ordinate and the recurrence interval is read on the abscissa. Again, it will be noted that the mean annual flood itself, yielding a dimensionless value of unity, has a recurrence interval of 2.33 years.

290

Another statistical approach to flood prediction is that of *stochastic simulation* which makes use of the fact that the magnitude of a runoff event and the frequency of its occurrence are related in a definable way, i.e., a statistical probability distribution, to generate sequential synthetic data from the statistical characteristics of the recorded historical pattern of runoff events. As far as the quality of the information is concerned the new data are no better than the data from which they were generated but the synthetic record has the outstanding advantage that it is longer than the historical record and thus more amenable to analysis. As Chow [29] pointed out, the concept of sequential generation of hydrological data is not new although it received renewed impetus through the work of Benson [16] [17] and Thomas and Fiering [145].

Two main approaches have been adopted in the sequential generation of runoff data. First it may be assumed that the data are purely random, in which case Monte Carlo methods are used [110] or more realistically it may be assumed that any runoff value is in part determined by the preceding runoff values at that point, in which case the process known as the Markov process or Markov chain is used [23]. Bonné [25] suggested that the Markov simulation process could be greatly improved where it is possible to incorporate representative precipitation data with the streamflow data for the drainage basin in question.

It has already been observed that in many countries, including Britain, the historical runoff record is short and this may lead to considerable errors when attempting to predict rare floods. For example, it is commonly found that the width of the 95 per cent confidence limits for an estimated 2000–10 000 year flood is of the same order of magnitude as the estimated rare flood magnitude itself [28]. This has been one of the factors underlying the growing tendency to base flood prediction, at least partially, on the physical analysis of historically recorded major storms and snowmelt periods. In some cases the concept of probable maximum precipitation (PMP) is used (see section 2.10), in which the assumption is that there is a physical upper limit to the depth of precipitation that can fall over a drainage basin, or the amount of snowmelt which can take place, and this has given rise to the concept of the *probable maximum flood,* i.e., the flood resulting from the PMP. Methods of estimating the PMP were discussed in section 2.10 and of estimating snow accumulation and melt in sections 2.12 and 2.13. The basic tool for converting the critical maximized meteorological conditions to the probable maximum flood (PMF) is the unit hydrograph method (see section 8.19) which allows the distribution in time of the estimated volume of runoff. Where the PMP concept is not used, a design flood for a given drainage basin may be estimated from the largest recorded storm in that region, often referred to as the *project storm.* This technique is clearly more satisfactory with a long run of precipitation data. Recently, however, doubt has been cast upon the statistical validity of estimating floods from extreme value frequency analysis of rainfall [2] [124].

8.19 Analytical methods of river forecasting

Although, as Nash and Sutcliffe [115] observed, the process linking rainfall and river flow is a deterministic one, in that it is governed by definite, known physical laws, it has so far proved impracticable to forecast river flow simply by applying these laws to the measured rainfall and boundary conditions within a catchment area, largely because of the complexity of the boundary conditions. Since there seems little point in applying exact physical laws to approximate boundary conditions, there has developed a traditional approach towards river forecasting which is largely empirical or analytical. This approach involves six basic steps [45]. (a) Collection of runoff and rainfall data, (b) calculation of areal precipitation over the catchment, (c) calculation of direct runoff, (d) conversion of direct runoff volume into hydrographs of discharge versus time, (e) determination of the change in shape of the hydrograph due to channel storage and the hydraulic characteristics of the channel as the water moves downstream, and (f) conversion of forecast discharge to stage. Step (a) is discussed in sections 2.6, 2.7, and 8.23–8.25, step (b) in section 2.8, and step (f) in section 8.23. The remaining steps (c)–(e) will be briefly discussed in the following paragraphs.

Calculation of direct runoff

Earlier discussions of the runoff process have indicated that during and after a storm, water reaches the stream channels by a variety of routes and at varying speeds. Such is the complexity of the process, however, that it is not possible to relate the speed of arrival of water in the channel to the route taken by that water in reaching the channel. One consequence of this is that traditional techniques of hydrograph analysis, intended to separate direct runoff from base runoff have been discredited while the time-based method suggested by Hewlett and Hibbert [65] is entirely arbitrary. Furthermore, the persistence of runoff for a long period after a storm event, and in the absence of further recharge, implies a very slow recession of runoff which makes it difficult to define even the gross volume of runoff associated with each storm event. The need to calculate the quick\flow resulting directly from a storm event, therefore represents a fundamental weakness in the traditional analytical approach to river flow forecasting and will undoubtedly lead to its eventual demise.

Until this happens, however, the volume of direct runoff will continue to be estimated by one of the well-tried methods. These include the traditional and current techniques of hydrograph analysis, as discussed in section 8.7, the infiltration approach of Horton [76] which has been shown to be generally incorrect in the light of modern concepts of the runoff process, and the use of rainfall-runoff relationships for the drainage basin usually incorporating an index of initial soil moisture conditions. One of the earliest and probably best-known attempts to forecast direct runoff volumes was the co-axial graphical correlation method developed by Linsley and others [97]. This method uses an antecedent precipitation index (API) combined with time of year to give an integrated index of soil moisture deficit at the beginning of rainfall. This soil moisture deficit index is corrected for storm duration and is then used with storm precipitation to calculate direct (storm) runoff as shown in Fig. 8.24. The API may be calculated on the basis of a logarithmic recession during periods of no precipitation thus

$$I_t = I_0 K^t \qquad (8.4)$$

where I_0 is the initial index, I_t is the index after t days have elapsed and K is a constant. The index for any day is, therefore, equal to that of the previous day

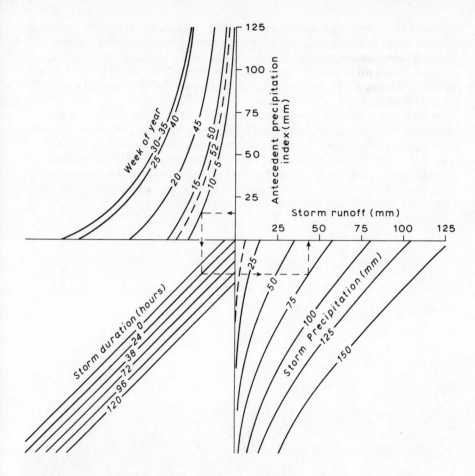

Fig. 8.24. *Coaxial diagram for estimating direct (storm) runoff. (From an original diagram by Kohler and Linsley [90]. With acknowledgment to NOAA.)*

multiplied by K. When precipitation is measured on a particular day the observed precipitation is added to the index. Values for the constant, K, are usually assumed to be in the range 0.80-0.98, although the choice is not usually critical since the calculation is used as an index of soil moisture deficit [51]. The usefulness of the API in estimating direct runoff can probably be explained less satisfactorily in simple infiltration terms, as has previously been the case, than in terms of the fact that conditions favouring a high API will also favour channel expansion and translatory flow.

Conversion of direct runoff volume into a time-distributed hydrograph

The distribution of direct runoff volume in time is normally attempted through the application of a unit hydrograph. This much-discussed technique, first suggested by Folse [44] and more clearly formulated by Sherman [133], was only slowly adopted by practising hydrologists, particularly in Britain. It has, however, provided a field in which an enormous amount of esoteric research continues to flourish. Ironically, the sophistications of recent unit hydrograph research have been made possible only by the advent of high-speed computers which, because of their great flexibility, have rendered the method obsolete. Linsley [96] summarized the situation most succinctly:

'Without decrying the tremendous impetus given to hydrology by the original concept or the very large amount of successful engineering design based on the unit hydrograph, it appears that the unit hydrograph is destined to be replaced in the future by more effective techniques'. Currently, however, the technique continues in widespread use and therefore merits consideration, although since the working procedures were well described by Chow [30] and by Gray and Wigham [50] only a brief description of the basic steps will be given here.

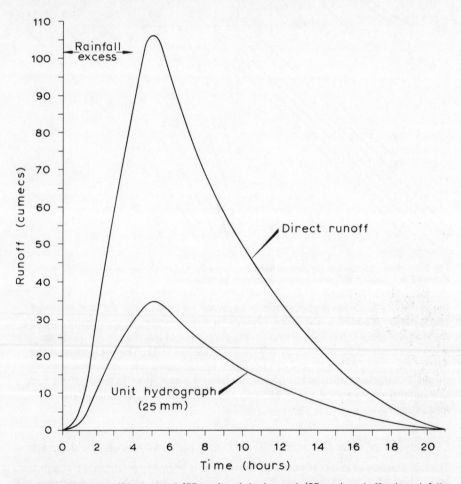

Fig. 8.25. *Direct runoff hydrograph (75 mm), unit hydrograph (25 mm), and effective rainfall duration (4 hours).*

294

The basis of the method proposed by Sherman [133] is that since a stream hydrograph reflects many of the physical characteristics of the catchment area, similar hydrographs will be produced by similar rainfalls occurring with comparable antecedent conditions. Thus, once a typical or unit hydrograph has been determined for certain, clearly defined conditions, it is possible to estimate runoff from a rainfall of any duration or intensity [133]. The unit hydrograph is the hydrograph of a unit volume (traditionally 1 inch or 25 mm) of direct runoff from the entire catchment area resulting from a short, uniform rainfall with an excess of unit duration. The duration of the unit rainfall will tend to vary with the size of the catchment from say, an hour to a day, or even longer but will always be of shorter duration than the period from the beginning of quickflow to the hydrograph peak. In other words, the unit rainfall provides the shortest time-base for the graph of direct runoff and one which, for practical purposes, is always the same regardless of storm intensity [133]. It follows, therefore, that for any observed unit rainfall, the ordinates of the runoff hydrograph will vary directly with the depth of runoff over the catchment [133].

The unit hydrograph is constructed by selecting a recorded hydrograph of a uniform isolated storm, preferably with a fairly large volume of runoff, and having separated out the base runoff, by dividing the discharge ordinates of the remaining direct runoff hydrograph according to the volume under the hydrograph to obtain the unit hydrograph, i.e., the hydrograph of 25 mm (or 1 inch). This is illustrated in Fig. 8.25 where the direct runoff hydrograph, which represents a runoff volume of 75 mm, has been divided by 3 to yield a unit hydrograph. It will be seen that in this example the unit storm was of four hours duration and so the derived unit hydrograph is referred to as a *4-hour unit hydrograph.* Different catchment characteristics may be expected to result in different unit hydrograph shapes, with steeply sloping terrain having a low storage potential yielding high, sharp-peaked hydrographs with a short time base, while flat terrain with a high storage potential will yield a flat, rounded peak with a long time base.

Bernard [19] introduced a refinement to the unit hydrograph technique in the form of the *distribution graph.* This has the same time-scale as the unit hydrograph, but the runoff resulting from a unit storm is plotted as a percentage of the total volume of runoff during a number of successive equal time units. Since all unit storms produce virtually identical distribution graphs for a given catchment area, the average distribution graph, once derived, can be used to construct the predicted hydrograph for any volume of direct runoff, provided that rational assumptions are made about the relationships between precipitation and catchment characteristics and, therefore, about the volume of precipitation excess available for direct runoff during the course of a storm.

It is in relation to this last condition that the main weaknesses of the unit hydrograph method have always been seen. It has been suggested that large errors may result from simplifying assumptions about variations of infiltration capacity, about the uniformity of precipitation over the catchment in both time and space [31] [59] and about the invariant linear response of the catchment to this simplified input of precipitation excess, especially since it is well known that not only does excess rainfall vary in time and area over a catchment but also that the response of a catchment to a given rainfall excess will change with time during the course of a storm and will depend, in part, on its responses to previous inputs of rainfall excess. In other words, the catchment is a non-linear, time variant system [141]. Such errors will tend to increase as the size of the catchment increases, since with larger areas, greater variations of permeability and rainfall distribution are to be expected [7]. Nevertheless, Wisler and Brater [160] considered that, for areas smaller than 8000 km^2, variations of areal rainfall patterns would not be significant. For substantially larger areas, it would be necessary to obtain unit hydrographs for the main tributaries, and then to route these to the main catchment outlet.

In the light of such considerations it is perhaps highly significant that the unit hydrograph technique works as well as it does over such a wide range of precipitation and catchment conditions. That it does so, may well be due to the fact that the runoff process is not as Sherman understood it, i.e., Hortonian, in the sense that direct runoff results when rainfall intensity exceeds infiltration capacity on a catchment-wide basis. Rather, the variable source area concept of Hewlett reflects a situation where, in fact, given prescribed antecedent conditions, it is likely that the quickflow-producing areas within a catchment will be more or less identical in size and that since infiltration capacity is effectively zero in these areas the total volume of quickflow and its time base will be closely similar for similar precipitation inputs. In other words, unit hydrograph theory is much more acceptable in the light of modern concepts of runoff formation than it was when related to the Horton runoff process.

Certainly, the unit hydrograph method has been one of the most dependable and most frequently used techniques in the forecasting of streamflow, and many attempts have been made to improve its accuracy and general applicability. Allowance needs to be made, for example, for storms whose duration considerably exceeds or falls below the duration of the unit storms used in deriving the initial unit and distribution graphs, and, for this purpose, the summation curve or *S-curve* is used. The S-curve is the hydrograph representing an infinite number of consecutive 25 mm runoffs resulting from rainfalls of a specified duration, and unit hydrographs for any rainfall duration, e.g., x-hours, may be obtained by subtracting two S-curves with their initial points displaced by x-hours. For example, in the case of the catchment whose 4-hour unit hydrograph is shown in Fig. 8.25, the 2-hour unit hydrograph would be derived by subtracting two 4-hour S-curves spaced 2 hours apart and multiplying the ordinates so derived by two. When the time interval between two S-curves is reduced almost to zero, the subtraction of the two curves results in the *instantaneous unit hydrograph,* which ideally represents the runoff generated in a catchment area by one infinitely short burst of rain, although, for practical purposes, this must be regarded as having a finite duration. By adding together the appropriate number of instantaneous unit hydrographs, the runoff generated by any storm may be determined, the main advantage of this method being that variations of rainfall intensity and of source area during the course of the storm can be considered.

Again, it may be necessary to determine unit hydrographs for catchments with few if any runoff records and, in such cases, close correlations between the physical characteristics of the catchment area and the resulting hydrographs are necessary. Snyder [138] was one of the earliest workers to derive *synthetic* unit hydrographs and found that the shape of the catchment and its 'lag', i.e., the time from the centre of the mass of rainfall to the hydrograph peak, were the main influencing characteristics. Much subsequent work has been done to investigate the main properties of the unit hydrograph, and to determine their correlation with catchment characteristics [49], [59], [114].

Downstream changes in hydrograph shape

In the case of large catchments, the total outflow hydrograph will normally be determined by summing the runoff contribution from each of the sub-catchments in which direct runoff has been generated. This involves the technique of *runoff routing* to determine the rate of movement and the change of shape of the flood waves moving downstream from each of the sub-catchments towards the main basin outlet. The movement of a flood wave in a stream channel is a highly complex problem of non-steady and usually non-uniform flow for not only does flow vary with time as the wave progresses downstream, but channel properties and the amounts of lateral inflow may also vary [50]. Even with the assistance of high-speed computers it becomes necessary to use approximate numerical techniques and simplifying assumptions.

Thus a flood wave may be considered to undergo simple translation and reservoir or pondage action as it moves downstream [45]. Figure 8.26 illustrates the change in shape of a flood wave moving down the Savannah River in Georgia. At first water is entering the channel reach faster than it is coming out at the downstream end resulting in the excess water being stored in the channel reach and thereby raising the water level there. Later the inflow into the reach drops below the rate of flow at the downstream end and the excess water stored between the two gauges comes

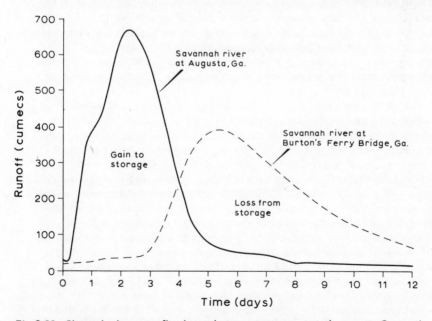

Fig. 8.26. *Change in shape of a flood wave between two gauging stations on the Savannah River in Georgia. (From an original diagram by Fox [45].)*

out of storage. Routing methods to determine the downstream change in shape of the flood wave require some knowledge of the river reach under consideration and particularly of the relationships between stage and storage or discharge and storage [95] and involve solution of the basic storage equation

$$\Delta S = \overline{I} - \overline{O} \qquad (8.5)$$

where ΔS is the change in storage, $\overline{I}$ is the average inflow into the reach, and $\overline{O}$ is the average outflow [45].

Numerous routing procedures have been developed and useful reviews were presented by Lawler [95] and Yevdjevich [164].

8.20 Catchment models

A fast-growing alternative to the traditional analytical method of runoff forecasting involves the construction of catchment models using digital computers. This development reflects the logical outcome of the increasing speed and flexibility of modern computers and the growing rejection by many hydrologists of the arbitrary separation of the quickflow and baseflow components of total runoff. As Nash and Sutcliffe [115] observed, if arbitrary hydrograph separation is rejected the steps by which rainfall is converted to effective rainfall and effective rainfall is converted to runoff can no longer be studied separately but must be treated simultaneously so that a model of the process of conversion of rainfall into runoff must be assumed.

The complexity of a catchment and of the runoff process itself inevitably mean that even the most rigorous model must involve considerable simplification and approximation. The parameters of the model may be determined from field measurement of hydrologic data or by means of successive optimization procedures continued until the model behaviour approximates most closely that of the catchment itself, as indicated by rainfall and runoff data, although care must be taken that the quest for 'curve-fitting' does not result in a model which has no inherent physical validity.

Undoubtedly, the most important pioneering work in this field was done by Crawford and Linsley [32]. This resulted in the development of the Stanford Watershed Model which seems likely to lead ultimately to a general hydrologic model [96]. The Stanford model, which has gone through a large number of development phases, is a simulation model of the hydrologic cycle which uses a moisture accounting procedure. A system of equations is used to keep a running tabulation of all moisture entering the catchment as precipitation, stored within the catchment, and leaving the catchment as runoff or evapotranspiration. All the water entering the catchment is accounted for until it evaporates, infiltrates to groundwater or enters a stream channel. The computer program then routes the runoff from the point of entry into the channel to some specified downstream location. One of the most useful features of the Stanford model, apart from its value in synthesizing historical hydrographs, is the way in which the values of the parameters which describe the physical characteristics of the catchment can be varied by a slight adjustment of the input data in order to study the effect of different catchment characteristics on runoff and other specific phases of the hydrologic cycle [111].

An infiltration model, first developed by Holtan [70] forms the core of another promising catchment model developed by the USDA Hydrograph Laboratory [71]. The program for this model, known as USDAHL-70, was written with the objective of preserving maximum clarity and versatility. Speed of calculation is sacrificed for ease of understanding and the program is constructed in distinct subroutines so that as new concepts are evolved the program may be relatively easily updated. The model is also programmed to simplify overlaying procedures so that it should be readily adaptable to a wide range of core size.

Many different models of the runoff process have been or are being developed but as Nash and Sutcliffe [115] observed, the results obtained are not always presented in a manner which facilitates comparisons between models, nor does there appear to be any general agreement on methods of model development and testing. Valuable work on the development of flow-forecasting models has been done at the UK Institute of Hydrology [104] [115] [116]. Development work continues with a view to developing satisfactory conceptual and stochastic models for a wide variety of uses.

(C) *Minimum flows*

The forecasting of the timing, magnitude, and probability of occurrence of periods of low runoff may be dealt with briefly since it tends to pose rather fewer problems than in the case of peak flows. This is largely because runoff during low flow periods comprises baseflow which changes more slowly, and often more predictably, than does quickflow. Again, several techniques are in common use although, for convenience, they can be considered under the headings *statistical methods* and the *hydrological approach.*

8.21 Statistical methods

Hazen [57] was one of the first to apply statistical methods to low flow forecasting. As with peak flows, statistical methods are based on the assumption that conditions recorded in the past will be repeated in the future so that the reliability of these methods has inevitably improved considerably since Hazen's initial work simply because of the availability of longer and more widespread flow records [77]. Forecasting procedures may be concerned with evaluating the minimum flow which is likely to occur in the most severe drought conditions but are more frequently concerned with estimating the probability of occurrence of low flows of a given magnitude and with the duration of flows below a certain threshold value.

An elementary but useful method, is to plot a runoff *duration curve,* either for the catchment whose low flow is to be predicted, or for adjacent catchments, if no runoff records are available for the area in question. These curves show the percentage of time that any given discharge is exceeded, and can normally be fairly confidently extrapolated at their lower ends. They are of use not only in studying the frequency of low flows, but also in reflecting the variability of the regime of the river. Thus, flat curves with high minimum values tend to be derived from rivers with a large baseflow component, which will probably show a slow but gradual decline to minimum flows, while a steep curve with low minimum values often

299

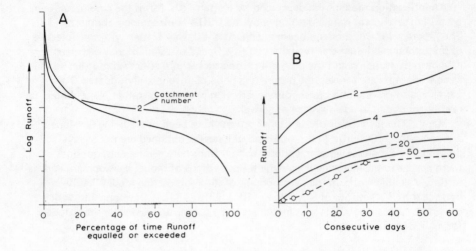

Fig. 8.27. *(A) Flow duration curves for contrasting catchments; (B) Curves for selected recurrence intervals showing the minimum flow occurring over periods of consecutive days on the Fransktown Branch of the Juniata River, Pa., 1930-48. (From original diagrams by (A) Walling [150] and (B) Hazen [58]. The latter is reprinted from* Journal American Water Works Association, *Vol. 48 by permission of the Association. Copyrighted 1956 by the American Water Works Association, Inc., 2 Park Avenue, New York, N.Y. 10016.)*

indicates a river with a large quickflow component, and in which the decline to minimum flows is likely to be more rapid. This contrast is illustrated in Fig. 8.27A where curve 1 represents a stream draining a predominantly clay area and curve 2 represents a stream draining a mixed clay, sandstone, and marl catchment.

Hudson and Hazen [77] observed that the weakness of the flow duration curve is that it deals only with discrete values of flow and reveals nothing about the sequence of low flows nor whether they occurred consecutively or were widely scattered in time. They suggested that a more valuable analysis results from determining the flows over a given period of consecutive days. This is illustrated in Fig. 8.27B where the broken line represents the lowest average flow of record over various duration periods whilst the remaining curves show the flows for various duration periods for recurrence intervals of 50, 20, 10, 4, and 2 years.

Another simple technique, which has already been discussed in relation to peak flows, is the derivation of equations relating rainfall to runoff for a given catchment. Again, a long run of reliable data is required, and even then the method is not suited to the prediction of short period flows although, in this respect, it is rather more reliable in the case of low flows than in the case of peak flows, especially when multiple regression analysis is incorporated. Thus Huff and Changnon [78] used rainfall-runoff relations for thirteen stations, together with a geomorphological factor, to estimate runoff volumes. The close correspondence between estimated and measured runoff suggests that the technique could usefully be used to calculate low flows having various recurrence intervals on ungauged streams and to estimate rare low flows on streams where there is a short run of flow data [28].

8.22 Hydrological approach

In many cases the problem is not to predict the frequency of occurrence of low flows of a given magnitude nor to estimate the most likely severe drought situation in a catchment but to provide a shorter period forecast, as early as possible during the year, of the amount of streamflow which will be available at various times during the growing season or of the minimum summer flow to which rates of abstraction for water supply purposes will have to be adjusted. In these situations low flow forecasts are frequently based upon the analysis of baseflow recessions. In other words the forecasts rest upon the basic hydrological principle that since low flows normally occur as a result of dry conditions, and particularly at the end of a long dry period during which evapotranspiration has exceeded rainfall, quickflow is non-existent so that low flows are the direct result of outflow from storage within the catchment.

In most large catchments groundwater storage will be the major source of dry-weather flows but it has been clearly demonstrated that in small, upland

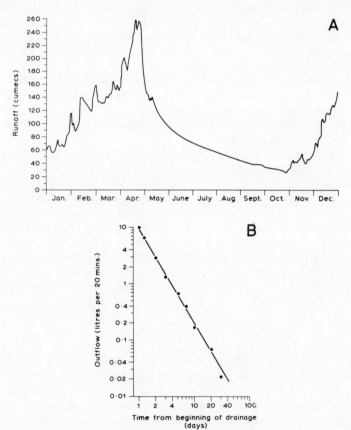

Fig. 8.28. *(A) Hydrograph of Lualaba River showing smooth baseflow recession and (B) recession of baseflow draining from the soil profile. (From original diagrams by Wisler and Brater [160] and by Hewlett [62].)*

catchments especially those having steep slopes, the recession of streamflow may be sustained entirely by the drainage of water from the soil profile [62] [64]. In either case, the decline of baseflow, as represented by the falling limbs of the stream hydrograph, follows an exponential exhaustion curve. In terms of runoff forecasting this is important in several ways.

First, the baseflow recession or depletion curve is easy to construct from existing stream records, since each individual section of falling hydrograph tends to form a displaced segment of one continuous curve which can be reassembled by juxtaposing the separate sections. In some cases the existence of a marked dry season, with no additions to runoff from rainfall or snowmelt, means that the stream hydrograph will describe an exponential curve without the need for relocating individual sections. The hydrograph of the Lualaba River in Fig. 8.28A serves to illustrate this point.

Secondly, since the baseflow depletion curve tends to plot as a straight line on semi-logarithmic paper, it is particularly easy to extrapolate with reasonable accuracy at the lower values. This is clearly illustrated in Fig. 8.28B which shows the declining outflow of water from the soil profile experiment of Hewlett [62]. Thus in the case of rivers used for irrigation or water supply purposes, provided that the slope of the recession curve can be established early in the year, reasonably accurate forecasts of the runoff in mid- to late summer may be made, assuming average rainfall and evapotranspiration conditions.

Thirdly, for catchments whose baseflow depletion characteristics can be rather precisely determined, it may be possible to predict late summer runoff from the appropriate baseflow equation without resorting to graphical techniques although, as Hall [54] demonstrated in an excellent review paper, only a relatively few workers have used this approach.

The hydrological approach may also be taken to include attempts to model the hydrological processes creating low flow conditions. Thus the Stanford Watershed Model [32] to which reference has already been made, is a comprehensive model of the entire catchment hydrologic cycle which is equally useful for studying low flows as it is for high flows. Other more restricted and, therefore, simpler models have been proposed as, for example, by Dooge [35] who provided what Ayers [10] referred to as a refreshing approach to the problem of baseflow synthesis by suggesting that groundwater systems consist of linear storage elements through which infiltrating rainfall may be routed to other storage, directly to the stream, or to the atmosphere by evapotranspiration. The discharge from a storage element may then be represented as

$$Q_n = C_0 R_n + C_1 R_{n-1} + C_2 Q_{n-1} \qquad (8.6)$$

where Q_n is the discharge during any time period, R_n is the recharge during the period, R_{n-1} is the recharge during the previous time period, Q_{n-1} is the discharge during the pervious time period and C_0, C_1, and C_2 are routing coefficients.

Final mention must be made of the fact that in some high latitude areas low flows are the result not of the excess of evapotranspiration over rainfall but of the freezing up of groundwater during the winter months. In such situations the prediction of runoff minima is a complex problem in which rainfall and temperature conditions assume a considerable importance [3].

302

The measurement of runoff

The reliability of many hydrological measurements, e.g., rainfall and evapotranspiration, depends upon the representativeness of the catch, in the case of rainfall, or the loss, in the case of evapotranspiration, from a very small sample area. Only a minute part of the total volume of rain falling on a catchment area, for example, is caught by even an extremely dense network of gauges, and yet large variations of rainfall over small distances have been recorded (see chapter 2). In the case of measurements of runoff, however, inaccuracies resulting from this type of sampling do not occur, since the total volume of streamflow which passes a given point may be measured. Methods of measuring runoff have been adequately discussed in the appropriate manuals, e.g., [27] and will only be summarized here. The three main approaches involve the velocity-area method, the use of control structures such as flumes and weirs, and dilution methods.

8.23 Velocity-area method

The velocity-area method comprises three related operations. The first involves the determination of the height or stage of the stream surface; the second the determination of the mean velocity of the water flowing in the stream channel; and the third the derivation of a known relationship between stage and the total volume of discharge.

Periodic observations of stream stage are normally made by reference to a graduated staff gauge while at most major gauging stations water levels are reproduced autographically or on tape by means of continuous recorders. The measurement of current velocity is normally accomplished by means of a current meter which may be located in the water on wading rods or on a cable suspended from a bridge, cableway or boat. Alternatively current velocities may be determined directly using deflecting vanes, pitometers or floats or may be estimated by means of one of the commonly used theoretical flow equations for the steady flow of water in open channels such as that proposed by Manning [105] or Chezy. Stream discharge is then obtained simply as the product of cross-sectional area and current velocity. If discharges are plotted as abscissa against the corresponding stages as ordinates, a smooth curve drawn through the plotted points defines the rating curve which can then be used in conjunction with a continuous record of stage to determine continuous discharge values.

8.24 Weirs and flumes

On streams where a physical obstruction in the channel is permissible, e.g., small tributaries and drainage channels which are not used for navigation, discharge may be measured by means of a weir or flume, thereby allowing a high level of accuracy to be achieved [130]. These are rigid structures whose cross-sectional area is closely defined and stable and which, therefore, act as a permanent control. Since the velocity of falling water (or indeed of any other substance) depends solely on the height of fall and the acceleration due to gravity (i.e., $V^2 = 2gh$), and since the acceleration due to gravity is constant, it follows that, in order to measure accurately the velocity of a stream, it is only necessary to cause the water to fall (as

303

over a weir or through a flume), and to measure the head of water by means of a continuous recorder at an appropriate point. The two main types of weir are, respectively, sharp-crested and broad-crested, the former being more commonly used where precise measurement is required. The critical depth or standing wave flume is so designed that a critical depth is produced in the throat section either by raising the floor of the channel in the form of a hump, or by contracting the sides, or by a combination of both. In so far as the basic theory of measurement is concerned, the only significant difference between weirs and flumes is that there is a loss of head (and of kinetic energy) over the weir, whereas there is a considerable recovery of head (preservation of kinetic energy) in the case of the flume.

8.25 Dilution gauging

A direct measurement of discharge may be made by injecting an aqueous solution of a tracer into a stream and then measuring its degree of dilution at an appropriate distance downstream. This technique, the principles of which were first reported more than a century ago [142], is mainly used in the calibration of existing gauging structures, such as weirs and flumes, but it can also be used to provide flow data on ungauged rivers. It is particularly suitable for use in the rapid, turbulent waters of mountain streams, beneath an ice cover, and in closed pipes [137], where conventional current metering is virtually impossible, but it has also been successfully used on lowland rivers [154].

Dilution gauging may be carried out either by injecting a tracer at a constant rate for a given period of time (constant-rate injection), or by injecting a known volume of tracer instantaneously (instantaneous injection or pulse integration). Each of these two methods has advantages and disadvantages although, in general, the constant-rate injection method is preferred. Detailed discussions of equipment and of field and analytical procedures were given by Smoot and Barnes [137] and the Water Research Association [154].

Various types of tracer may be used including chemical salts, dyes, and radioactive materials. Chemical salts were, until recently, the most commonly used tracer agent [137], especially lithium, potassium, magnesium, chloride and dichromate ions [154]. Sodium dichromate was selected by the Water Research Association as the most suitable tracer for routing gauging purposes [154]. With the development of commercially available fluorescent dyes and fluorometers which can detect these dyes at very low concentrations, dye dilution has increased in popularity with extensive use of four organic dyes of the rhodamine family, i.e., fluorescein, Pontacyl Pink, rhodamine B, and rhodamine WT [137]. The use of radioisotopes in general was extensively reviewed by Ellis [40], while Florkowski and others [43] specifically discussed the use of tritiated water as a tracer for flow gauging in turbulent rivers.

References

1. ALEXANDER, G. N., Using the probability of storm transposition for estimating the frequency of rare floods, *J. Hydrol.*, 1: 46-57, 1963.

2. ALEXANDER, G. N., A. KAROLY, and A. B. SUSTS, Equivalent distributions with application to rainfall as an upper bound to flood distributions, *J. Hydrol.,* 9: 322-344, 1969.

3. AMERICAN SOCIETY OF CIVIL ENGINEERS, *Hydrology Handbook,* New York, 1949.

4. AMERICAN SOCIETY OF CIVIL ENGINEERS, Report of Committee on Effects of urban development on flood discharge, Presented at ASCE (Hydraulics Division) Conf., Vicksburg, Miss., Aug. 1964. Quoted by Espey and others [41].

5. AMERMAN, C. R., The use of unit-source watershed data for runoff prediction, *WRR,* 1: 499-508, 1965.

6. ANDERSON, D. G., Effects of urban development on floods in northern Virginia, *USGS Open-file report,* 26 pp., 1968.

7. ANDREWS, F. M., Some aspects of the hydrology of the Thames basin, *Proc. ICE,* 21: 55-90, 1962.

8. ARNETT, R. R., Environmental factors affecting the speed and volume of topsoil interflow, in Gregory, K. J. and D. E. Walling (Eds.), *Fluvial Processes in Instrumented Watersheds* IBG. Spec. Publ. No. 6, pp. 7-22, 1974.

9. AYERS, H. D., Effect of agricultural land management on winter runoff in the Guelph, Ontario region, *Research Watersheds: Proc. Hydrol. Sympos.,* 4: 167-182, Univ. of Guelph, Ontario, 1965.

10. AYERS, H. D., Basin yield, Section X in Gray, D. M. (Ed.), *Handbook on the Principles of Hydrology,* National Research Council of Canada, Ottawa, 1970.

11. BAKER, D. G., Prediction of spring runoff, *WRR,* 8: 966-972, 1972.

12. BARNES, B. S. in discussion of Meyer, O. H. Analysis of runoff characteristics, *Trans. ASCE,* 103: 83-141, 1938.

13. BARNES, B. S., The structure of discharge-recession curves, *Trans. AGU,* 20: 721-725, 1939.

14. BAUER, W. J., Urban hydrology, *The Progress of Hydrology,* Univ. of Illinois, Urbana, Illinois, 2: 605-637, 1969.

15. BECKINSALE, R. P., River regimes, in Chorley, R. J. (Ed.), *Water, Earth and Man,* Methuen, London, pp. 455-471, 1969.

16. BENSON, M. A., Characteristics of frequency curves based on a theoretical 1000-year record, in Dalrymple, T. (Ed.), Flood frequency analyses, Manual of Hydrology, part 3, Flood-flow techniques, *USGS Wat. Sup. Pap.,* 1543-A: 57-74, 1960.

17. BENSON, M. A., Plotting positions and economics of engineering planning, *Proc. ASCE, J. Hydraulics Div.,* 88: 57-71, 1962.

18. BENSON, M. A., Uniform flood frequency estimating methods for Federal agencies, *WRR,* 4: 891-908, 1968.

19. BERNARD, M. M., An approach to determinate stream flow, *Trans. ASCE,* 100: 347-395, 1934.

20. BERNDT, H. W., Inducing snow accumulation on mountain grassland watersheds, *J. Soil and Water Cons.,* 19: 196-198, 1964.

21. BETSON, R. P., What is watershed runoff?, *JGR,* 69: 1541-1551, 1964.

22. BETSON, R. P., and J. B. MARIUS, Source areas of storm runoff, *WRR,* 5: 574-582, 1969.

23. BHARUCHA-REID, A. T., *Elements of the Theory of Markov Processes and their Applications,* McGraw-Hill, New York, 1960.

24. BLEASDALE, A., A. G. BOULTON, J. INESON, and F. LAW, Study and assessment of water resources, in *Conservation of Water Resources,* Institution of Civil Engineers, pp. 121-136, London, 1963.

25. BONNÉ, J., Stochastic simulation of monthly streamflow by a multiple regression model utilizing precipitation data, *J. Hydrol.,* 12: 285-310, 1971.

26. BRATER, E. F., The unit hydrograph principle applied to small watersheds, *Trans. ASCE,* 105: 1154-1178, 1940.

27. BRITISH STANDARDS INSTITUTION, *Methods of measurement of liquid flow in open channels,* BS 3680, The Institution, London, 1965.

28. BRUCE, J. P., and R. H. CLARK, *Introduction to Hydrometeorology,* Pergamon, Oxford, 1966.

29. CHOW, V. T., Statistical and probability analysis of hydrologic data, Part IV. Sequential

generation of hydrologic information, in Chow, V. T. (Ed.)., *Handbook of Applied Hydrology,* McGraw-Hill, New York, 1964.

30. CHOW, V. T., Runoff, Section 14 in Chow, V. T. (Ed.). *Handbook of Applied Hydrology,* McGraw-Hill, New York, 1964.

31. COLE, G., An application of the regional analysis of flood flows, in *River Flood Hydrology,* The Institution of Civil Engineers, London, 1966.

32. CRAWFORD, N. H., and R. K. LINSLEY, Digital simulation in hydrology: Stanford watershed model IV, *Dept. Civil Eng., Tech. Rept.,* 39, 210 pp. 1966.

33. DALRYMPLE, T., Flood frequency analysis: Manual of Hydrology, part 3, Flood flow techniques, *USGS Wat. Sup. Pap.,* 1543-A 80 pp., 1960.

34. DINGMAN, S. L., Characteristics of summer runoff from a small watershed in Central Alaska, *WRR,* **2**: 751-754, 1966.

35. DOOGE, J. C. I., The routing of groundwater recharge through typical elements of linear storage, *Gen. Ass. Helsinki, IASH Publ.,* 52: 286-300, 1960.

36. DRAGOUN, F. J., Effects of cultivation and grass on surface runoff, *WRR,* **5**: 1078-1083, 1969.

37. DUNNE, T., and R. D. BLACK, An experimental investigation of runoff processes in permeable soils, *WRR,* **6**: 478-490, 1970.

38. DUNNE, T., and R. D. BLACK, Partial area contributions to storm runoff in a small New England watershed, *WRR,* **6**: 1296-1311, 1970.

39. DUNNE, T., and R. D. BLACK, Runoff processes during snowmelt, *WRR,* **7**: 1160-1172, 1971.

40. ELLIS, W. R., A review of radioisotope methods of stream gauging, *J. Hydrol.,* **5**: 233-257, 1967.

41. ESPEY, W. H., C. W. MORGAN, and F. D. MASCH, A study of some effects of urbanization on storm runoff from a small watershed, *Texas Water Devel. Board Rept.,* 23, 110 pp., 1966.

42. EYRE, S. R., *Vegetation and Soils,* Arnold, London, 1963.

43. FLORKOWSKI, T., T. G. DAVIS, B. WALLANDER, and D. R. L. PRABHAKAR, The measurement of high discharges in turbulent rivers using tritium tracer, *J. Hydrol.,* **8**: 249-264, 1969.

44. FOLSE, J. A., A new method of estimating streamflow based upon a new evaporation formula, part II: A new method of estimating streamflow, *Carnegie Inst. of Washington, Publ.* 400, 1929.

45. FOX, W. E., Methods of river forecasting, *Proc. Conf. Hydrologic Activities in the South Carolina Region,* pp. 41-75, Clemson Univ., Clemson, 1965.

46. FREEZE, R. A., Role of subsurface flow in generating surface runoff 1. Baseflow contributions to channel flow, *WRR,* **8**: 609-623, 1972.

47. FREEZE, R. A., Role of subsurface flow in generating surface runoff 2. Upstream source areas, *WRR,* **8**: 1272-1283, 1972.

48. GARNIER, B. J., Maps of the water balance in West Africa, *Bulletin de l'Institut Français d'Afrique Noire,* **32**: 709-722, 1960.

49. GRAY, D. M., Interrelationships of watershed characteristics *JGR,* **66**: 1215-1223, 1961.

50. GRAY, D. M., and J. M. WIGHAM, Peak flow-rainfall events, Section VIII in Gray, D. M. (Ed.), *Handbook on the Principles of Hydrology,* National Research Council of Canada, Ottawa, 1970.

51. GRAY, D. M., D. I. NORUM, and J. M. WIGHAM, Infiltration and the physics of flow of water through porous media, Section V in Gray, D. M. (Ed.), *Handbook on the Principles of Hydrology,* National Research Council of Canada, Ottawa, 1970.

52. GREGORY, K. J., and D. E. WALLING, The variation of drainage density within a catchment, *Bull. IASH,* **13** (2): 61-68, 1968.

53. GUILCHER, A., *Precis d'Hydrologie, Marine et continentale,* Masson, Paris, 1965.

54. HALL, F. R., Baseflow recessions—A review, *WRR,* **4**: 973-983, 1968.

55. HAMLIN, M. J., The correlation of rainfall and run-off for water resources and flood prediction studies, Duplicated notes prepared for a Short Course in Hydrology at Loughborough College of Technology., Dept. of Civil Engineering, 5-16 October, 1964.

56. HARROLD, L. L., D. L. BRAKENSIEK, J. L. McGUINNESS, C. R. AMERMAN, and F. R. DREIBELBIS, Influence of land use and treatment on the hydrology of small watersheds

at Coshocton, Ohio, 1938-1957, *USDA Tech. Bull.* 1256, 194 pp., (USGPO), 1962.

57. HAZEN, A., Storage to be provided in impounding reservoirs for municipal water supply, *Trans. ASCE,* **77**: 1539-1667, 1914.

58. HAZEN, R., Economics of streamflow regulation, *Amer. Water Works Assoc.,* **48**: 761-767, 1956.

59. HENDERSON, F. M., Some properties of the unit hydrograph, *JGR,* **68**: 4785-4793, 1963.

60. HERTZLER, R. A., Engineering aspects of the influence of forests on mountain streams, *Civil Eng.,* **9**: 487-489, 1939.

61. HEWLETT, J. D., Watershed management, in *USFS Southeast, Forest Expt. Sta. Rept.,* pp. 61-66, 1961.

62. HEWLETT, J. D., Soil moisture as a source of baseflow from steep mountain watersheds, *USFS, Southeast. Forest Expt. Sta. Paper* 132, 11 pp., 1961.

63. HEWLETT, J. D., and J. D. HELVEY, Effects of forest clear-felling on the storm hydrograph, *WRR,* **6**: 768-782, 1970.

64. HEWLETT, J. D., and A. R. HIBBERT, Moisture and Energy conditions within a sloping soil mass during drainage, *JGR,* **68**: 1081-1087, 1963.

65. HEWLETT, J. D., and A. R. HIBBERT, Factors affecting the response of small watersheds to precipitation in humid areas, in Sopper, W. E. and H. W. Lull (Eds.), *Forest Hydrology,* Pergamon, Oxford, pp. 275-290, 1967.

66. HEWLETT, J. D., and W. L. NUTTER, *An Outline of Forest Hydrology,* Univ. of Georgia Press, Athens, 1969.

67. HEWLETT, J. D., and W. L. NUTTER, The varying source area of streamflow from upland basins, *Proc. Sympos. on Watershed Management,* pp. 65-83, ASCE, New York, 1970.

68. HIBBERT, A. R., Forest treatment effects on water yield, in Sopper, W. E. and H. W. Lull (Eds.), *Forest Hydrology,* Pergamon, Oxford, pp. 527-543, 1967.

69. HIBBERT, A. R., and G. B. CUNNINGHAM, Streamflow data processing opportunities and application, in Sopper, W. E. and H. W. Lull (Eds.)., *Forest Hydrology,* Pergamon, Oxford, pp. 725-736, 1967.

70. HOLTAN, H. N., A concept for infiltration estimates in watershed engineering, *USDA, Agric. Res. Serv.,* ARS 41-51, 25 pp., 1961.

71. HOLTAN, H. N., and N. C. LOPEZ, USDAHL-70 model of watershed hydrology, *USDA, Agric. Res. Serv., Tech. Bull.,* 1435, 84 pp., 1971.

72. HOOVER, M. D., and C. R. HURSH, Influence of topography and soil depth on runoff from forest land, *Trans. AGU,* **24**: 692-698, 1943.

73. HORNBECK, J. W., Storm flow from hardwood-forested and cleared watersheds in New Hampshire, *WRR,* **9**: 346-354, 1973.

74. HORTON, J. H., and R. H. HAWKINS, Flow path of rain from the soil surface to the water table, *Soil Science,* **100**: 377-383, 1965.

75. HORTON, R. E., Discussion of report of committee on yield of drainage-areas, *J. New England Water Works Assoc.,* 538-542, 1914.

76. HORTON, R. E., The role of infiltration in the hydrologic cycle, *Trans. AGU,* **14**: 446-460, 1933.

77. HUDSON, H. E., and R. HAZEN, Droughts and low streamflow, Section 18 in Chow, V.T. (Ed.), *Handbook of Applied Hydrology,* McGraw-Hill, New York, 1964.

78. HUFF, F. A., and S. A. CHANGNON, Relation between precipitation, drought and low streamflow, *Gen. Ass. Berkeley, IASH Publ.,* **63**: 167-180, 1963.

79. HURSH, C. R., Storm-water and absorption, Contribution to Report of the Committee on Absorption and Transpiration, 1935-36, *Trans. AGU,* **17**: 296-302, 1936.

80. HURSH, C. R., Report of the sub committee on subsurface flow, *Trans. AGU,* **25**: 743-746, 1944.

81. HURSH, C. R., and E. F. BRATER, Separating storm-hydrographs from small drainage areas into surface- and subsurface-flow, *Trans. AGU,* **22**: 863-870, 1941.

82. INSTITUTE OF HYDROLOGY, *Research 1971-72,* 67 pp., The Institute, Wallingford, 1972.

83. INSTITUTE OF HYDROLOGY, *Research 1972-73,* 66 pp., The Institute, Wallingford, 1973.

84. JAMIESON. D. G., and C. R. AMERMAN, Quick-return subsurface flow, *J. Hydrol.*, 8: 122-136, 1969.
85. JOHNSON, M. L., Research on Sleepers River at Danville, Vermont, Paper 7184, IRI, *J. Irrig. and Drainage Div., Proc. ASCE,* pp. 67-88, 1970.
86. JONES, A., Soil piping and stream channel initiation, *WRR,* 7: 602-610, 1971.
87. KIRKBY, M. J., Infiltration, throughflow and overland flow, in Chorley, R. J. (Ed.), *Water, Earth and Man,* Methuen, London, pp. 215-227, 1969.
88. KIRKBY, M. J., and R. J. CHORLEY, Throughflow, overland flow and erosion, *Bull. IASH,* 12 (3): 5-21, 1967.
89. KIRKBY, M. J., and D. R. WEYMAN, Measurements of contributing area in very small drainage basins, *Univ. of Bristol, Geography Seminar Papers, Series B,* No. 3, 12 pp. 1973.
90. KOHLER, M. A., and R. K. LINSLEY, Predicting the runoff from storm rainfall, *U.S. Weather Bur., Res. Paper,* 34, 1951.
91. KUICHLING, E., The relation between the rainfall and the discharge of sewers in populous districts, *Trans. ASCE,* 20: 1-56, 1889.
92. KULANDAISWAMY, V. C., and S. SEETHARAMAN, A note on Barnes' method of hydrograph separation, *J. Hydrol.* 9: 222-229, 1969.
93. KUNKLE, G. R., A hydrogeologic study of the groundwater reservoirs contributing base runoff to Four Mile Creek, east-central Iowa, *USGS Wat. Sup. Pap.,* 1839-0, 1968.
94. LANGBEIN, W. B. and others, Topographic characteristics of drainage basins, *USGS Wat. Sup. Pap.,* 968-C: 125-157, 1947.
95. LAWLER, E. A., Flood routing, Section 25-II in Chow, V. T. (Ed.), *Handbook of Applied Hydrology,* McGraw-Hill, New York, 1964.
96. LINSLEY, R. K., The relation between rainfall and runoff, *J. Hydrol.,* 5: 297-311, 1967.
97. LINSLEY, R. K., M. A. KOHLER, and J. L. H. PAULHUS, *Applied Hydrology,* McGraw-Hill, New York, 1949.
97A. LINSLEY, R. K., M. A. KOHLER, and J. L. H. PAULHUS, *Hydrology for Engineers,* McGraw-Hill, New York, 1958.
98. LINTON, D. L., River flow in Great Britain, 1955-56, *Nature,* 183: 714-716, 1959.
99. LLOYD-DAVIS, D. E., The elimination of storm water from sewerage systems, *Min. Proc. ICE,* 164: 41-67, 1906.
100. LULL, H. W., and W. E. SOPPER, Prediction of average annual and seasonal streamflow of physiographic units in the northeast, in Sopper, W. E. and H. W. Lull (Eds.), *Forest Hydrology,* Pergamon, Oxford, pp. 507-522, 1967.
101. LVOVITCH, M. I., Streamflow formation factors, *IASH Gen. Ass. of Toronto, Proc.,* 3: 122-132, 1958.
102. LVOVITCH, M. I., The global water balance, *Trans. AGU,* 54: 28-42, 1973.
103. McCANN, S. B., P. J. HOWARTH, and J. G. COGLEY, Fluvial processes in a periglacial environment, *Trans. IBG,* 55: 69-82, 1972.
104. MANDEVILLE, A. N., P. E. O'CONNELL, J. V. SUTCLIFFE, and J. E. NASH, River flow forecasting through conceptual models: Part III—The Ray catchment at Grendon Underwood, *J. Hydrol,* 11: 109-128, 1970.
105. MANNING, R., On the flow of water in open channels and pipes, *Trans. Inst. Civ. Engrg. Ireland,* 20: 161-207, 1889.
106. MARTINELLI, M., Watershed management in the Rocky Mountain alpine and subalpine zones, *USFS Res. Note* RM-36, 7 pp., 1964.
107. MELENTIJEVICH, M., Estimation of flood flows using mathematical statistics, *Floods and their computation, Studies and reports in hydrology* No. 3: 164-174, UNESCO-IASH-WMO, 1969.
108. MEYER, A. F., *The Elements of Hydrology,* John Wiley, New York, 1928.
109. MEYER, A. F., Introduction to Runoff, Chapter 11A in Meinzer, O. E. (Ed.), *Hydrology,* McGraw-Hill, New York, 1942.
110. MEYER, H. A. (Ed.), *Symposium on Monte Carlo Methods,* Wiley, New York, 1956.
111. MILLER, C. F., Evaluation of runoff coefficients from small natural drainage areas, *Univ. of Kentucky, Water Res. Institute, Res. Rept.,* 14, 112 pp., 1968.
112. MULVANEY, T. J., On the use of self-registering rain and flood gauges in making observations of the relations of rainfall and of flood discharges in a given catchment, *Trans. Inst. Civ. Engrs. Ir.* (Dublin), 4: 18, 1851.

113. NAKAMURA, R., Runoff analysis by electrical conductance of water, *J. Hydrol.,* **14**: 197-212, 1971.

114. NASH, J. E., A unit hydrograph study, with particular reference to British catchments, *Proc. ICE,* **17**: 249-282, 1960.

115. NASH, J. E., and J. V. SUTCLIFFE, River flow forecasting through conceptual models, Part I—A discussion of principles, *J. Hydrol.,* **10**: 282-290, 1970.

116. O'CONNELL, P. E., J. E. NASH, and J. P. FARRELL, River flow forecasting through conceptual models, Part II—The Brosna catchment at Ferbane, *J. Hydrol.,* **10**: 317-329, 1970.

117. PARDÉ, M., Hydrologie fluviale des Iles Britanniques, *Annales de Géographie,* **48**: 369-384, 1939.

118. PARDÉ, M., *Fleuves et Rivières* (3rd Ed.), Armand Colin, Paris, 1955.

119. PINDER, G. F., and J. F. JONES, Determination of the groundwater component of peak discharge from the chemistry of total runoff, *WRR,* **5**: 438-445, 1969.

120. RAGAN, R. M., An experimental investigation of partial area contributions, *Gen. Ass. of Berne, IASH Publ.* **76**: 241-249, 1967.

121. RAGAN, R. M., Role of basin physiography on the runoff from small watersheds, *Vermont Resources Research Centre, Rept.* 17, 25 pp., Univ. of Vermont, 1967.

122. RAWITZ, E., E. T. ENGMAN, and G. D. CLINE, Use of the mass balance method for examining the role of soils in controlling watershed performance *WRR,* **6**: 1115-1123, 1970.

123. REEVE, W. T. N., and D. T. EDMONDS, Zambezi River flood hydrology and its effect on design and operation of Kariba Dam, in *River Flood Hydrology,* The Institution of Civil Engineers, London, 1966.

124. REICH, B. M., Flood series compared to rainfall extremes, *WRR,* **6**: 1655-1667, 1970.

125. REICH, B. M., Land surface form in flood hydrology, *Proc. Binghamton Geomorph. Sympos.,* **1**: 49-68, 1970.

126. ROCHE, M., *Hydrologie de Surface,* Gauthier-Villars, Paris, 1963.

127. RODDA, J. C., The flood hydrograph, in Chorley, R. J. (Ed.), *Water, Earth and Man* Methuen, London, pp. 405-418, 1969.

128. RODDA, J. C., The significance of characteristics of basin rainfall and morphometry in a study of floods in the United Kingdom, *Floods and their computation, Studies and reports in hydrology,* No. 3: 834-845, UNESCO-IASH-WMO, 1969.

129. ROGERS, W. F., New concept in hydrograph analysis, *WRR,* **8**: 973-981, 1972.

130. ROTHACHER, J., and N. MINER, Accuracy of measurement of runoff from experimental watersheds, in Sopper, W. E. and H. W. Lull (Eds.), *Forest Hydrology,* Pergamon, Oxford, pp. 705-714, 1967.

131. SATTERLUND, D. R., Forest types and potential runoff, in Sopper, W. E. and H. W. Lull (Eds.), *Forest Hydrology,* Pergamon, Oxford, pp. 497-505, 1967

132. SAVINI, J., and J. C. KAMMERER, Urban growth and the water regimen, *USGS Wat. Sup. Pap.,* 1591-A, 43 pp., 1961.

133. SHERMAN, L. K., Streamflow from rainfall by unit-graph method, *Eng. News-Record,* **108**: 501-505, 1932.

134. SMITH, K., Some thermal characteristics of two rivers in the Pennine area of northern England, *J. Hydrol.,* **6**: 405-416, 1968.

135. SMITH, K., Some features of snow-melt recession in the upper Tees basin, *Water and Water Engg.,* **75**: 345-346, 1971.

136. SMITH, K., *Water in Britain,* Macmillan, London, 1972.

137. SMOOT, G. F., and H. H. BARNES, Recent developments in measuring techniques for flow in open channels, *The Progress of Hydrology,* University of Illinois, Urbana, Illinois, **1**: 194-225, 1969.

138. SNYDER, F. F., Synthetic unit-graphs, *Trans. AGU,* **19**: 447-454, 1938.

139. SNYDER, W. M., Hydrograph analysis by the method of least squares, *Proc. ASCE,* **81** (793), 1955.

140. SNYDER, W. M., Subsurface implications from surface hydrograph analysis, Proc. 2nd Seepage Symposium, Phoenix, Arizona, 1968, *USDA Agric. Res. Service Publ.* 41-147: 35-45, 1969.

141. SNYDER, W. M., W. C. MILLS, and J. C. STEPHENS, A method of derivation of

nonconstant watershed response functions, *WRR*, **6**: 261-274, 1970.

142. SPENCER, E. A., and J. S. TUDHOPE, A literature survey of the salt-dilution method of flow measurement, *J. Inst. Water Engrs.*, **12**: 127-138, 1958.

143. TENNESSEEE VALLEY AUTHORITY, *Hydrology of small watersheds in relation to various crop covers and soil characteristics,* Tennessee Valley Authority in co-operation with N. Carolina State College of Agriculture and Engineering, 1960.

144. TENNESSEE VALLEY AUTHORITY, Bradshaw Creek—Elk River: A pilot study in area-stream factor correlation, *Office of Tributary Area Develt. Research Paper, 4,* 64 pp., 1964.

145. THOMAS, H. A., and M. B. FIERING, Mathematical synthesis of streamflow sequences for the analysis of river basins by simulation, chap. 12 in Maas, A., M. M. Huffschmidt, R. Dorfman, H. A. Thomas, S. A. Marglin, and G. M. Fair (Eds.), *Design of Water-Resource Systems,* Harvard Univ. Press, 1962.

146. THORNTHWAITE, C. W., and J. R. MATHER, Instructions and tables for computing potential evapotranspiration and the water balance, *Publ. in Climat.,* **10**: 185-311, 1957.

147. TOLER, L. G., Use of specific conductance to distinguish two base-flow components in Econfina Creek, Florida, *USGS Prof. Pap.,* 525-C, 206-208, 1965.

148. TSUKAMATO, Y., An experiment on subsurface flow, *J. Jap. Soc. Forestry,* **43**: 61-68, 1961.

149. VORONKOV, P. P., Hydrochemical bases for segregating local runoff and a method of separating its discharge hydrograph, *Meteorologiya i Gidrologiya,* **8**: 21-28, 1963.

150. WALLING, D. E., Streamflow from instrumented catchments in south-east Devon, in Gregory, K. J. and W. Ravenhill (Eds.), *Exeter Essays in Geography,* Univ. of Exeter, pp. 55-81, 1971.

151. WARD, R. C., Some runoff characteristics of British rivers, *J. Hydrol.,* **6**: 358-372, 1968.

152. WARD, R. C., *Small Watershed Experiments,* 254 pp., Univ. of Hull, 1971.

153. WARD, R. C., Estimating streamflow using Thornthwaite's climatic water-balance, *Weather,* Feb. pp. 73-84, 1972.

154. WATER RESEARCH ASSOCIATION, River flow measurement by dilution gauging, *Water Research Assoc. Tech. Pap.,* 74, 85 pp., 1970.

155. WEYMAN, D. R., Runoff process, contributing area and streamflow in a small upland catchment, in Gregory, K. J. and D. E. Walling (Eds.), *Fluvial Processes in Instrumented Watersheds* IBG, Spec. Publ. No. 6, pp. 33-43, 1974.

156. WHIPKEY, R. Z., Subsurface stormflow from forested slopes, *Bull. IASH,* **10** (2): 74-85, 1965.

157. WHIPKEY, R. Z., Theory and mechanics of subsurface stormflow, in Sopper, W. E. and H. W. Lull (Eds.), *Forest Hydrology,* Pergamon, Oxford, pp. 255-260, 1967.

158. WHIPKEY, R. Z., Storm runoff from forested catchments by subsurface routes, *Floods and their computation, Studies and reports in hydrology,* No. 3: 773-779, UNESCO-IASH-WMO, 1969.

159. WHITE, E. L., and B. M. REICH, Behaviour of annual floods in limestone basins in Pennsylvania, *J. Hydrol.,* **10**: 193-198, 1970.

160. WISLER, C. O., and E. F. BRATER, *Hydrology,* 2nd Ed., Wiley, New York, 1959.

161. WOLF, P. O., Forecast and records of floods in Glen Cannich in 1947, *J. Inst. Water Eng.,* **6**: 298-324, 1952.

162. WOLF, P. O., Comparison of methods of flood estimation, in *River Flood Hydrology,* The Institution of Civil Engineers, London, 1966.

163. WOODRUFF, J. F., and J. D. HEWLETT, Predicting and mapping the average hydrologic response for the Eastern United States, *WRR,* **6**: 1312-1326, 1970.

164. YEVDJEVICH, V. M., Bibliography and discussion of flood-routing methods and unsteady flow in open channels, *USGS Wat. Sup. Pap.,* 1690, 1964.

165. ZAVODCHIKOV, A. B., Computation of spring high water hydrographs using genetic formula of runoff, *Soviet Hydrol.,* **5**: 464-476, 1965.

166. ZELENHASIC, E., Theoretical probability distributions for flood peaks, *Colorado State Univ., Hydrology Papers,* No. 42, 35 pp., 1970.

9. The Drainage Basin

9.1 The need for synthesis

The success of environmental studies, including geography, depends in large measure on the ability to synthesize as well as to analyse. There is a need to 'put back together' those elements and factors which have previously been discussed separately in order to understand properly the workings of the natural environment. The essential unity of the hydrological cycle was introduced in chapter 1 and has been emphasized over and again throughout this book. However, the importance of recognizing it in an age which shows such an interest in ecosystems, ecology, and environment, is such that a final, synthesizing chapter seems appropriate.

Just as the geographer attempts a synthesis within the framework of 'the region' so does the hydrologist perform a similar exercise within the context of the drainage basin. This provides the almost ideal natural unit over which hydrological processes are integrated and for which a water balance may be constructed to show the disposal of precipitation into a number of subsequent forms, e.g., interception, soil moisture and groundwater storages, evapotranspiration, and runoff. It is also the area over which fluvial geomorphic processes operate and for which an energy balance can be constructed whereby the precipitation input is equated with an erosional output of water and gravity-moved load. Again, biologists, ecologists, and biogeographers have turned to the drainage basin as an ideal unit in which to develop the ecosystem approach and finally, it is by definition the unit for river basin planning by hydrologists, economists, and sociologists, where the intimate ties between physical, economic, and human resources needs not only to be clearly shown but positively utilized.

The unity of the drainage basin is most clearly seen in the delicate balance of processes, geomorphological and climatological, biogeographical and hydrological, which are at work within it. But such is the complexity of the natural system that it is almost impossible to illustrate this unity simply and concisely, except by reference to specific examples. (Much the same problem is faced by the geographer attempting to write regional geography.) This final chapter will be concerned with specific examples illustrating the geomorphological and hydrological unity of the drainage basin and with particular reference to the development of interest in small catchment studies and in catchment modelling.

311

The geomorphological unity of the drainage basin

9.2 Introduction

The geomorphological unity of the drainage basin was perhaps most succinctly expressed by W. M. Davis in 1899:

> Although the river and the hill-side waste sheet do not resemble each other at first sight, they are only the extreme members of a continuous series; and when this generalization is appreciated, one may fairly extend the 'river' all over its basin and up to its very divides. Ordinarily treated, the river is like the veins of a leaf; broadly viewed, it is like the entire leaf [22].

This quotation emphasizes that geomorphologically and to a large extent hydrologically the river and its drainage basin are synonymous, that it is meaningless to regard the water flowing in the river channel as distinct from the remaining water within the drainage basin and that geomorphologically water is at work within the drainage basin from the moment of impact of a raindrop on the soil surface to its final exit from the basin in the main, trunk stream.

This theme was developed in chapter 8 where it was emphasized that runoff is a basin-wide process which occurs *everywhere* within the drainage basin both *on* the surface (although usually over only limited areas of the basin) and particularly *beneath* the surface as either interflow or groundwater flow. Since this movement of water is the prime agent in creating landforms in fluvially eroded areas and is influenced by a wide variety of factors, including climate, vegetation cover, and human activity, it will be evident that expressions of the surface shape of the drainage basin and of the nature and disposition of its channel network may be used as indices both of the landforms which have been developed and of the hydrological processes operating to modify those landforms.

Intuitively it would seem clear that basin and channel network characteristics are strongly associated. This is not to say that the two are causally related in a systematic way, i.e., basin plan does not invariably determine channel network arrangement or vice versa. Indeed, often both characteristics are in large measure determined by other factors such as geological structure or vegetation cover. There is thus no logical order in which the two should be discussed and for convenience in the following sections basin characteristics will be considered before channel characteristics.

9.3 Basin plan

The planform of a drainage basin may be simply characterized by reference to its maximum axial length, L, its perimeter length, P, and its area, A, as illustrated in Fig. 9.1. Natural drainage basins are unique and there is thus a virtually infinite variety of basin sizes and shapes. For convenience this infinite variety has often been categorized into a number of simple geometric shapes. We thus speak, for example, of circular, rectangular, triangular, pear-shaped, or ovoid basins. If all drainage basins were geometrically similar in plan then the ratio of basin area to the square of basin length would be a constant although, in fact, a wide variety of natural and digitally simulated drainage basin data indicate a tendency towards elongation of the larger catchments [28].

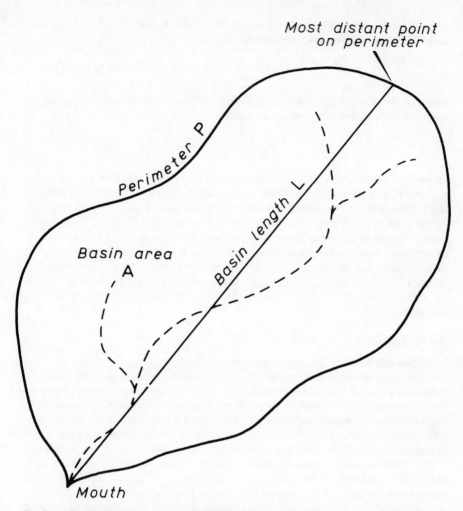

Fig. 9.1. *Commonly used drainage basin parameters.*

Basin shape or plan may be numerically described in a number of ways. Thus Horton [49] used a dimensionless *form factor* (R_f) such that

$$R_f = \frac{A}{L^2} \qquad (9.1)$$

A compact basin will yield a value close to unity, while a long, narrow basin will yield a low value, and generally a lower peak runoff total than a similar sized area with a high form factor, since a heavy rainfall is less likely to fall simultaneously over the entire area. The *compactness coefficient* (C_c) was first proposed by Gravelius in 1914 [37] and is

$$C_c = \frac{P}{P_c} \qquad (9.2)$$

313

where P_c is the perimeter length of a circle whose area equals that of the drainage basin. Miller [71] used a *circularity ratio* (R_c)

$$R_c = \frac{A}{A_c} \qquad (9.3)$$

where A_c is the area of a circle having the same perimeter length as the drainage basin, while Schumm [88] used an *elongation ratio* (R_e)

$$R_e = \frac{D_c}{L} \qquad (9.4)$$

where D_c is the diameter of a circle having the same area as the drainage basin. C_c, R_c, and R_e all have a theoretical minimum value of unity for a completely circular basin. These and other shape indices have been criticized because they do not relate to the ovoid or pear-shape which has often been proposed as the ideal drainage-basin shape. This led Chorley and others [17] to suggest the lemniscate loop as a more appropriate model. Others have derived even more sophisticated morphometric indices and in this context the work of Anderson [4] seems most promising but, as Wilcock [110] cautioned, the search for magic numbers which can be used to sum up all the important features of a segment of reality is likely to meet with only limited success.

9.4 Basin relief

It is equally difficult to derive a single number which meaningfully quantifies slope and relief over an entire drainage basin. Some methods represent only very simple approximations, so that *maximum basin relief* is the difference in height between the basin mouth and the highest point on the basin perimeter while this value, divided by the horizontal distance over which it is measured yields the *relief ratio* [88]. The relief ratio thus measures the overall slope of a drainage basin and provides an index of the intensity of erosion processes operating on the basin slopes [100]. Other methods, perhaps yielding more representative values, make use of the grid squares or grid intersections on a drainage basin map as a basis for sampling the relevant characteristics. For example, drainage basin slope may be derived by determining the frequency distribution of slope normal to the contours at a number of grid intersections. The resulting graph can then be used both as a visual aid and also to determine the mean and median slopes [64]. More simply, Nash [74] used a value of slope in parts per 10 000 which was defined as the mean of the slopes obtained at the intersection of a grid of about 100 points. Another excellent grid square technique for evaluating mean slope and aspect was presented by Foyster [34]. Finally, a common approach to the problem is through the *hypsographic curve,* which shows the cumulative percentage of the area of a drainage basin above or below a given height. This method can give only a partial picture, but may serve to indicate the existence of significant flat areas, such as erosion surfaces or structural benches, which may considerably affect the speed and direction of movement of water into the stream channels, as in the case of the Lynmouth flood of 1952 [24]. However, Aronovici [6] argued that while the hypsographic curve may be of interest to the geomorphologist '. . . it does not appear to be a meaningful parameter in watershed hydrology'.

314

It has been shown that a significant index of the average slope steepness over a drainage basin is provided by the *maximum valleyside slope* measured at intervals along the valley sides between the divides and the adjacent stream channels [100].

Stream channel slope may similarly be expressed in a number of ways including the now outmoded attempts to fit mathematical curves to the longitudinal profiles. Commonly used measures include the average slope of individual channel segments, the simple gross slope of the main channel, which is derived by dividing the difference in elevation between the head and the mouth by the channel length, and the mean slope of the entire channel network which is obtained by averaging the gradients of all channels draining at least 10 per cent of the total drainage basin area [16].

Strahler [98] observed a close quantitative relationship between valleyside slopes and the slope of the adjacent stream channel over a wide range of geographical conditions (see Fig. 9.2). Strahler suggested that steep slopes contribute large quantities of coarse debris to the stream channels, necessitating steep channel gradients to enable streamflow to transport that debris as bed load, while gentle slopes yield smaller quantities of fine debris which in turn requires correspondingly low channel gradients for its transport.

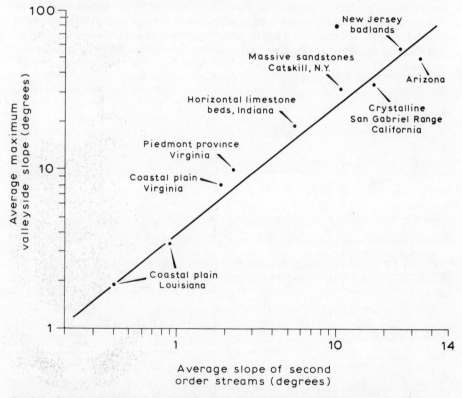

Fig. 9.2. *Relation between valleyside slope and adjacent channel slope in regions of widely different relief, climate, and rock type. (From an original diagram by A. N. Strahler,* American Journal of Science, *248, 1950.)*

9.5 Channel networks

Even a brief perusal of the relevant literature indicates that many of the discussions of drainage basin morphometry and of its relations to channel networks and drainage basin hydrology have been at a fairly naive level. Many of the relationships which have been 'discovered' are essentially self-evident and amount to little more than statements that little streams feed large streams and that little streams drain smaller basins and have lower discharges than do large streams. In terms of channel networks, geomorphologists have long recognized that ordered patterns exist (see Fig. 9.3), based upon either the relative orientation of individual streams in the same network, e.g., dendritic, trellis, etc., or the relative orientations of individual networks, e.g., radial, parallel, etc. [93]. However, R. E. Horton was one of the first to attempt to quantify and systematize channel networks in such a way that small differences in network structure could be characterized and their hydrological significance demonstrated. His useful contribution was subsequently elaborated and modified by a number of workers including A. N. Strahler.

Horton [50] developed a system of stream ordering, or numerical ranking of the channels within a channel network, that could be used to calculate a number of dimensionless network parameters such as the bifurcation ratio and the stream length ratio. This type of network analysis, often referred to as 'Horton analysis' or 'Horton-Strahler analysis' is now a basic technique in the description and analysis of channel network structure. The stream ordering system most frequently used is the modification suggested by Strahler [99] in which the smallest fingertip tributaries are designated order 1. Where two first-order channels join, a channel of order 2 is

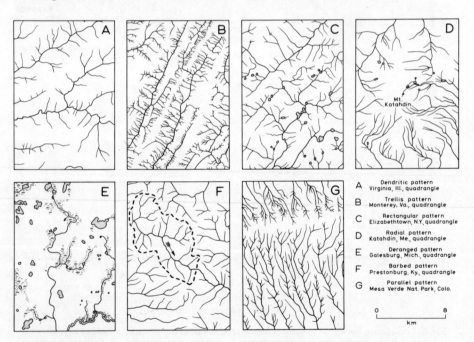

A   Dendritic pattern
Virginia, Ill., quadrangle

B   Trellis pattern
Monterey, Va., quadrangle

C   Rectangular pattern
Elizabethtown, N.Y., quadrangle

D   Radial pattern
Katahdin, Me., quadrangle

E   Deranged pattern
Galesburg, Mich., quadrangle

F   Barbed pattern
Prestonburg, Ky., quadrangle

G   Parallel pattern
Mesa Verde Nat. Park, Colo.

Fig. 9.3. *Some common drainage patterns. (From original diagrams by Thornbury [104].)*

316

formed: where two second-order channels join a channel of order 3 is formed and so on. In this way the main channel at the mouth of the drainage basin has the highest order. Melton [68] subsequently explained the mathematical concepts involved. The main disadvantage of the Strahler system is that it violates the distributive law, since the entry of a lower-order tributary (e.g., the confluence of an order 1 and order 2 stream) does not necessarily increase the order of the main stream. Shreve [91] overcame this by dividing the network into separate segments or links at each junction and allowing each segment to reflect the order of the tributaries comprising it. Thus two first-order streams combine to form a second-order segment, a first-order stream and a second-order segment combine to form a third-order segment, two third-order segments combine to form a sixth-order segment, and so on. Many other topologic refinements have subsequently been proposed, c.f. Jarvis [53] but, as Chorley [16] observed, the simpler unambiguous Strahler system is now firmly established as the normal method.

Horton [50] proposed three drainage basin 'laws'. The *law of stream numbers* states that the number of stream segments of each order form an inverse geometric sequence with order number and can be described by the equation

$$\ln N_u = C - Du \qquad (9.5)$$

where u is stream order, N_u is the number of uth order streams, and C and D are constants. The *law of stream lengths* states that the mean lengths of stream segments of each of the successive orders of a drainage basin tend to approximate a direct geometric sequence in which the first term is the average length of segments of the first order and can be described by the equation

$$\ln L_u = E - Fu \qquad (9.6)$$

where u is stream order, L_u is the average length of uth order streams, and E and F are constants. Finally, the *law of stream slopes* states that the mean slopes of stream segments of each of the successive orders of a drainage basin tend to approximate an inverse geometric sequence with order number and can be described by the equation

$$\ln S_u = G - Hu \qquad (9.7)$$

where u is stream order, S_u is the average slope of uth order streams, and G and H are constants. Horton did not actually state a law of stream areas although he inferred the relationship between drainage basin area and stream order. It was, in fact, Schumm [88] who expressed this relationship in a *law of stream areas* which states that the mean area of drainage basins of each order tend closely to approximate a direct geometric sequence in which the first term is the mean area of the first-order basins. This may be described by the equation

$$\ln A_u = J - Ku \qquad (9.8)$$

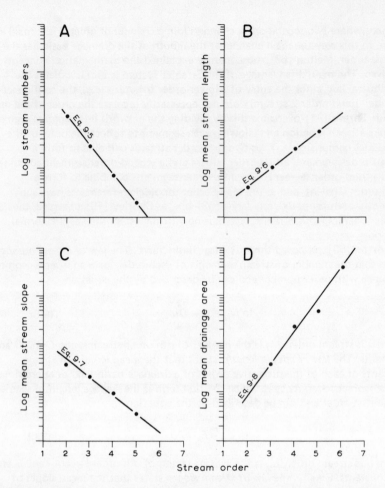

Fig. 9.4. *Horton-Strahler relations for the Big Sandy River basin in Kentucky. (From an original diagram by C. T. Yang,* Water Resources Res., **7**, *316, Fig. 2, 1971. © American Geophysical Union.)*

where u is stream order, A_u is the mean area of uth order drainage basins, and J and K are constants.

Equations (9.5) to (9.8) plot as straight lines on semilog paper and are shown for the Big Sandy River basin in Kentucky in Fig. 9.4. The slope of each of the lines defines a dimensionless ratio. Thus from eq. (9.5) can be derived the *bifurcation ratio* of stream number (Rb)

$$Rb = N_u/N_{u+1} \qquad (9.9)$$

The *stream length ratio* (Rl) can be derived from eq. (9.6)

$$Rl = L_u/L_{u+1} \qquad (9.10)$$

and the *stream slope (or concavity) ratio (Rs)* from eq. (9.7)

$$Rs = S_u/S_{u+1} \qquad (9.11)$$

Finally, the *drainage area ratio (Ra)* is derived from eq. (9.8) so that

$$Ra = A_u/A_{u+1} \qquad (9.12)$$

Further derivations from these basic equations have yielded the *law of stream fall* [33] [116] which states that in a river basin that has reached its dynamic equilibrium condition, the ratio of the average fall between any two different order streams in the same river basin is unity and the *law of least time rate of energy expenditure* [116] which states that during the evolution towards its equilibrium condition, a natural stream chooses its course of flow in such a manner that the time rate of potential energy expenditure per unit mass of water along this course is a minimum. In addition Ghosh and Scheidegger [35] proposed a *law of link lengths* which states that link lengths (distances along the stream between successive tributary junctions) tend to increase geometrically with stream order.

A considerable amount of field measurement and cartographic analysis has been invested in verifying these basic Horton laws and their subsequent derivatives and Strahler [100] referred to many of the important contributions. It might, indeed, be argued that too much effort has been devoted to fitting the data to these laws and not enough to discovering why particular drainage networks deviate from the norm. In this respect acknowledgement should be paid to the work of Gregory [38], who showed that the law of stream numbers applies best to active stream systems mapped in the field, rather than to a valley network which may include 'fossil' elements from an earlier period of erosion, and to the work of Eyles [31] who suggested that in some areas recent rejuvenation might have disproportionately increased the numbers of low order streams and resulted in a tendency for Rb to be inversely related to stream order. It is significant to note, as did Strahler [100], that verification of Horton's laws of stream numbers and lengths supports the notion that there is a preservation of geometrical similarity in basins of increasing order, i.e., that a third-order basin will tend to be geometrically similar to the second-order basins which lie within it, and a second-order basin similar to its component first-order basins. In fact, however, attention has already been drawn in section 9.3 to the view, supported by Hack [39], that in general larger drainage basins are more elongated than smaller ones.

Smart [93] referred to Horton's original view that the bifurcation and stream length ratios would be functions of a number of environmental factors, including topography, geology, and climate, and could therefore be used to distinguish channel networks of widely differing structure, but noted that such expectations have not been confirmed, although such Horton parameters have proved useful for characterizing individual networks. Shreve [92] suggested that this failure could be explained partly by the random nature of network topology and link lengths, particularly in the case of channel networks developed in the absence of geological controls, and partly by the fact that calculation of the various Horton ratios involves the summation of large numbers of stream lengths and areas and this tends to average out many of the details responsible for differences in network structure.

319

It is not surprising, therefore, that many stream networks randomly generated on a computer yield stream number relationships which satisfy Horton's law.

However, investigations into the use of Horton parameters as indices of the changing structure of channel networks continues. The channel network within a drainage basin will normally adjust to changes in the energy input to that basin. Some aspects of the adjustment to varying precipitation were examined by Williams and Fowler [112].

In addition to the analysis of channel-network structure it is also frequently desirable to have numerical information on *drainage density* which may reflect a variety of geologic, climatic, and hydrologic controls. Discussions in the preceding chapter emphasized that normally, over virtually the entire drainage basin, all precipitation will infiltrate the soil surface and that only over limited, specifically defined areas, e.g., alongside streams, is overland flow likely to be generated. It seems clear, therefore, that this type of overland flow *results from* the presence of the streams and their adjacent 'wetlands'. This is a logical corollary of the Hewlett runoff model discussed in section 8.4 for if it is normal for precipitation to infiltrate rather than to flow initially over the ground surface, the subsequent appearance of the water at the surface as channel flow presumably reflects the inability of the subsurface transmission of water to keep pace with the delivery of water by means of infiltration. Thus where the subsurface material is sufficiently permeable there will be no surface flow, where it is totally impermeable there will be no infiltration and where, as is normally the case, permeability decreases with depth the resulting lateral flow of water at comparatively shallow depths results in water-logging and overland flow on the lower slopes. A number of workers, including Bunting [13] and Woodruff and Gergel [113] have demonstrated that concentrated subsurface flow may result in concentrated subsurface solution and consequently the formation of shallow surface valleys.

Viewed in these terms drainage density is a direct reflection of the relationship between the input of precipitation to the surface of a drainage basin and the transmission of infiltrated water through the subsurface materials. It is thus an extremely complex function of precipitation, infiltration and permeability characteristics and just as the Horton runoff model was shown in chapter 8 to apply only in a special set of conditions, so the related Horton erosion model, postulating rill formation by overland flow at a certain critical distance from the drainage basin divides, may be considered a special case which applies only to areas of abnormally low infiltration capacity, such as the badlands in which it was first developed.

Horton [49] [50] defined drainage density (D) as the average length of stream channel (L) per unit area within a drainage basin of area A. Thus

$$D = \Sigma L/A \qquad (9.13)$$

Broadly speaking, drainage density reflects the closeness of spacing of channels within a drainage network and is equally important as a hydrological and geomorphological index since it will clearly be closely related to the character of runoff-response to precipitation and to the topographic features of the drainage basin.

The inverse of drainage density was used by Schumm [88] and termed the *constant of channel maintenance (C)*:

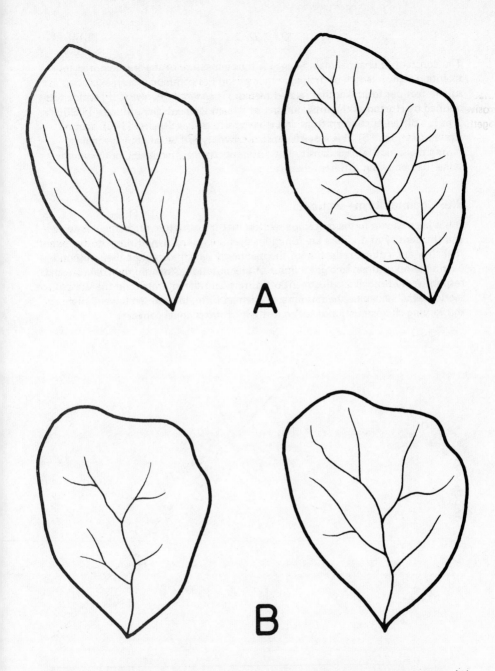

Fig. 9.5. *An illustration of the difference between drainage density and channel frequency: (A) The two basins have the same drainage density but different channel frequency (B) The two basins have the same channel frequency but different drainage density. (From an original diagram by A. N. Strahler in Chow, V. T. (Ed.),* Handbook of Applied Hydrology, *McGraw-Hill, New York, 1964.)*

321

$$C = 1/D \qquad (9.14)$$

The constant of channel maintenance is thus a measure of the area required to maintain a given length of drainage channel and has commonly been expressed as square feet per foot. Another related measure is *channel frequency* (*F*) which was defined by Horton as the total number of stream segments per unit area [100]. This is a different concept from drainage density and as Strahler [100] demonstrated (see Fig. 9.5), it is easy to construct two hypothetical drainage basins having the same drainage density but different channel frequency, or having the same frequency but different density.

9.6 Channel form—sectional

It has been shown in the preceding section that the structure of channel networks and the density of drainage are functions of drainage basin hydrology to the extent that they are a direct reflection of the manner in which a drainage basin responds, over a period of time, to a given input of precipitation. An even more delicate and responsive hydrological adjustment of form is considered in this and the subsequent section which examine the response of channel form, in both section and plan, to the varying characteristics of the runoff which the channel conveys.

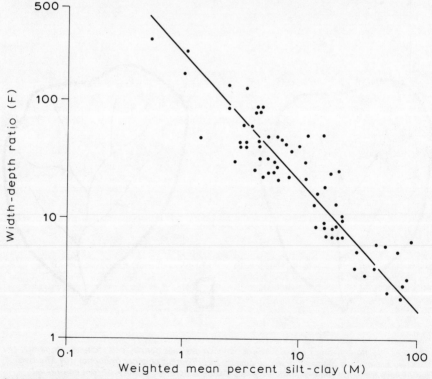

Fig. 9.6. *Relation between width/depth ratio and weighted mean per cent silt-clay. (From an original diagram by Schumm [89]. With acknowledgment to the US Geological Survey.)*

322

An input of precipitation to a drainage basin may result in both an increase in the discharge of water and debris in the stream channels and also in a change in the form of the channels themselves. These variables are closely interrelated and according to Leopold and others [62] the form of the stream channel at a given location is a function of stream discharge, the quantity and character of the debris moved through the channel, and the character or composition of the materials comprising the bed and banks of the channel.

Thus, observing that stream channels formed in predominantly sandy materials were wider and shallower than those formed in predominantly silty materials, Schumm [89] calculated the width/depth ratio (F) for a large number of locations in the northern Great Plains and related this to the weighted mean percentage of silt-clay (less than 0.074 mm diameter) forming the channel banks and bed, as

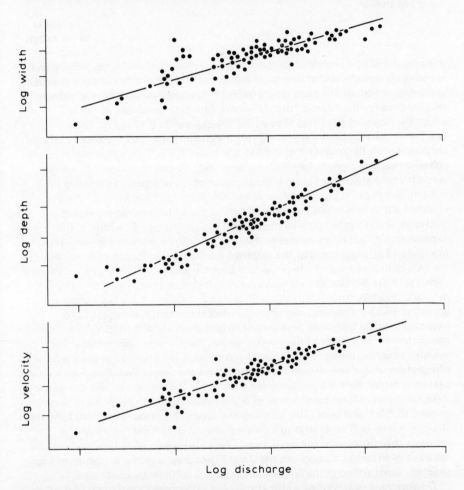

Fig. 9.7. *At-a-station changes of width, mean depth, and mean velocity with discharge. Powder River at Locate, Montana. (From an original diagram by Leopold and Maddock [60]. With acknowledgment to the US Geological Survey.)*

323

shown in Fig. 9.6. It will be noted that as the bed and bank materials contain progressively more silt-clay and become progressively more cohesive and resistant to erosion, the channels become relatively narrower and deeper, the actual relationship being

$$F = M^{-1.08} \tag{9.15}$$

where M is the weighted percentage of silt-clay in the channel perimeter.

Again, in an important pioneering study of natural streams, Leopold and Maddock [60] demonstrated that channel characteristics constitute an interdependent system of *hydraulic geometry* whereby depth, width, and velocity of flow at a given cross-section can be related to discharge by simple power functions so that

$$w = aQ^b, \quad d = cQ^f, \quad v = kQ^m \tag{9.16}$$

where w is width, d is mean depth, v is mean velocity, Q is discharge, and a, c, k, b, f, and m are numerical coefficients. It will be apparent that since $Q = wdv$ the exponents in eqs. (9.16) must satisfy $b+f+m = 1$. Examples of these *at-a-station* relationships for the Powder River at Locate, Montana are shown in Fig. 9.7 in which the slopes of the fitted lines define the exponents b, f, and m. These exponents describe both the geometry of the channel and the resistance to erosion associated with the character of the bed and banks [62]. Thus a channel incised in cohesive material tends to be relatively deep and box-like so that width changes very little with discharge (i.e., b is small), most of the adjustment occurring in velocity and depth (i.e., m and f are large). On the other hand in wide, shallow channels cut in non-cohesive materials most of the adjustment to increasing discharge occurs in the width of the channel (i.e., b is large). In addition, the exponents b, f, and m are related to the transportation load of the stream. Thus Morisawa [73] suggested that the sediment load would partly determine whether the exponent b or f was the larger, whilst Leopold and others [62] showed that the more rapid the increase of sediment load with increasing discharge the higher is the m/f ratio. Furthermore, since stream discharge and sediment load at a given station are influenced by drainage basin characteristics above that station it is to be expected that the hydraulic geometry of stream channels will, in part, reflect the morphometric properties of the entire drainage basin as was demonstrated in a valuable study in Illinois [96]. Leopold and others [62] studied the geographical distribution of b, f, and m over the United States and found that although no clear pattern emerges there is a tendency for channels in the humid east and the wetter mountain areas to have lower values of b than channels in the semi-arid southwest or parts of the High Plains. This confirms the suggestion made earlier that ephemeral streams in semi-arid areas tend to increase more rapidly in width as discharge increases than do streams in humid areas. There also appeared to be a tendency for values of m to be low throughout the Great Plains area, possibly as a result of high sediment loads although the limited data were not sufficiently conclusive.

Downstream variations of width, depth, and velocity with variations of discharge constitute further elements of the hydraulic geometry of stream channels provided that such variations are related to the same discharge. In Fig. 9.8 downstream

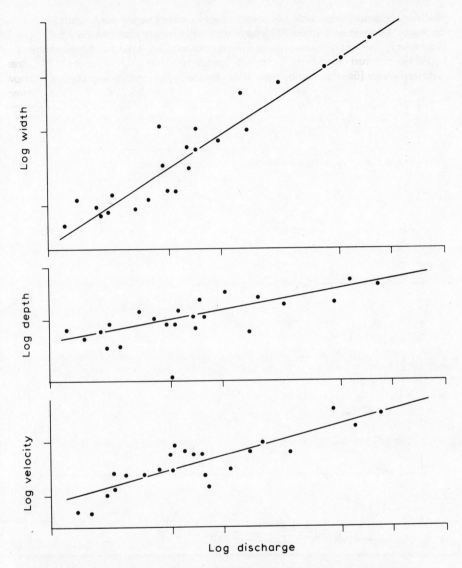

Fig. 9.8. *Changes of width, mean depth, and mean velocity in relation to mean annual discharge as discharge increases downstream, Powder River system, Wyoming and Montana. (From an original diagram by Leopold and Maddock [60]. With acknowledgment to the US Geological Survey.)*

variations of width, depth, and velocity are related to mean annual discharge for the Powder River and its tributaries in Wyoming and Montana. Mean annual discharge is frequently used because the relevant data are readily available. It has been argued that the most meaningful discharge for this purpose is the one which forms or maintains the channel [62]. Since channel form responds in some measure to virtually the complete range of discharge, the calculation of channel-forming discharge is almost impossible but is often approximated by the use of bankfull discharge. Leopold and others [62] suggested that in many rivers the bankfull discharge is one that has a recurrence interval of about 1.5 years but although this suggestion has been confirmed by some subsequent workers, c.f. Woodyer [114], it has been severely questioned by others, c.f. Kennedy [56] and in any case, as

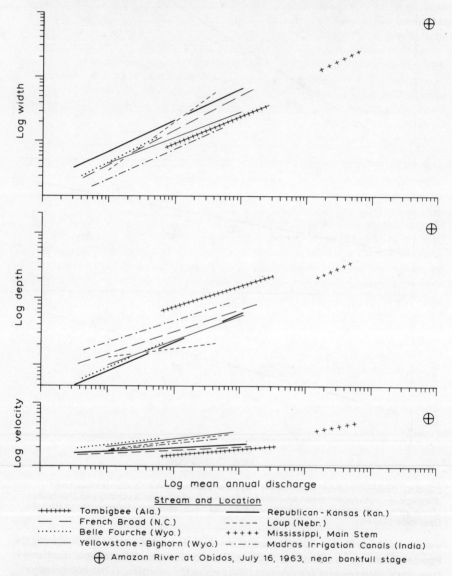

Fig. 9.9. *Changes of width, mean depth, and mean velocity in relation to mean annual discharge as discharge increases downstream in various river systems. (From* Fluvial Processes in Geomorphology *by Luna B. Leopold, M. Gordon Wolman, and John P. Miller, W. H. Freeman and Company. Copyright ©1964.)*

Nixon [76] observed, the assessment of bankfull discharge is not easy without intimate knowledge of the river concerned. Some of the most valuable British work was carried out by Harvey [41] who suggested that streams with a dominant quickflow component appear to be adjusted to the 1-2 year flood but that baseflow streams appear to be adjusted to rarer floods. Harvey concluded that channel capacity must ultimately be governed by a balance between the erosive forces associated with high discharges and the aggradational processes, together with vegetational growth, associated with lower discharges.

In Fig. 9.9 downstream variations of width, depth, and velocity in relation to mean annual discharge, as discharge increases downstream, are compared for various river systems. The inclusion of the single available measurement for the Amazon suggests that the general relations of the hydraulic geometry extend even to the world's largest river [62]. There is clearly a considerable similarity in the slope of the lines among the various river systems, even though the intercepts vary, apart from the Loup River which was included to show the variance that may exist [62]. The discharge of most rivers increases downstream and could theoretically be accommodated by a number of variations in hydraulic geometry. Thus, for a given frequency of discharge, the channel could increase in width while depth and velocity remained constant, or velocity could increase while width and depth remained constant. Each combination would involve accompanying changes in channel slope and thus lead to different longitudinal profiles. In fact, however, as Leopold and others [62] observed and as Fig. 9.9 confirms, the values of the exponents are rather conservative with width usually increasing most consistently as the square root of the discharge ($b \cong 0.5$) and mean velocity tending to increase slightly downstream in most rivers ($m \cong 0.05 - 0.1$).

Adjustments in channel cross-section cannot be considered in isolation since they are closely related to the *long-profile* of the river which Leopold and others [62] considered to be a function of the following variables:

Discharge	Q
Load (delivered to the channel)	G
Size of debris	D
Flow resistance	n
Velocity	v
Width	w
Depth	d
Slope	s

It has already been shown that discharge tends to increase downstream, as also does width of channel, and although less consistently, depth of channel. Velocity may either increase very slightly or remain more or less constant in a downstream direction [58]. (It is almost unbelievable that, twenty years or more after publication of evidence to the contrary, the quaint notion still persists in some quarters that streams which flow rapidly in their mountain courses slow down when they reach the gentler gradients of the lowlands.) The downstream increases in width, depth, and discharge result in a downstream increase in channel efficiency (i.e., n decreases downstream) and since this is not accompanied by a corresponding increase in velocity it tends to be compensated by a downstream reduction of slope.

This reduction of channel slope downstream inevitably yields the normal longitudinal profile which is concave to the sky.

However, flow resistance within the channel is also affected by both the total quantity and particle size of the debris load. Thus, all other factors being equal, an increase in total load must be accompanied by an increase in channel slope, and an increase in the particle size of the bed load will also result in an increase in channel slope, with the result that coarse debris appears in streams having steep slopes [62]. Finally, as Hack [39] showed, the change of channel slope downstream, i.e., the concavity of the long-profile, is closely related to the change of debris size downstream—the more rapid the downstream decrease in bed material size, the more concave is the longitudinal profile.

The complexity of such interrelationships as those discussed briefly above emphasizes the considerable difficulty of generalizing about the form of longitudinal profiles. Since changes in discharge can result in changes in the width and depth of the channel as well as in adjustments of slope, and since changes in load are not wholly reflected in changes of slope, it is clear that although long profiles are commonly envisaged in two dimensions the problem is, in fact, multi-dimensional and the conclusion is inevitable that no long-profile need ever become an entirely smooth, concave-upward curve. It also follows that much of the voluminous early literature on graded profiles, which considered only the two-dimensional form, is rendered largely meaningless. More relevant discussions of graded profiles have concentrated on the distribution and transformation of energy within river systems [59] [86] [116], an aspect which will be discussed in more detail in the ensuing section which is concerned with the plan-form of stream channels.

9.7 Channel form—plan

A similar complex interrelationship between hydrological and geomorphological variables is evident when the plan-form of stream channels is considered. The basic channel plan-forms or patterns that have been recognized are meandering, braided, and straight, although since natural streams are seldom straight over any but short distances only the meandering and braided patterns will be discussed here. In fact, it is found that even in a straight channel the line of maximum depth meanders from near one bank to the other [62] so that straight channels can be considered as an initial, unstable form. This is confirmed by those examples of artificially straightened channels which rapidly develop meander characteristics and by a consideration of energy relationships.

Meandering channels

There is a long history of discussion and attempted explanation of meandering which, like that concerned with graded profiles, is rendered largely valueless for the identical reason that the meander form has been seen too often as a two-dimensional rather than as a multi-dimensional problem. As Yang [117] observed, a satisfactory explanation of meandering must account for the fact that meandering channels occur in association with a wide range of variation of stream discharge and turbulence, debris load, geology, and location and size of channel,

328

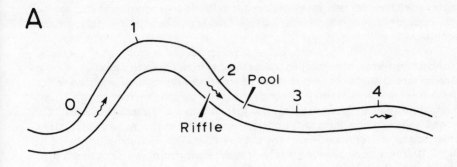

A

0

1

2   Pool

3   4

Riffle

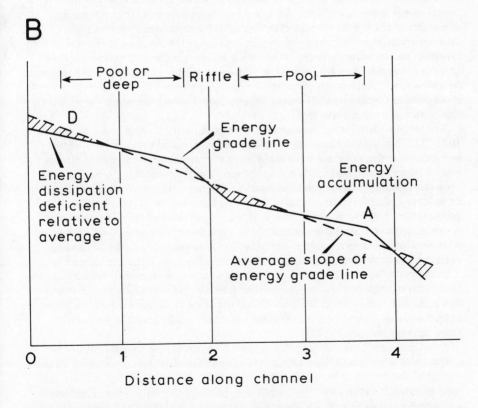

B

Pool or
deep | Riffle | Pool

D

Energy
grade line

Energy
dissipation
deficient
relative to
average

Energy
accumulation

A

Average slope of
energy grade line

O   1   2   3   4

Distance along channel

Fig. 9.10. *(A) Plan view of hypothetical reach of stream channel showing location of pools and riffles and (B) long profile of total energy (the energy grade line) for the same reach. (From* Fluvial Processes in Geomorphology *by Luna B. Leopold, M. Gordon Wolman, and John P. Miller. W. H. Freeman and Company. Copyright © 1964.)*

that meandering channels always follow a smooth sinuous course, and that they may occur on ice in the absence of a debris load as well as on alluvial material and bedrock.

The numerous theories and hypotheses which have been proposed to explain channel meandering have invoked causes ranging from the earth's rotation, through random local disturbances and patterns of erosion, sedimentation, and flow within the channel, to considerations of energy dissipation within the river system. Thus the first really plausible explanation was that advanced by Leopold and others [62] in which the sectional form of the channel was related to energy considerations in order to account for meandering. The basic relationships are summarized in Fig. 9.10. It had been observed that the regular downstream variations of depth, known as pools and riffles, are associated with changes of channel cross-section as the line of deepest water swings across the channel over the riffle reach and approaches closest to the banks in the intervening pools. Leopold and others [62] showed that in such situations the rate of energy dissipation, compared with the average energy grade line, is very unequal being greatest over the riffle and least in the deeper and smoother pools. Meandering represents a mechanism by means of which energy losses along a channel reach may be regularized. Thus the energy excess in the pools is compensated by being partly consumed in the creation of channel bends, thereby causing an increase in energy loss because of changes in flow direction and increased internal interference and friction between water particles. Only in a meandering reach, therefore, is the slope of the energy grade line likely to be uniform and thereby achieve what Leopold and others [62] considered to be the most probable distribution of energy namely, one in which there is uniform energy loss in each unit of channel length.

These ideas were further developed and refined in major contributions by Yang [97] [117] which simultaneously explain meandering and the concavity of the long profile and relate these to variables such as channel slope and width, stream discharge, and suspended solids concentration. Yang showed that the law of least time rate of energy expenditure (see section 9.5) not only forms the basis of a complete explanation of meandering but is the basic law which governs fluvial processes. In brief, Yang argued that a natural stream seldom follows a straight course in either the lateral or vertical plane. Lateral deviations result in meandering while vertical deviations result in a concave longitudinal profile with undulating deeps and shallows such as pools and riffles, and dunes and antidunes. Both the meander forms and the bed forms are important to the resistance to flow which, in turn, determines the rate of energy expenditure and the regulation of unit stream power, so that according to the law of least time rate of energy expenditure, 'the smooth sinuous meandering course is the only course that a natural unbraided stable channel can take' [117].

Yang concluded that the overall time rate of potential energy expenditure per unit mass of water should decrease in the downstream direction, causing an increase in channel width and a decrease in channel slope (i.e., a concave long profile). He further noted, however, that the overall time rate of potential energy expenditure per unit mass of water will also depend on factors such as discharge, sediment concentration, valley slope, and geological constraints. Thus an increase in sediment concentration will result in an increase in channel slope which in turn causes an increase in the time rate of potential energy expenditure per unit mass of water and a similar effect results from an increase in stream discharge. In effect, meandering is another form of geometrical adjustment, similar to those discussed in section 9.6, which in combination result in a downstream decrease in the rate of potential energy expenditure per unit mass of water. Since adjustments to changing energy conditions are almost certain to be effected by more than one variable, it is almost inevitable that there will be a strong relationship between aspects of meandering and other features of hydraulic geometry. Figure 9.11, for example, shows relationships between meander wavelength and channel width (A) and between wavelength and stream discharge (B). However, it will be clear from the brief discussion in this section that no *one* element in the hydraulic geometry of a stream system can be held accountable for meander wavelength. It then follows that simple

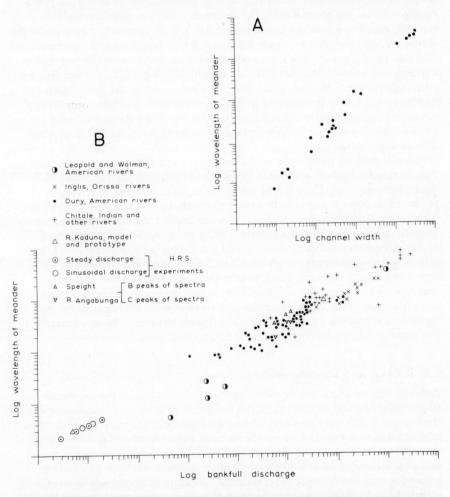

Fig. 9.11. *Meander wavelength as a function of (A) channel width and (B) bankfull discharge. (From original diagrams by Leopold and Wolman [61], with acknowledgment to the US Geological Survey, and by Ackers and Charlton [1].)*

correlations between meander wavelength and lithology, channel width, or discharge are simply too naive and that many of the edifices which have been built upon such foundations (e.g., Dury's palaeohydrological explanation of misfit meanders [26] [27]) are really castles in the air. It also follows that more sophisticated statistical analyses of the variables which influence meandering, which have become fashionable in recent years, are unlikely to advance our understanding of the complex, inter-related processes which are in operation.

Braided channels

Like meandering, braiding appears to be a mechanism by which a stream regularizes its energy relationships and achieves a stable channel pattern by separating into a number of channels which are divided by islands or bars. Braided channels are normally associated with non-cohesive materials such as sands and gravels where the shifting of large amounts of mobile bed material is responsible for bar formation at high discharges. Indeed, deposition and sediment transport are clearly essential to braiding. It is equally clear, however, as Leopold and others [62] pointed out, that with stable banks deposition would reduce the water section and increase the rate of flow and therefore the capacity for sediment transport. In other words, if bars are to become sufficiently stable to divert streamflow the banks must be sufficiently erodible so that they, rather than the incipient bars, are removed as the flow pattern changes.

Braided channels have a geometry which results in a greater resistance to flow than in straight channels and this is compensated for by steeper bed slopes [40]. Again, however, interrelations between variables are so complex that it seems unlikely that channel slope can be explained simply by reference to one variable. Thus, in addition, the type of material in which braided channels typically develop are often associated with steeper slopes. Furthermore, as Leopold and others [62] observed, because braided reaches are wide and shallow with unstable banks, the rate of sediment transport per unit width of channel may be relatively low, and deposition in such reaches is characteristic. Braided channels, then, may be associated with aggradation which, in turn, is associated with increased channel slope. However, much of the deposition which takes place in braided channels is localized and transitory and there is much evidence to support the contention that, like meandering, 'braiding is a valid equilibrium form' [62].

9.8 Erosion and sedimentation

Although erosion and sedimentation have been referred to in preceding discussions of drainage basin and channel form, it is important at this stage to emphasize that the manner in which the products of drainage-basin decomposition reach the stream channels is yet another example of the complex hydrological and geomorphological interrelations involved in the total unity of the drainage basin. Evidence of the erosional activity of water is found in stream channels in the form of dissolved, suspended, and bed material and is reflected throughout the drainage basin in slope angles, the varying depth and composition of the weathered mantle, and in the gradual or rapid downslope movement of slope material. It has, unfortunately, become customary to separate considerations of erosion-sedimentation processes in the stream channels from those within the remainder of the basin. It is now recognized, however, that such a separation is quite unrealistic and that just as there is a continuum of water movement from all parts of the basin to the stream channels so too is there a continuous movement of weathered material from all parts of the drainage basin towards and through the stream channels. In addition, the processes of detachment, transportation, and deposition of material may take place at any point within the drainage basin system. It is, therefore, encouraging to see that some of the more recent work on drainage-basin geomophology, c.f. [94], has recognized the essential unity between drainage basin slopes and stream channels.

Erosion comprises the processes of weathering and transportation of materials by water, gravity, ice, and wind, although it is only with the first two that we are concerned here. A distinction is often made between *natural erosion,* operating under the influence of normal geological factors, and man-induced or *accelerated erosion* such as that resulting from widespread burning of vegetation, deforestation, or overgrazing. It is likely, however, that such a distinction conceals more than it reveals. Thus, there is no longer any part of the earth's surface where erosion is not affected to a greater or lesser degree by human activity and in any case 'natural erosion' operates over an enormous range of intensities from that resulting from the gentlest rain to that resulting from a catastrophic storm having a recurrence interval of hundreds or even thousands of years. There is, in fact, considerable discussion on the relative importance of rare events in geomorphology, accompanied (almost by definition) by a sparsity of data. Leopold and others [62] suggested that events of moderate frequency, rather than catastrophic events, account for a large percentage of the total sediment removed from a drainage basin and although this has been supported by some subsequent studies of actual events, c.f. [66] there is by no means universal agreement. For all these reasons no distinction between natural and accelerated erosion will be attempted here.

The main types of water or water-induced erosion occur when water moves over or through the slope material, when the slope material itself moves and finally, during channel flow. Erosion at the soil surface will be most effective in the absence of a vegetation cover and may result from raindrop impact as well as from overland flow. Ellison [30] demonstrated the considerable kinetic energy exerted by rainfall, which may be sufficient to kick soil particles 60 cm vertically into the air and up to 130 cm horizontally, and showed that during a single storm on a bare soil up to 25 kg/m^2 of soil may be detached and splashed into the air. Clearly, with a flat surface there will be little, if any, net removal of material from this cause, although net erosion will increase markedly with slope. If overland flow builds up to a sufficient depth and velocity it too may result in the detachment and transport of soil particles. It was observed in chapter 8 that this process, which formed the core of the Horton erosion model, normally occurs only over limited areas of a drainage basin and is considerably less important than was once thought. Work by Young and Mutchler [118], simulating soil movement on irregular slopes, showed that most soil loss results from raindrop-splash-induced movement to a rill system and then from channel transport through the rill system.

Subsurface erosion occurs mainly by solution but also by the throughwashing of fine colloidal material and is extremely important. Indeed, solution alone accounts for more than 50 per cent of the material carried by the world's rivers to the oceans

[43]. The role of groundwater in this respect is well known, particularly in limestone areas, and a valuable review of its geomorphological effects was given by Williams [111]. In non-limestone areas groundwater movement has been held responsible for the formation of seepage steps several metres in height [105] and for the formation of surface rills and similar depressions as a result of solution-induced subsurface collapse [13] [113]. Much less is known about the geomorphological effects of interflow although, clearly, solution does occur. This was demonstrated by Johnson and others [55] who related the movement of water through the soil and subsoil layers in a mountain drainage basin in New Hampshire with variations of stream water chemistry. Since fine material is leached vertically through the soil profile by percolating water it seems almost certain that there will be a similar lateral movement of fines as a result of interflow. Apart from Arnett's valuable work [5] in the North Yorkshire Moors, however, there is little direct supporting field evidence.

An important element in the denudation of a drainage basin is the gravity-induced mass movement of slope material which may take many forms ranging from the rapid movements, such as rock falls, landslides, and mudflows, to the almost imperceptible movements associated with soil creep. In all cases water, particularly in the form of soil moisture, plays an important role in determining the stability of the slope material through its effects on shear strength. Briefly, the negative pressure or suction of soil moisture in unsaturated conditions helps to draw the solid particles of the slope material closer together and thereby increase its shear strength. Thus moist material normally has a higher shear strength than completely dry material. When, however, the slope material becomes saturated a positive soil-water pressure develops which tends to push the individual solid particles apart and which thereby reduces shear strength [14]. Many natural landslides appear to be associated with the development of positive soil-water pressure [106]. It follows, then, that slope material which neither completely dries out nor becomes saturated will always possess negative soil-water pressures and may thus stand at a steeper angle than that dictated simply by the shear strength of the material itself, while slope material which frequently becomes saturated, as in most humid areas, will stand at correspondingly lower angles [14]. Soil moisture also plays an important role in the slower, but no less important, process of soil creep both through its effects on shear strength and also on the swelling and shrinking of soils during phases of wetting and drying.

Finally, channel erosion occurs as water impinges upon the bed and particularly the banks of a stream and occasionally the floodplain itself may be scoured by overbank flows. Most of the material carried as bed load and in suspension through a channel network is derived from bank erosion, the bank material having been initially eroded and transported by the processes which have already been discussed. Most of the material carried in solution is derived from the broader area of the drainage basin as a whole. The total load of a stream thus provides an indication of the amount of erosion taking place within its drainage basin but the picture is severely complicated by the patterns of temporary and semi-permanent deposition of eroded material both on the surface of the drainage basin, e.g., in depressions, on the floodplain, at breaks of slope and at fence lines, and also in the channels themselves. The geomorphological interpretation of stream load is thus extremely difficult and attempts to extrapolate long-term geomorphological trends from short runs of stream load data are quite unrealistic and almost certainly doomed to failure. However, understanding of the hydraulic relationships between streamflow characteristics and the transportation of bed load and suspended solids in particular has improved considerably in recent years, a major contribution having been made by H. A. Einstein, c.f. [29].

The hydrological unity of the drainage basin

9.9 Introduction

The preceding sections (9.2—9.8) have clearly illustrated the geomorphological unity of the drainage basin and have indicated that, in areas of fluvial erosion, that unity is achieved largely through the medium of moving water. It is now proposed to discuss briefly the hydrological unity of the drainage basin, paying particular attention to the complex interrelationships between the various components of the hydrological cycle and the implications of these interrelationships for the management and control of drainage basin water resources. The basic theme of this discussion was concisely stated by Wicht [109]:

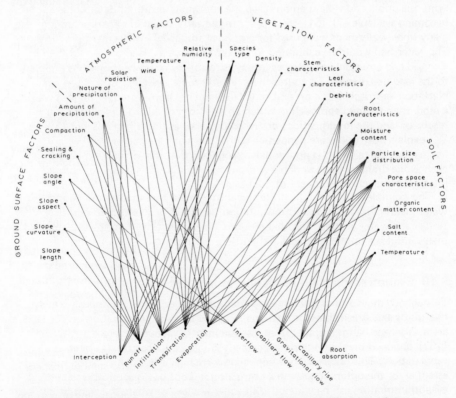

Fig. 9.12. *Simplified diagram of relations between some of the main biophysical factors influencing the circulation of water in a drainage basin. (From an original diagram by R. L. Wright,* Geoforum, *10, 1972.)*

The catchment of a natural stream . . . is an ecosystem in which the living organisms and their non-living environment are reciprocally related in an energy cycle. . . If the dynamics of water in the catchments are to be adequately understood, so that they can be managed to improve water yields, the ecosystem must be analysed into its component parts and the interactions between these investigated to gain a holistic conception of the catchment.

Some aspects of the complex interrelationships involved are illustrated in Fig. 9.12 which, for the sake of simplicity, is restricted to include only six of the more important factors from each of four main groups, i.e., ground surface, atmosphere, vegetation, and soil. This selection represents less than half the biophysical factors influencing water circulation which were identified by Wright [115] and completely excludes, for example, consideration of groundwater-surface water movements. However, the interconnections are even more complex than those indicated in the diagram because of the interdependent nature of links between many processes and between many factors.

As in the earlier part of this chapter, discussion will proceed by reference to specific examples. The majority of these have been drawn from the field of small watershed or drainage basin experiments which have played an increasingly important role in hydrological investigations in recent years. A substantial review of many such experiments was provided by Ward [108] and a number of recent symposia publications have emphasized the enormous wealth of data that are becoming available [2] [51] [52] [75]. In the early stages of experimentation many small watersheds were used for empirical studies of management practices. The watershed would be calibrated for a number of years, land use would then be changed and any resulting changes in, say, streamflow could be (hopefully) related to the land-use changes. Of course there are still some examples of this type of black-box approach at the present time but there has been a gradual trend towards a more sophisticated approach in which an attempt is made to *understand* the results of land-use modifications on hydrology and, as stated by Holscher [46], to '. . . explain *how* things are happening within the watershed and *why* they are happening'. Certainly, it is now widely recognized that it is virtually impossible to change one drainage-basin variable without setting in motion a train of complicated interlocking reactions throughout the basin.

The examples chosen to illustrate this theme relate first to attempts to increase water yields by reducing evapotranspiration, secondly to attempts to optimize timber and water yields in forested mountain areas and finally, to investigations of man's influence on the chemical and nutrient balances of drainage basins.

9.10 Evapotranspiration reduction to increase water yields

Throughout the world as a whole evapotranspiration accounts for approximately two-thirds and streamflow for the remaining one-third of precipitation on the land areas. Drainage basins are, therefore, predominantly vapour loss areas and Wicht [109] observed that '. . . their management should be judged on its effects on vaporization'. This observation is not valid where high evapotranspiration values are essential for the optimum growth of agricultural crops but is applicable where evapotranspiration can be equated with water wastage or where the transpiring vegetation has little if any commercial value. Where it is demonstrably desirable to reduce evapotranspiration losses the two main approaches have normally been

either to remove or replace the vegetation cover or to apply transpiration
suppressants to the existing vegetation cover. A third possibility, involving
simultaneous weather and vegetation modification, was referred to by Satterlund
[85] who argued on theoretical grounds that there would be periods when the
combined practice would yield an increase in streamflow when neither weather
modification nor vegetation modification, applied separately, would be effective in
so doing. Only the first two approaches, however, will be considered in detail in this
section.

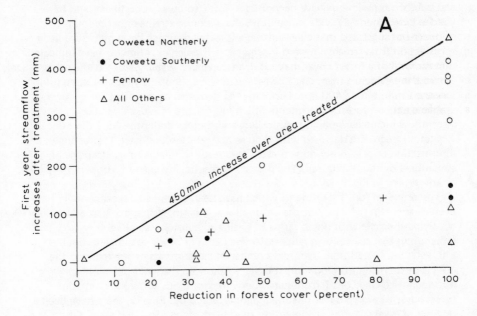

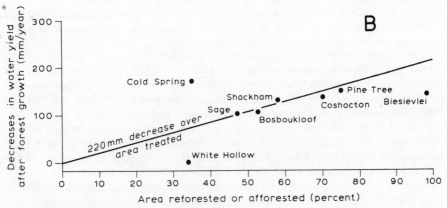

Fig. 9.13. *Effects of (A) reduction of forest cover and (B) increase of forest cover on
streamflow in selected basins. (From original diagrams by A. R. Hibbert in* Forest Hydrology, ©
1967, Pergamon Press Ltd)

Removal/Replacement of vegetation cover

There have been many 'black box' attempts to alter the vegetation cover of drainage basins in order to improve water yields. Investigations of the effects have frequently involved the use of paired catchments where precipitation and streamflow from both are observed for an initial, undisturbed calibration period. The vegetation cover is then removed or replaced on one of the catchments and the resulting streamflow is compared with the streamflow which statistical analysis indicates would have occurred in the absence of any change. These 'calibrate, cut, and publish' experiments have been used in forested areas in particular and have yielded a bewilderingly wide range of results. At one extreme experimental evidence has suggested that clear-cutting a forested drainage basin may have little or no effect on total streamflow and indeed some Russian experiments have indicated that removal of a forest cover may result in decreased water yields. At the other extreme the evidence from one Coweeta watershed showed a first-year streamflow increase of more than 400 mm. Hibbert [44] provided an excellent review of the available data on forest experiments which indicated that most first-year streamflow increases were 300 mm or less and that generally speaking the effect of vegetation treatment declined with time as revegetation occurred. Interestingly, the afforestation of previously non-forested area resulted in an average decrease of streamflow of about 220 mm. The pattern of results discussed by Hibbert [44] is clearly shown in Fig. 9.13.

The principal forest treatments which have been examined experimentally have been clear cutting of a part or the entire drainage basin, strip cutting in which clear cut strips alternate with uncut strips of similar dimension, selective thinning based either upon age, species or quality, selective patch or block cutting and finally, afforestation. In addition, a limited number of experiments have examined the effects of changing from one forest cover to another, the most commonly documented change being from hardwoods to coniferous softwoods, or from a forest cover to a grass cover or similar agricultural crop. Finally, there have been a number of investigations of vegetation treatments in non-forested areas. Of these, some of the most important examples have been concerned with attempts to reduce erosion and flood runoff [25] [65] or to improve water yields in semi-arid areas by converting water-wasting chaparral scrub to grass [45].

Apart from the very wide range of results, to which reference has already been made, two other major points emerge from extensive study of many drainage basin experiments. First, problems of experimental control and of data accuracy are crucial because the experimenter has virtually no control over a natural drainage basin, however carefully his experiment is designed, and because all the basic hydrological measurements of precipitation, streamflow, evapotranspiration, and storage of soil moisture and groundwater are subject to considerable error and uncertainty [108]. Indeed Lee [57] considered that, in relation to the normal type of paired catchment experiment, combined uncertainties in comparisons of theoretical estimates and observed values of streamflow or water losses are usually '. . . greater than the difference between the values compared'. One implication of this is that techniques of data analysis and interpretation must be rigorously conceived and applied although, even then, Lee's pessimism was foreshadowed by that of Sharp and others [90] who concluded, after extensive study, that the quantitative evaluation of the effects of land treatment upon streamflow could not be satisfactorily demonstrated by statistical analyses given the type and quality of data then generally available. A most valuable review of methods of detecting hydrological changes was presented by Snyder [95].

Secondly, consideration of drainage basin experiments reveals a widespread lack of understanding of the hydrological mechanisms and processes involved in the basin water circulation. Thus, for example, it is possible that the apparently discrepant Russian results [72] [80] indicating that streamflow from forested areas is greater than that from non-forested areas, may be explicable in terms of the processes of snow accumulation, storage and melt in specific geographical conditions. Again, in the 'normal' situation where removal of forest cover results in an increase of streamflow, the effect may be explained largely in terms of reduced transpiration or reduced interception losses. Inevitably both factors make a contribution but it seems clear from the evidence that in some cases the transpiration effect dominates and in others the change in interception characteristics is more important [101]. The implications are significant since if the interception effect dominates the greatest improvements in water yield would be obtained by treating the vegetation cover over the entire area of the drainage basin while, if the transpiration effect dominates, proportionately greater improvements in water yield might be obtained by removing that vegetation with the readiest access to water supplies. Indeed, a number of workers have shown that the removal of streamline, riparian vegetation may result in substantial increases in streamflow [67] [82] [84].

Further reference to Fig. 9.12 will indicate that, quite apart from its obvious effects on transpiration and interception, removal or replacement of a vegetation cover may have other repercussions. Many of these are not fully understood and some of them may not even be recognized in normal circumstances. For example, the replacement of vegetation having a marked litter layer may, as a result of the reduction in water detention previously associated with the litter, cause an increase in overland flow [79] or an increase in the amount of infiltration. Again, an increase in total infiltration resulting from reduced surface detention, reduced transpiration, reduced interception loss, or a combination of these factors may be associated with either an increased vertical drainage of water through the soil profile to underlying groundwater and a corresponding increase in groundwater flow, or in increased lateral movement in the form of interflow. Since, as was demonstrated in section 8.4, the hydrograph recessions of some streams may be most appropriately explained in terms of the drainage of water from the soil profile rather than by the exhaustion of groundwater storage, it is frequently difficult to determine the precise reasons for improved low flows following the removal of forest vegetation unless detailed soil moisture and groundwater data are available. In any event, increased total infiltration will almost certainly be associated with wetter soils and with corresponding reductions in available soil moisture storage. This, in turn, encourages the more rapid expansion of variable source areas during precipitation thereby resulting in increased volumes of quickflow and in higher instantaneous flood peaks [42] [47].

As might be expected, vegetation removal or replacement is likely to have a marked effect on water quality. Experiments have demonstrated that increased sediment yields frequently result from vegetation manipulation and from associated disturbance to the soil profile [48] [78] [81]. Less obviously, perhaps, dramatic vegetation changes such as clearcutting have been shown to cause changes in stream-water temperatures sufficient to modify or damage fish and other aquatic life [11] [12] [63]. Daily and monthly mean maximum water temperatures are increased when vegetation removal exposes the stream surface to direct insolation which is the principal source of energy for the warming process.

Transpiration suppressants

As an alternative to vegetation replacement or removal high evapotranspiration losses may be reduced by the application of transpiration suppressants or antitranspirants to the existing vegetation cover. This approach is still at a comparatively early stage of development but could prove to be a particularly useful means of improving water yields in areas of high aesthetic, recreational, or wildlife value where forest clearance or other vegetation disturbance would be objectionable [10].

From the extensive discussion of evapotranspiration in chapter 5 it will be clear that, to a large extent, the successful application of this technique will depend upon a correct understanding of the transpiration process and particularly of the plant physiological factors which affect it. A useful review was provided by Davenport and others [21]. In terms of foliar sprays the three main possibilities are firstly, the use of reflecting materials that decrease the heat load on the leaf, secondly the application of film-forming materials that slow down the escape of water vapour from leaves by decreasing their permeability, and thirdly the use of substances which increase stomatal resistance either by closing or by narrowing the stomatal apertures. At the present time the most promising approach appears to be through the modification of stomatal aperture, using either phenylmercuric acetate (PMA) or decenylsuccinic acid. Both have been demonstrated as effective for a range of plants under controlled conditions. For large natural areas, aerial spraying represents the only economically viable application technique but has the disadvantage that transpiration suppressant is delivered mainly to the upper leaf surface whereas the stomata are normally located in the underside of the leaf. Where stomata are more amenably distributed, as in red pine, impressive reductions in soil water depletion have been recorded after the application of antitranspirants [107]. As Davenport and others [21] pointed out, however, whether such reductions in evapotranspiration losses will be hydrologically significant in increasing streamflow will depend on the nature of the drainage basin and particularly on soil depth and water holding capacity, plant characteristics such as stomatal distribution, foliar growth, and rooting depth, and climatic characteristics such as the amount and distribution of precipitation. All these, together with errors in the measurement of soil moisture and streamflow, make it difficult to evaluate the possible benefits of applying antitranspirants to natural drainage basins.

9.11 Optimizing water and timber yields in forested mountain catchments

The problem of optimizing water and timber yields from forested, high-altitude catchments, where most winter precipitation falls as snow and where little snowmelt occurs during the winter period, provides a further example of the interdependence of hydrological variables within a drainage basin. In this case, however, the situation is even more complex than those discussed in section 9.10 because of the need to consider the additional processes of snow accumulation, storage, and melt. In many areas, and certainly in North America, these upland drainage basins contain some of the most important reserves of exploitable timber and are also the major water-supply catchments. Optimization procedures are therefore essential if the development of one resource is not to adversely affect the development of the other. Ideally, management practices would aim to achieve the highest timber yields, using economically viable harvesting methods, consistent with the highest water yields distributed as uniformly as possible in time. In view of the marked seasonality of snow accumulation and melt this involves the manipulation of vegetation cover and snow accumulation in such a way that melting is spread over an extended period of time. It will become apparent that such optimization is not possible without a detailed understanding of the energy balance and water balance relationships of snowmelt and of the way in which the energy and water balances are affected by changing geographical conditions, including factors of climate, vegetation, soils, topography, and hydrology.

A forest cover interacts with snow precipitation through the process of interception and through its effects on snow accumulation, evaporation and melting. Earlier discussions (sections 3.7 and 4.6) indicated that evaporation from snow, including snow intercepted by the vegetation cover, is small and hydrologically insignificant. Patterns of snow accumulation in forested areas will be affected by the falling, blowing, and sliding of intercepted snow and by the aerodynamic effects of the vegetation cover. Clearly, over an extended area total snow accumulation will not differ materially between forested and non-forested areas but locally, variations in the depth of accumulated snow may have important implications for snowmelt and for subsequent runoff. In terms of optimal forest management a particularly significant possibility is that of encouraging the maximum accumulation of snow in areas of minimum snowmelt. As Costin and others [20] observed, the retarding effect of forest cover on the disappearance of snow has long been recognized and was discussed by J. E. Church as early as 1912 [19]. Subsequently, the main debate has been over the relative importance of locally increased snow accumulation or retarded snowmelt in explaining the persistence of snow and the associated prolongation of the spring runoff period.

An excellent summary of forest effects upon *snow accumulation* was provided by Jeffrey [54]. It is known that less snow accumulates beneath dense forest canopies than beneath more open stands, that less accumulates beneath forest canopies than in small clearings in the stand and that the maximum difference in snow accumulation between forest and small clearing occurs when the opening has a diameter or width equivalent to the height of the adjacent trees. The differences in accumulation are presumably attributable to aerodynamic effects, i.e., reduced wind velocities across the clearings, and also to the effects of differential interception and evaporation, although the relative importance of these three factors is still not totally clear. It is not possible yet, as both Miller [70] and Jeffrey [54] emphasized, to generalize either about the magnitude of the difference in snow accumulation between forest and clearing or, by implication, about the quantitative effects on snow accumulation of specific forest treatments.

Forest effects upon *snowmelt* occur through modifications to both energy and water balances and may differ considerably between deciduous hardwoods and conifers. A large percentage of incoming solar radiation is absorbed by the forest canopy with a corresponding reduction in the amount of short-wave radiation transmitted to the snowpack on the forest floor. Some effects of vegetation type and canopy density are illustrated in Fig. 9.14. The decreasing transmission of radiation with increasing canopy cover is clearly shown in Fig. 9.14A while Fig. 9.14B shows the marked fall in transmission accompanying leaf development in a birch stand and the much smaller, but statistically significant, decrease with the advent of new growth on spruce. The reduction by absorption of short-wave radiation is partly offset by the long-wave radiation exchange between the forest canopy and the snow surface. A large radiation flux towards the snowpack will occur when foliar temperatures are markedly higher than those of the snow surface. The degree of compensation between reduction in short-wave flux and increase in long-wave flux to the snow surface, resulting from the presence of a forest canopy, depends largely on the canopy density as is shown in Fig. 9.15. Net short-wave gain and net long-wave loss decrease progressively with increasing canopy cover while net long-wave gain increases. As a result the net radiation gain to the snow surface decreases up to a canopy coverage of about 20 per cent and increases thereafter, although with total canopy coverage the net radiation gain at the snow surface is little more than 50 per cent of that in the open (canopy coverage = 0). Small clearings within a forest tend to benefit, in net radiation terms, from the surrounding forest, particularly in terms of the net gain of longwave radiation.

Apart from radiation other components of the energy balance affect patterns of snowmelt in forested areas. Thus sensible heat is transferred from the vegetation to

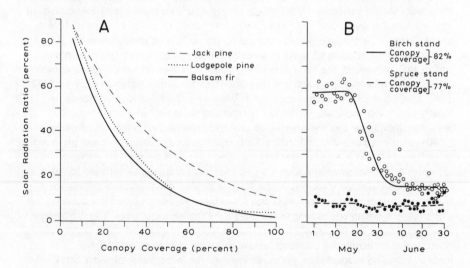

Fig. 9.14. *Reduced transmission of solar radiation through forest canopies. (A) Relations between canopy coverage and relative transmissions of solar radiation in stands of balsam fir, jack pine, and lodgepole pine. (B) Relative transmission of solar radiation through a deciduous and coniferous canopy, May-June 1966. (From original diagrams by W. W. Jeffrey in Gray, D. M. (Ed.),* Handbook on the Principles of Hydrology, *Canadian National IHD Committee, Ottawa 1970 and by C. E. Schomaker,* Forest Science, **14:** *31-38 1968.)*

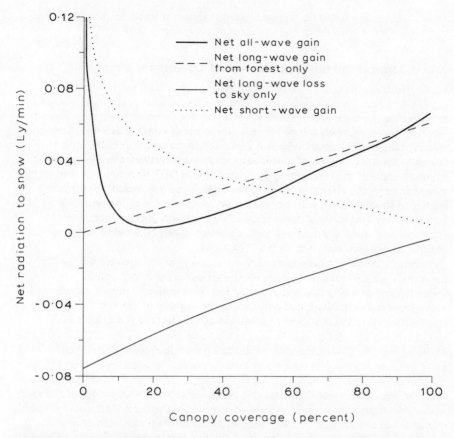

Fig. 9.15. *Calculated net gains and losses of radiation to the snowpack in relation to canopy coverage for spruce-fir forest in Oregon. (From an original diagram by W. E. Reifsnyder and H. W. Lull,* USDA Tech. Bull., *1344, 1965. With acknowledgment to the US Department of Agriculture.)*

the surrounding air and thence to the snowpack while with a downward vapour pressure gradient, or in conditions favouring advection, the transfer of moisture to the snowpack results in the addition of latent heat.

Valuable work on the total heat budget of snowpacks in forested mountain areas was presented by Miller [69] who suggested that forest snowmelt is generally about 75 per cent of that in the open, while Anderson [3] suggested the figure of 64 per cent for dense forest. Because small clearings in the forest receive convected heat and longwave radiation from the surrounding trees, as well as an increased solar radiation component because of the absence of canopy cover, melt rates are greater than in the closed forest [54]. The persistence of snow in small forest clearings must therefore be a function not of lower melt rates but of greater snow accumulation [54].

In summary, therefore, considerations of water- and energy-balances in high-altitude forested areas confirm the long-held conclusion, noted by Costin and others [20], that a forest honeycombed with numerous small openings or cleared strips is the most efficient in accumulating and conserving snow.

9.12 Chemical and nutrient balances of agricultural areas

Man's effects on the quality of both surface and subsurface waters result from a variety of causes. Thus air pollution increases the acidity of precipitation water, changes in agricultural and urban land use affect the volume and routing of water reaching stream channels and direct pollution occurs in both urban and agricultural areas. One aspect of human influence which is now causing particular concern relates to the increased use of pesticides and chemical fertilizers in agriculture. Although some of the long-lasting pesticides such as DDT have been banned in many countries the effects of their former widespread use, including pesticide loading of stream waters, will continue to be felt in the future. Agricultural fertilizers may lead to the deterioration of water quality through nutrient enrichment of streams. As Jeffrey [54] observed, about 30.5 billion kilogrammes of fertilizer are used each year in the USA alone.

Recent evidence of high nitrate concentrations in drinking-water supplies in some chalk areas of eastern England illustrated our present incomplete understanding of subsurface water circulation, even in comparatively simple hydrogeological conditions, and emphasized the urgency of the many ongoing investigations into the transport of solutes in both saturated and unsaturated conditions. Initial reaction was to lay the blame on the greatly increased application of nitrogenous fertilizers, the assumption being that downward leaching and percolation of nitrates in solution was responsible for the abnormally high concentrations found in wells tapping the chalk groundwater. Follow-up studies, however, have been inconclusive on this point and recent investigations of the rates of downward percolation in high-yield aquifers in southern England, to which some reference has already been made in section 7.8, and which indicated downward movement of less than 1 m per year, have thrown even more serious doubts on this explanation.

It is clear that, before the effects of man's interventions can be properly assessed, a number of areas need further investigation. For example, how much of the total dissolved matter in stream water is contributed directly by precipitation? All precipitation contains some dissolved matter although the concentrations and the actual elements in solution vary with location, e.g., sodium and chloride contents decrease inland, and with season as well as with the nature and intensity of precipitation. Maximum concentrations of the major constituents in precipitation may approximate values found in natural streams [62] although individual investigations have yielded widely varying results. Thus in coastal forest areas of Oregon it was found that the role of precipitation in nitrogen cycling was minor in comparison with the much greater amounts of nitrogen cycled in litterfall [102], while in a New Hampshire forest it was found that, although effluent nitrate levels were affected by the forest biota, the input of nitrate and ammonium in precipitation exceeded the discharge of these constituents [32]. Other aspects of the natural cycling of nutrients in a wide variety of climatic, vegetation, and soil conditions, together with details of the transfer processes involved in the movement of solutes through porous media, also need further investigation. An excellent

344

theoretical discussion of the latter problem was provided by Bresler [9] and comparisons with field data indicated that solute displacement can now be predicted quantitatively.

Even with improved understanding of the natural movement of solutes, however, the problem of interpreting man's influence remains a complex one. For example, measurements of nutrients and pesticides in 43 tile-drain systems in the San Joaquin Valley of California showed that nitrogen concentrations ranged from a maximum of 400 to a minimum of 2 mg/l while seasonal variations were as high as 600 per cent [36]. These variations were attributed to complex interrelationships between the physiographic position of the tile systems, the soil series, and agricultural management practices.

A valuable piece of work in this field was reported by Taylor and others [103] and involved measurements of nitrogen, phosphate, and potassium concentrations in streams draining woodland and farmland basins in Ohio. The analyses emphasized that total nutrient losses cannot be calculated meaningfully unless both hydrological and chemical data are available. It was found that nutrient losses from farmland were significantly higher than those from woodland and were very erratic in nature, reflecting variations in streamflow and concentrations. For example, 76 per cent of the nitrate-nitrogen carried off the farmland in one year was lost in a single month when high nitrate concentrations fortuitously coincided with a period of high flow. In contrast, phosphate concentrations at that particular time were not unusually high. Comparison of the nutrients added to and lost from the farmland (see Fig. 9.16) revealed no relationship between the two but indicated large differences between the three nutrients, losses of nitrogen and potassium being

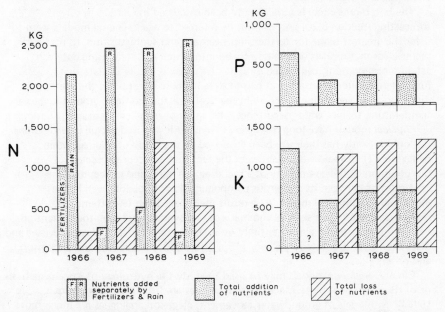

Fig. 9.16. *Annual additions and losses of nutrients in runoff water from 123 ha of farmland at Coshocton, Ohio. (Based on data in Taylor and others [103].)*

always much larger than losses of phosphate. Referring particularly to the nitrogen budget, it will be observed that this was dominated by the input rainfall which greatly exceeded the fertilizer applied. However, since no data were available for the amount of nitrogen fixed by legumes, the amount removed in harvested crops or that returned to the air as gaseous byproducts of denitrification, the complete nitrogen budget could not be calculated. Thus the effect of fertilizer nitrogen on the nitrogen content of the streams could not be calculated, although it was clear that the loss in stream water was but a small fraction of the total turnover.

9.13 Models

It is hoped that the foregoing discussions have made clear that the processes of analysis and synthesis, which are essential to a proper understanding of the hydrological unity of a drainage basin and to a study of man's hydrological impact, have been greatly advanced by the development of small experimental watersheds. Equally clearly, although much has been achieved there are still many basic hydrological problems which remain unsolved. Since the hydrological investigation of actual drainage basins is both time-consuming and protracted (commonly requiring data runs of up to 50 years) an alternative has been sought and in this respect the most promising approach involves the use of drainage basin models. These can be used to increase our understanding of fundamental hydrological processes, to reconstruct past occurrences and predict future conditions and to assess man's influence on the hydrological cycle. Models have been classified in many ways but for present purposes it will be convenient to differentiate physical, analogue, and digital models.

The best *digital* models have already been discussed in relation to runoff forecasting (section 8.20) and will not be referred to again. Digital models seem to offer the greatest scope for further improvement and sophistication. In particular, it seems clear that models will be developed into which remote sensing data, e.g., satellite observations, can be fed directly. This would finally obviate the lumped-parameter approach and permit detailed consideration of the spatial distribution of vegetation cover, soil type, soil moisture content, ground surface temperature, surface water occurrence, and so on over the entire drainage basin.

Physical models have long been used in hydraulic and hydrologic investigations but only recently has their use been extended to the study of drainage basin behaviour. Their main weakness lies in the fact that an accurate, scaled reproduction of the complexity of a real drainage basin and of the inter-relationships between its numerous components is not possible even in the most complex physical model. As a result there have been few attempts to reproduce more than a skeletal summary of real systems and even then attention has been concentrated almost entirely on surface characteristics. Useful reviews and discussion were presented by Chery [15], Chow [18], Black [8], and Dickinson and others [23].

Finally, *analogue* models may be used to study the hydrology of a drainage basin or of its component parts. Mechanical analogues are little used apart from Hele-Shaw models of groundwater movement; electrical analogues are much more common although there are very few examples in which they have been used to simulate a complete drainage basin system. Riley and others [83] discussed a

346

preliminary analogue model of the surface hydrological system within a sub-basin of the Walnut Gulch experimental watershed in Arizona. O'Donnell [77] referred to the Kansas River model as an operational example of elaborate analogue models of entire drainage basin systems.

Epilogue

The complexity of the interrelationships and interdependence between the various components of drainage basin hydrology and geomorphology have been referred to frequently in this chapter in order to stress the essential unity of the drainage basin and the dangers of an elemental approach to the study of its hydrology. Ultimately, however, the conclusion is inescapable that, for practical purposes, this very complexity poses a virtually insurmountable problem and it is the fundamental truth of this conclusion which has provided the ultimate justification for the 'geographical' approach adopted in this book. The contrast between the approach to hydrology of the civil engineer and the geographer (or indeed of some other environmental or earth scientists) provides a simple illustration.

The civil engineer normally requires a working, numerical solution to a specific hydrological problem and must frequently base his conclusions upon data which, from the point of view of both quality and duration, the scientist would regard as inadequate. Inevitably then the engineer resorts to empiricism and coefficients, to simplifications and generalizations of systems and processes. Eagleson [28] referred to the 'idealized' systems of the civil engineer and contrasted these with the 'natural' systems of the environmental scientist. Thus in the idealized drainage basin the input precipitation produces streamflow by passing through a system whose physical structure is assumed to be unrelated to the nature of the input, whereas in the natural system, as has been emphasized, both the initial impact of precipitation on the surface of the drainage basin and the subsequent movement of water through the basin to its outlet change the structure of the system itself and result in an output from the basin not only of water but also of water-borne material in the form of dissolved, suspended, and bed load. In the same way the individual hydrological processes operating within the drainage basin, e.g., precipitation, interception, evapotranspiration, soil moisture, and groundwater movement and storage, and the runoff process itself, must be idealized by the engineer if numerical solutions are to be readily obtained.

The geographer, on the other hand, is primarily interested in how the landscape works and in man's interactions with it or, as in the more explicit definition of Bird [7], with '. . . the changing spatial patterns and relationships of terrestrial phenomena viewed as the world of man. . .' The geographer thus recognizes that water is but one of the terrestrial phenomena in the total, complex interacting ecosystem in which he is really interested. This implies that much of the geographer's hydrological endeavour is directed, not towards the solution of a specific hydrological problem but towards a more complete understanding of the total landscape. It is therefore much more important that the geographer should have a detailed understanding of hydrological processes and systems than that he should be able, through simplification, to generate approximate numbers.

Recent trends in physical geography pose great dangers for if the geographer abandons the contemplation of natural systems and processes, for which he is ideally qualified, and attempts to ape the role of the engineer, for which he is desperately ill-qualified, he will rapidly find himself superfluous in fields such as hydrology. It is hoped that this book, which has throughout stressed the understanding of processes and basic principles rather than the ability to obtain quick answers, has illustrated the potentially valuable contribution of the 'geographical' approach in hydrology.

References

1. ACKERS, P., and F. G. CHARLTON, Meander geometry arising from varying flows, *J. Hydrol.*, **11**: 230-252, 1970.
2. AMERICAN WATER RESOURCES ASSOCIATION, *Watersheds in Transition,* National Sympos. on Watershed Hydrology, Colorado State Univ., Ft. Collins, 1972.
3. ANDERSON, H. W., Forest cover effects on snowpack accumulation and melt, Central Sierra snow laboratory, *Trans. AGU,* **37**: 307-312, 1956.
4. ANDERSON, M. G., Measure of three-dimensional drainage basin form, *WRR,* **9**: 378-383, 1973.
5. ARNETT, R. R., Private communication, 1973.
6. ARONOVICI, V. S., The area-elevation ratio curve as a parameter in watershed analysis, *J. Soil and Water Cons.,* **21**: 226-228, 1966.
7. BIRD, J. H., Desiderata for a definition; or is geography what geographers do? *Area,* **5**: 201-203, 1973.
8. BLACK, P. E., Runoff from watershed models, *WRR,* **6**: 465-477, 1970.
9. BRESLER, E., Simultaneous transport of solutes and water under transient unsaturated flow conditions, *WRR,* **9**: 975-986, 1973.
10. BROOKS, K. N., and D. B. THORUD, Transpiration of Ponderosa pine and Douglas fir after treatment with Phenylmercuric Acetate, *WRR,* **6**: 957-959, 1970.
11. BROWN, G. W., and J. T. KRYGIER, Changing water temperatures in small mountain streams, *J. Soil and Water Cons.,* **22**: 242-244, 1967.
12. BROWN, G. W., and J. T. KRYGIER, Effects of clear-cutting on stream temperature, *WRR,* **6**: 1133-1139, 1970.
13. BUNTING, B. T., The role of seepage moisture in soil formation, slope development, and stream initiation, *Amer. J. Sci.,* **254**: 503-518, 1961.
14. CARSON, M. A., Soil moisture, in Chorley, R. J. (Ed.), *Water, Earth and Man,* pp. 185-195, Methuen, London, 1969.
15. CHERY, D. L., A review of rainfall-runoff physical models as developed by dimensionless analysis and other methods, *WRR,* **3**: 881-889, 1967.
16. CHORLEY, R. J., The drainage basin as the fundamental geomorphic unit, in Chorley, R. J. (Ed.), *Water, Earth and Man,* pp. 77-99, Methuen, London, 1969.
17. CHORLEY, R. J., D. E. G. MALM, and H. A. POGORZELSKI, A new standard for estimating drainage basin shape, *Amer. J. Sci.,* **255**: 138-141, 1957.
18. CHOW, V. T., Laboratory study of watershed hydrology, *Illinois Univ., Civ. Eng. Studies,* Hydraulic Eng. Series No. 14, 1967.
19. CHURCH, J. E., The conservation of snow, *Sci. Amer. Suppl.* **74**: 152, 1912.
20. COSTIN, A. B., L. W. GAY, D. J. WIMBUSH, and D. KERR, Studies in catchment hydrology in the Australian Alps, III. Preliminary snow investigation, *CSIRO, Div. Plant. Ind., Tech. Paper,* 15, 1961.
21. DAVENPORT, D. C., R. M. HAGAN, and P. E. MARTIN, Antitranspirants research and its possible application in hydrology, *WRR,* **5**: 735-743, 1969.
22. DAVIS, W. M., The geographical cycle, *Geog. J.,* **14**: 481-504, 1899.
23. DICKINSON, W. T., M. E. HOLLAND, and G. L. SMITH, An experimental rainfall-runoff facility, *Colorado State Univ., Hydrol. Papers,* No. 25, 1967.
24. DOBBIE, C. H., and P. O. WOLF, The Lynmouth flood of August, 1952, *Proc. ICE,* **2**: 522-588, 1953.
25. DRAGOUN, F. J., Effects of cultivation and grass on surface runoff, *WRR,* **5**: 1078-1083, 1969.

26. DURY, G. H., General theory of meandering valleys, *USGS Prof. Pap.,* 452, 1965.
27. DURY, G. H., Relation of morphometry to runoff frequency, in Chorley, R. J. (Ed.), *Water, Earth and Man,* pp. 419-430, Methuen, London, 1969.
28. EAGLESON, P. S., *Dynamic Hydrology,* McGraw-Hill, New York, 462 pp., 1970.
29. EINSTEIN, H. A., River sedimentation, in Chow, V. T. (Ed.), *Handbook of Applied Hydrology,* Section 17-II, McGraw-Hill, New York, 1964.
30. ELLISON, W. D., Some effects of raindrops and surface flow on soil erosion and infiltration, *Trans. AGU,* 26: 415-429, 1945.
31. EYLES, R. J., Stream net ratios in west Malaysia, *Bull. Amer. Geol. Soc.,* 79: 701-712, 1968.
32. FISHER, D. W., A. W. GAMBELL, G. E. LIKENS, and F. H. BORMANN, Atmospheric contributions to water quality of streams in the Hubbard Brook Experimental Forest, New Hampshire, *WRR,* 4: 1115-1126, 1968.
33. FOK, Y. S., Law of stream relief in Horton's stream morphological system, *WRR,* 7: 201-203, 1971.
34. FOYSTER, A. M., Application of the grid square technique to mapping of evapotranspiration, *J. Hydrol.,* 19: 205-226, 1973.
35. GHOSH, A. K., and A. E. SCHEIDEGGER, Dependence of stream link lengths and drainage areas on stream order, *WRR,* 6: 336-340, 1970.
36. GLANDON, L. R., and L. A. BECK, Monitoring nutrients and pesticides in subsurface agricultural drainage, Abstract in *Trans. AGU,* 50: 612, 1969.
37. GRAVELIUS, H., *Flusskunde,* 1, Berlin and Leipzig, 1914.
38. GREGORY, K. J., Dry valleys and the composition of the drainage net, *J. Hydrol.,* 4: 327-340, 1966.
39. HACK, J. T., Studies of longitudinal stream profiles in Virginia and Maryland, *USGS Prof. Paper,* 294-B, 1957.
40. HAGGETT, P., and R. J. CHORLEY, *Network Analysis in Geography,* Arnold, London, 1969.
41. HARVEY, A. M., Channel capacity and the adjustment of streams to hydrologic regime, *J. Hydrol.,* 8: 82-98, 1969.
42. HEWLETT, J. D., and J. D. HELVEY, Effects of forest clear-felling on the storm hydrograph, *WRR,* 6: 768-782, 1970.
43. HEWLETT, J. D., and W. L. NUTTER, *An Outline of Forest Hydrology,* Univ. of Georgia Press, Athens, Ga., 137 pp., 1969.
44. HIBBERT, A. R., Forest treatment effects on water yield, in Sopper, W. E. and H. W. Lull (Eds.), *Forest Hydrology,* pp. 527-543, Pergamon, Oxford, 1967.
45. HIBBERT, A. R., Increases in streamflow after converting chaparral to grass, *WRR,* 7: 71-80, 1971.
46. HOLSCHER, G. E., Forest hydrology research in the United States, in Sopper, W. E. and H. W. Lull, (Eds.), *Forest Hydrology,* pp. 99-103, Pergamon, Oxford, 1967.
47. HORNBECK, J. W., Storm flow from hardwood-forested and cleared watersheds in New Hampshire, *WRR,* 9: 346-354, 1973.
48. HORNBECK, J. W., and K. G. REINHART, Water quality and soil erosion as affected by logging in steep terrain, *J. Soil and Water Cons.,* 19: 23-27, 1964.
49. HORTON, R. E., Drainage basin characteristics, *Trans. AGU,* 13: 350-361, 1932
50. HORTON, R. E., Erosional development of streams and their drainage basins: Hydrophysical approach to quantitative morphology, *Bull. Amer. Geol. Soc.,* 56: 275-330, 1945.
51. INTERNATIONAL ASSOCIATION OF SCIENTIFIC HYDROLOGY, *Representative and Experimental Basins,* Sympos. of Budapest, *IASH Publ.* 66, 2 vols., 1965.
52. INTERNATIONAL ASSOCIATION OF SCIENTIFIC HYDROLOGY, *Results of Research on Representative and Experimental Basins,* Sympos. of Wellington, *IASH, Publ.* 96, 1970.
53. JARVIS, R. S., New measure of the topologic structure of dendritic drainage networks, *WRR,* 8: 1265-1271, 1972.
54. JEFFREY, W. W., Hydrology of land use, Sect 13 in Gray, D. M. (Ed.), *Handbook on the Principles of Hydrology,* Canadian National Committee for the IHD, Ottawa, 1970.
55. JOHNSON, N. M., G. E. LIKENS, F. H. BORMANN, D. W. FISHER, and R. S. PIERCE,

Working model for the variation in stream water chemistry at the Hubbard Brook Experimental Forest, New Hampshire, *WRR,* **5**: 1353-1363, 1969.

56. KENNEDY, B. A., 'Bankfull' discharge and meander forms, *Area,* **4**: 209-212, 1972.

57. LEE, R., Theoretical estimates versus forest water yield, *WRR,* **6**: 1327-1334, 1970.

58. LEOPOLD, L. B., Downstream change of velocity in rivers, *Amer. J. Sci.,* **251**: 606-624, 1953.

59. LEOPOLD, L. B., and W. B. LANGBEIN, The concept of entropy in landscape evolution, *USGS Prof. Paper,* 500-A, 1962.

60. LEOPOLD, L. B., and T. MADDOCK, The hydraulic geometry of stream channels and some physiographic implications, *USGS Prof. Paper,* **252**, 1953.

61. LEOPOLD, L. B., and M. G. WOLMAN, River channel patterns: braided meandering and straight, *USGS Prof. Paper,* 282-B: 39-84, 1957.

62. LEOPOLD, L. B., M. G. WOLMAN, and J. P. MILLER, *Fluvial Processes in Geomorphology,* Freeman, San Francisco, 522 pp., 1964.

63. LEVNO, A., and J. ROTHACHER, Increases in maximum stream temperatures after logging in old-growth Douglas-fir watersheds, *USFS Res. Note* PNW-65, 1967.

64. LINSLEY, R. K., M. A. KOHLER, and J. L. H. PAULHUS, *Applied Hydrology,* McGraw-Hill, New York, 1949.

65. McGUINESS, J. L., L. L. HARROLD, and F. R. DREIBELBIS, Some effects of land use and treatment on small single crop watersheds, *J. Soil and Water Cons.,* **15**: 65-69, 1960.

66. McPHERSON, H. J., and W. F. RANNIE, Geomorphic effects of the May 1967 flood in Graburn watershed, Cypress Hills, Alberta, Canada, *J. Hydrol.,* **9**: 307-321, 1969.

67. MEIMAN, J. R., and J. S. HORTON, Watershed management research from desert to alpine, *Proc. Soc. Amer. Foresters,* **70**: 93-96, 1964.

68. MELTON, M. A., A derivation of Strahler's channel-ordering system, *J. Geol.,* **67**: 345-346, 1959.

69. MILLER, D. H., Snow cover and Climate in the Sierra Nevada, *Publ. in Geog.,* **11**: 218 pp., Univ. of California, Berkeley, 1955.

70. MILLER, D. H., Transport of intercepted snow from trees during snow storms, *USFS Res. Pap.,* PSW-33, 1966.

71. MILLER, V. C., A quantitative geomorphic study of drainage basin characteristics in the Clinch Mountain area, Virginia and Tennessee, *Tech. Rept.* 3, Dept. of Geology, Columbia Univ., New York, 1953.

72. MOLCHANOV, A. A., *The Hydrological Role of Forest,* Israel Program for Scientific Translations, Jerusalem, 1963.

73. MORISAWA, M., *Streams: their dynamics and morphology,* McGraw-Hill, New York, 175 pp., 1968.

74. NASH, J. E., and B. L. SHAW, Flood frequency as a function of catchment characteristics, in *River Flood Hydrology,* The Institution of Civil Engineers, London, 1966.

75. NATIONAL RESEARCH COUNCIL OF CANADA, *Research Watersheds,* Proc. Hydrology Sympos. No. 4, Univ. of Guelph, Ontario, 1965

76. NIXON, M., A study of the bankfull discharge of rivers in England and Wales, *Proc. ICE,* **12**: 157-174, 1959.

77. O'DONNELL, T., Mathematical modelling, Paper presented to NERC Hydrol. Cttee., 3rd Ann. Representative Meeting, 11 June 1970.

78. PACKER, P. E., and H. F. HAUPT, The influence of roads on water quality characteristics, *Proc. Soc. Amer. Foresters,* **71**: 112-115, 1965.

79. PILSBURY, A. F., R. E. PELISHEK, J. F. OSBORN, and T. E. SZUSZKIEWICZ, Effects of vegetation manipulation on the disposition of precipitation on chaparral-covered watersheds, *JGR,* **67**: 695-702, 1962.

80. RAKHMANOV, V. V., *Role of Forests in Water Conservation,* Israel Program for Scientific Translations, Jerusalem, 1966.

81. REINHART, K. G., and G. R. TRIMBLE, Forest cutting and increased water yield, *J. Amer. Water Works Assoc.,* **54**: 1464-1472, 1962.

82. RICH, L. R., Water yields resulting from treatment applied to mixed conifer watersheds, *Proc. Ann. Arizona Watershed Sympos.,* **9**: 12-15, 1965.

83. RILEY, J. P., D. G. CHADWICK, and E. K. ISRAELSEN, Application of an electronic

analog computer for the simulation of hydrologic events on a southwest watershed, *Utah State Univ., Wat. Res. Lab. Rept.,* PRWG 38-1, 1967.

84. ROWE, P. B., Streamflow increases after removing woodland-riparian vegetation from a southern Californian watershed, *J. For.,* **61**: 365-70, 1963.

85. SATTERLUND, D. R., Combined weather and vegetation modification promises synergistic streamflow response, *J. Hydrol.,* **9**: 155-166, 1969.

86. SCHEIDEGGER, A. E., Some implications of statistical mechanics in geomorphology, *Bull. IASH,* **9**: 12-16, 1964.

87. SCHOMAKER, C. E., Solar radiation measurements under a spruce and a birch canopy during May and June, *Forest Science,* **14**: 31-38, 1968.

88. SCHUMM, S. A., Evolution of drainage systems and slopes in badlands at Perth Amboy, New Jersey, *Bull. Amer. Geol. Soc.,* **67**: 597-646, 1956.

89. SCHUMM, S. A., The shape of alluvial channels in relation to sediment type, *USGS Prof. Paper,* 353-B: 17-30, 1960.

90. SHARP, A. L., A. E. GIBBS, and W. J. OWEN, Development of a procedure for estimating the effects of land and watershed treatment on streamflow, *USDA Tech. Bull.,* 1352, 1966.

91. SHREVE, R. L., Statistical law of stream numbers, *J. Geol.* **74**: 17-37, 1966.

92. SHREVE, R. L., Stream lengths and basin areas in topologically random channel networks, *J. Geol.,* **77**: 397-414, 1969.

93. SMART, J. S., Quantitative characterization of channel network structure, *WRR,* **8**: 1487-1496, 1972.

94. SMITH, T. R., and F. P. BRETHERTON, Stability and the conservation of mass in drainage basin evolution, *WRR,* **8**: 1506-1529, 1972.

95. SNYDER, W. M., Summary and evaluation of methods for detecting hydrologic changes, *J. Hydrol.,* **12**: 311-338, 1971.

96. STALL, J. B., and Y. S. FOK, Hydraulic geometry of Illinois streams, *Univ. of Illinois Water Resources Center, Research Rept.,* 15, 1968.

97. STALL, J. B., and C. T. YANG, Hydraulic geometry and low flow regimen, *Univ. of Illinois, Water Resources Center Research Rept.,* 54, 31 pp., 1972.

98. STRAHLER, A. N., Equilibrium theory of erosional slopes approached by frequency distribution analysis, *Amer. J. Sci.,* **248**: 673-696, 800-814, 1950.

99. STRAHLER, A. N., Hypsometric analysis of erosional topography, *Bull. Amer. Geol. Soc.,* **63**: 1117-1142, 1952.

100. STRAHLER, A. N., Quantitative geomorphology of drainage basins and channel networks, Section 4-II in Chow, V. T. (Ed.), *Handbook of Applied Hydrology,* McGraw-Hill, New York, 1964.

101. SWANK, W. T., and N. H. MINER, Conversion of hardwood-covered watersheds to white pine reduces water yield, *WRR,* **4**: 947-954, 1968.

102. TARRANT, R. F., K. C. LU, W. B. BOLLEN, and C. S. CHEN, Nutrient cycling by throughfall and stemflow precipitation in three coastal Oregon forest types, *USFS Res. Pap.,* PNW-54, 1968.

103. TAYLOR, A. W., W. M. EDWARDS, and E. C. SIMPSON, Nutrients in streams draining woodland and farmland near Coshocton, Ohio, *WRR,* **7**: 81-89, 1971.

104. THORNBURY, W. D., *Principles of Geomorphology* (2nd Ed.), Wiley, New York, 1969.

105. TUCKFIELD, C. G., Seepage steps in the New Forest, Hampshire, England, *WRR,* **9**: 367-377, 1973.

106. VARGAS, M., and E. PICHLER, Residual soil and rock slides in Santos (Brazil), *Proc. 4th Internat. Conf. Soil Mechanics,* **2**: 394-398, 1957.

107. WAGGONER, P. E., and B. A. BRAVDO, Stomata and the hydrologic cycle, *Proc. Nat. Acad. Sci.,* **57**: 1096-1102, 1967.

108. WARD, R. C., *Small Watershed Experiments,* Univ. of Hull Occ. Papers in Geography No. 18, 254 pp., 1971.

109. WICHT, C. L., The influence of vegetation in South African mountain catchments on water supplies, *South Afr. J. Sci.,* **67**: 201-209, 1971.

110. WILCOCK, A. A., Drainage basin shapes, *Area,* **5**: 231-234, 1973.

111. WILLIAMS, P. W., The geomorphic effects of groundwater, in Chorley, R. J. (Ed.), *Water, Earth and Man,* pp. 269-284, Methuen, London, 1969.

112. WILLIAMS, R. E., and P. M. FOWLER, A preliminary report on an empirical analysis of drainage network adjustment to precipitation input, *J. Hydrol.,* **8**: 227-238, 1969.
113. WOODRUFF, J. F., and T. J. GERGEL, On the origin and headward extension of first-order channels, *J. Geog.,* **68**: 100-105, 1969.
114. WOODYER, K. D., Bankfull frequency in rivers, *J. Hydrol.,* **6**: 114-142, 1968.
115. WRIGHT, R. L., Some perspectives in environmental research for agricultural land-use planning in developing countries, *Geoforum,* **10**: 15-33, 1972.
116. YANG, C. T., Potential energy and stream morphology, *WRR,* **7**: 311-322, 1971.
117. YANG, C. T., On river meanders, *J. Hydrol.,* **13**: 231-253, 1971.
118. YOUNG, R. A., and C. K. MUTCHLER, Soil movement on irregular slopes, *WRR,* **5**: 1084-1089, 1969.

Reference Abbreviations

The following is a list of the abbreviations most commonly used in the alphabetical lists of chapter references:

Bull. Amer. Geol. Soc.	Bulletin of the American Geological Society
Bull. Amer. Met. Soc.	Bulletin of the American Meteorological Society
Bull. IASH	Bulletin of the International Association of Scientific Hydrology (now the International Association of Hydrological Sciences)
Geog. J.	Geographical Journal
IASH Publ.	International Association of Scientific Hydrology Publication (now the International Association of Hydrological Sciences)
J. Appl. Met.	Journal of Applied Meteorology
JGR	Journal of Geophysical Research
J. Hydrol.	Journal of Hydrology
J. Soil and Water Cons.	Journal of Soil and Water Conservation
J. Soil Sci.	Journal of Soil Science
Met. Mag.	Meteorological Magazine
Mon. Wea. Rev.	Monthly Weather Review
Proc. ASCE	Proceedings American Society of Civil Engineers
Proc. ICE	Proceedings Institution Civil Engineers
Proc. SSSA	Proceedings Soil Science Society of America
Publ. in Climat.	Publications in Climatology (New Jersey)
QJRMS	Quarterly Journal of the Royal Meteorological Society
Soviet Hydrol.	Soviet Hydrology
Trans. AGU	Transactions American Geophysical Union
Trans. ASAE	Transactions American Society of Agricultural Engineers
Trans. ASCE	Transactions American Society of Civil Engineers
Trans. IBG	Transactions Institute of British Geographers
USDA	United States Department of Agriculture
USFS	United States Forest Service
USGS Circ.	United States Geological Survey Circular
USGS Prof. Pap.	United States Geological Survey Professional Papers
USGS Wat. Sup. Pap.	United States Geological Survey Water Supply Papers
WRR	Water Resources Research

Simplified Metric Conversion Tables

DISTANCE*

Inches		Millimetres		Feet		Metres
0.039	1	25.4		3.281	1	0.305
0.079	2	50.8		6.562	2	0.610
0.118	3	76.2		9.842	3	0.914
0.158	4	101.6		13.123	4	1.219
0.197	5	127.0		16.404	5	1.524
0.236	6	152.4		19.685	6	1.829
0.276	7	177.8		22.966	7	2.134
0.315	8	203.2		26.246	8	2.438
0.354	9	228.6		29.527	9	2.743

Yards		Metres		Miles		Kilometres
1.094	1	0.914		0.621	1	1.609
2.187	2	1.829		1.243	2	3.219
3.281	3	2.743		1.864	3	4.828
4.375	4	3.658		2.486	4	6.437
5.468	5	4.572		3.107	5	8.047
6.562	6	5.486		3.728	6	9.656
7.656	7	6.401		4.350	7	11.265
8.750	8	7.316		4.971	8	12.875
9.843	9	8.230		5.592	9	14.484

VOLUME*

Cu. feet		Cu. metres		Gallons (imp)		Litres
35.315	1	0.028		0.220	1	4.544
70.629	2	0.057		0.440	2	9.087
105.943	3	0.085		0.660	3	13.631
141.258	4	0.113		0.880	4	18.174
176.572	5	0.142		1.101	5	22.718
211.887	6	0.170		1.321	6	27.262
247.201	7	0.198		1.541	7	31.805
282.516	8	0.227		1.761	8	36.349
317.830	9	0.255		1.981	9	40.892

AREA*

Sq. feet		Sq. metres
10.764	1	0.093
21.528	2	0.186
32.292	3	0.279
43.056	4	0.372
53.819	5	0.465
64.583	6	0.557
75.347	7	0.650
86.111	8	0.743
96.875	9	0.836

Sq. yards		Sq. metres
1.196	1	0.836
2.392	2	1.672
3.588	3	2.508
4.784	4	3.345
5.980	5	4.181
7.176	6	5.016
8.372	7	5.853
9.568	8	6.690
10.764	9	7.526

TEMPERATURE
(Centigrade to Fahrenheit)

°C	0	1	2	3	4	5	6	7	8	9
+40	104.0	105.8	107.6	109.4	111.2	113.0	114.8	116.6	118.4	120.2
+30	86.0	87.8	89.6	91.4	93.2	95.0	96.8	98.6	100.4	102.2
+20	68.0	69.8	71.6	73.4	75.2	77.0	78.8	80.6	82.4	84.2
+10	50.0	51.8	53.6	55.4	57.2	59.0	60.8	62.6	64.4	66.2
+ 0	32.0	33.8	35.6	37.4	39.2	41.0	42.8	44.6	46.4	48.2
− 0	32.0	30.2	28.4	26.6	24.8	23.0	21.2	19.4	17.6	15.8
−10	14.0	12.2	10.4	8.6	6.8	5.0	3.2	1.4	−0.4	−2.2
−20	−4.0	−5.8	−7.6	−9.4	−11.2	−13.0	−14.8	−16.6	−18.4	−20.2
−30	−22.0	−23.8	−25.6	−27.4	−29.2	−31.0	−32.8	−34.6	−36.4	−38.2
−40	−40.0	−41.8	−43.6	−45.4	−47.2	−49.0	−50.8	−52.6	−54.4	−56.2

*In the Distance, Area, and Volume conversion tables the figures in the central columns may be read as either the metric or the imperial unit, e.g., 1 cubic foot = 0.028 cubic metre; or 1 cubic metre = 35.315 cubic feet.

Index

McCann, S.B., 263
McClean, W.N., 2
McCulloch, J.S.G., 116
McEwan, G.F., 86
McGavin, R.E., 124
McGuinness, J.L., 35, 110, 118
McIlroy, I.C., 120
McKay, G.A., 37
McMillan, W.D., 66
Maddock, T., 323, 324, 325
Maher, F.J., 123
Makkink, G.F., 105
Manley, G., 46
Manning, R., 2, 303
Man's effect (*see* Human influence)
Mansfield, T.A., 104
Mariotte, E., 2, 244
Marius, J.B., 253
Markov chain, 291
Martinelli, M., 272
Mass movement, of slope material, 334
Massive spaces, 185
Mather, J.R., 77, 144
Matric potential, 137
Maxey, G.B., 214, 228
Maximization of storms, 45
Maximum basin relief, 314
Maximum capillary capacity, 142
Maximum valleyside slope, 315
Maximum water capacity, 142
Mead, D.W., 3
Mean travel time, 264
Meandering channels, 328-332
Measurement:
    actual evapotranspiration, 118-124
    free-water evaporation, 86, 87
    groundwater, 230
    interception, 56-59
    moisture flux, 122-124
    potential evapotranspiration, 110-118
    precipitation, 30-39
    runoff, 303, 304
    soil evaporation, 87, 88
    soil moisture content, 143, 144
    stemflow, 58
    throughfall, 57, 58
Meidner, H., 104
Meiman, J.R., 49
Meinzer, O.E., 4, 185, 202, 203, 206, 214,
    216
Melton, M.A., 317
Meriam, R.A., 56
Meteoric water, 183
Meteorological influences:
    on free water evaporation, 72-75
    on interception loss, 55, 56
    on PE, 99, 100
    on runoff, 263-265

Meyboom, P., 121, 197, 217, 218
Meyer, A.F., 286
Mifflin, M.D., 228
Miller, A.A., 73, 267
Miller, D.H., 64, 341, 343
Miller, D.W., 20, 90
Miller, E.E., 172
Miller, J.F., 44
Miller, J.P., 326, 329
Miller, V.C., 314
Miner, N.H., 59
Minimum flow forecasting, 299-302
Models, 346, 347
    catchment, 298, 299
    distributed system, 13
    lumped-system, 13
Mohrmann, J.C.J., 91
Moisture equivalent, 142
Moisture flux, in calculation of evapotrans-
    piration, 122-124
Moisture retention, 134-143
Moisture supply, effect on actual evapotrans-
    piration of, 106, 107
Monte Carlo methods, 291
Monteith, J.L., 119
Morgan, W.A., 117
Morin, K.V., 114
Movement, of groundwater, 204-222
Mulvaney, T.J., 286
Musgrave, G.W., 168
Mutchler, C.K., 333

Nace, R.L., 5, 7, 16, 189
Nagel, J.F., 65
Nakamura, R., 262
Nash, J.E., 259, 291, 298, 299, 314
Natural erosion, 333
Net interception loss, 54
Net radiometer, 84, 85
Networks, raingauge, 35-39
Neutral stresses, in confined aquifers, 203
Neutron probe, 143
Nicholass, C.A., 38
Nicholson, H.H., 149
Nielsen, D.R., 153, 156
Nine-hole gauge, 35
Nixon, M., 327
Non-capillary porosity, effect on infiltration
    of, 172
Non-cyclic variations of precipitation, 22, 23
Non-linear systems, 13
Norum, D.I., 145, 172, 173, 175
Nutrient balance, of agricultural areas, 344-
    346
Nutter, W.L., 96, 97, 242, 247, 254, 261

O'Donnell, T., 347
Original interstices, 185

Snow:
    accumulation of, forest effects upon, 341
    amount of, 46-48
    distribution of, 46
    effect of on infiltration, 168
    evaporation from, 82
    interception of, 63, 64
    water equivalent of, 47
Snow courses, 48
Snow pillow, 48
Snow sampler, 48
Snowmelt, 48, 49
    as a component of runoff, 241, 242
    effects of forest on, 341-344
    forecasting of runoff from, 285
Snowpack, heat budget of, 343
Snyder, F.F., 266, 296
Snyder, W.M., 259, 260, 339
Soil characteristics, effects of on runoff, 268,
        269
Soil colour, effect of on evaporation, 80
Soil evaporation, 78-82
    measurement of, 87, 88
Soil moisture:
    definition, 132
    effect of on evaporation, 78-80
    effect of on PE, 104-106
    energy relationships, 136-138
    in calculation of evapotranspiration, 122
    measurement of, 143, 144
    movement, 144-166
        after infiltration, 155-160
        during infiltration, 151-154
        and the soil profile, 166
        upward movement, 160-165
    storage, 134-144
Soil moisture characteristics, 138-142
Soil moisture 'constants', 142, 143
Soil moisture gradient, effect of on infiltra-
        tion, 169, 172
Soil moisture potential, 136
Soil moisture suction, 137
Soil profile:
    effect of on infiltration, 172, 173
    and moisture movement, 166
Soil surface, effect of on infiltration, 167
Soil zone, 132
Solar evaporation, 72
Solar radiation, effect of on evaporation, 72
Sopper, W.E., 271
SPAC, 97
Specific retention, 191, 192
Specific water capacity, 140
Specific yield, 192
    measurement of, 230
Splash, effect of on raingauge catch, 33, 34
Spongy leaf cells, 95
Springs, 198, 199

Stanford watershed model, 298, 302
Stark, N., 123
Statistical methods:
    of forecasting minimum flows, 299, 300
    of predicting peak flows, 286-291
Stemflow, 54
    measurement of, 58
Stevenson, T., 35
Stigter, C.J., 103, 104
Stochastic hydrology, 1
Stochastic processes, 13, 14
Stochastic simulation, 291
Stochastic variations, of precipitation, 24, 25
Stol, P.T., 38
Stomata, 95
Stomatal control, of transpiration, 103
Storage, of groundwater, 189-204
Storage changes:
    confined aquifers, 203, 204
    unconfined aquifers, 200-203
Storage coefficient, 199
Storm hydrograph, 245
Storm patterns, 19-22
Storm seepage, 240
Strahler, A.N., 315, 316, 317,
        319, 321
Stream areas, law of, 317, 318
Stream fall, law of, 319
Stream length ratio, 318
Stream lengths, law of, 317
Stream numbers, law of, 317
Stream ordering, 316
Stream slope ratio, 319
Stream slopes, law of, 317
Stream types, 242-244
Streamlines:
    definition, 210
    refraction of, 210, 211, 215
Stringfield, V.T., 221
Strip cutting, 338
Structure, effect of on runoff, 268
Subpermafrost groundwater, 229
Subsurface storm flow, 240
Subsurface water, zones of, 132, 133
Suction, of soil moisture, 137
Summation curve, 296
Suprapermafrost groundwater, 230
Surface storage, 67
Surface tension, effect of on soil moisture
        retention, 135
Sutcliffe, J.V., 259, 291, 298, 299
Sutcliffe, R.C., 5
Swank, W.T., 59
Swartzendruber, D., 146
Swenson, F.A., 228
Swinbank, W.C., 123
Symons, G.J., 2
Synthetic unit hydrographs, 296

Computer Typesetting by Print Origination,
Bootle, Merseyside L20 6NS.
Reproduced photolitho in Great Britain by
J. W. Arrowsmith Ltd, Bristol